STATUS OF
MOLTEN SALT REACTOR
TECHNOLOGY

The following States are Members of the International Atomic Energy Agency:

AFGHANISTAN
ALBANIA
ALGERIA
ANGOLA
ANTIGUA AND BARBUDA
ARGENTINA
ARMENIA
AUSTRALIA
AUSTRIA
AZERBAIJAN
BAHAMAS
BAHRAIN
BANGLADESH
BARBADOS
BELARUS
BELGIUM
BELIZE
BENIN
BOLIVIA, PLURINATIONAL STATE OF
BOSNIA AND HERZEGOVINA
BOTSWANA
BRAZIL
BRUNEI DARUSSALAM
BULGARIA
BURKINA FASO
BURUNDI
CABO VERDE
CAMBODIA
CAMEROON
CANADA
CENTRAL AFRICAN REPUBLIC
CHAD
CHILE
CHINA
COLOMBIA
COMOROS
CONGO
COSTA RICA
CÔTE D'IVOIRE
CROATIA
CUBA
CYPRUS
CZECH REPUBLIC
DEMOCRATIC REPUBLIC OF THE CONGO
DENMARK
DJIBOUTI
DOMINICA
DOMINICAN REPUBLIC
ECUADOR
EGYPT
EL SALVADOR
ERITREA
ESTONIA
ESWATINI
ETHIOPIA
FIJI
FINLAND
FRANCE
GABON
GAMBIA
GEORGIA
GERMANY
GHANA
GREECE
GRENADA
GUATEMALA
GUYANA
HAITI
HOLY SEE
HONDURAS
HUNGARY
ICELAND
INDIA
INDONESIA
IRAN, ISLAMIC REPUBLIC OF
IRAQ
IRELAND
ISRAEL
ITALY
JAMAICA
JAPAN
JORDAN
KAZAKHSTAN
KENYA
KOREA, REPUBLIC OF
KUWAIT
KYRGYZSTAN
LAO PEOPLE'S DEMOCRATIC REPUBLIC
LATVIA
LEBANON
LESOTHO
LIBERIA
LIBYA
LIECHTENSTEIN
LITHUANIA
LUXEMBOURG
MADAGASCAR
MALAWI
MALAYSIA
MALI
MALTA
MARSHALL ISLANDS
MAURITANIA
MAURITIUS
MEXICO
MONACO
MONGOLIA
MONTENEGRO
MOROCCO
MOZAMBIQUE
MYANMAR
NAMIBIA
NEPAL
NETHERLANDS
NEW ZEALAND
NICARAGUA
NIGER
NIGERIA
NORTH MACEDONIA
NORWAY
OMAN
PAKISTAN
PALAU
PANAMA
PAPUA NEW GUINEA
PARAGUAY
PERU
PHILIPPINES
POLAND
PORTUGAL
QATAR
REPUBLIC OF MOLDOVA
ROMANIA
RUSSIAN FEDERATION
RWANDA
SAINT KITTS AND NEVIS
SAINT LUCIA
SAINT VINCENT AND THE GRENADINES
SAMOA
SAN MARINO
SAUDI ARABIA
SENEGAL
SERBIA
SEYCHELLES
SIERRA LEONE
SINGAPORE
SLOVAKIA
SLOVENIA
SOUTH AFRICA
SPAIN
SRI LANKA
SUDAN
SWEDEN
SWITZERLAND
SYRIAN ARAB REPUBLIC
TAJIKISTAN
THAILAND
TOGO
TONGA
TRINIDAD AND TOBAGO
TUNISIA
TÜRKİYE
TURKMENISTAN
UGANDA
UKRAINE
UNITED ARAB EMIRATES
UNITED KINGDOM OF GREAT BRITAIN AND NORTHERN IRELAND
UNITED REPUBLIC OF TANZANIA
UNITED STATES OF AMERICA
URUGUAY
UZBEKISTAN
VANUATU
VENEZUELA, BOLIVARIAN REPUBLIC OF
VIET NAM
YEMEN
ZAMBIA
ZIMBABWE

The Agency's Statute was approved on 23 October 1956 by the Conference on the Statute of the IAEA held at United Nations Headquarters, New York; it entered into force on 29 July 1957. The Headquarters of the Agency are situated in Vienna. Its principal objective is "to accelerate and enlarge the contribution of atomic energy to peace, health and prosperity throughout the world".

TECHNICAL REPORTS SERIES No. 489

STATUS OF MOLTEN SALT REACTOR TECHNOLOGY

INTERNATIONAL ATOMIC ENERGY AGENCY
VIENNA, 2023

COPYRIGHT NOTICE

Marketing and Sales Unit, Publishing Section
International Atomic Energy Agency
Vienna International Centre
PO Box 100
1400 Vienna, Austria
fax: +43 1 26007 22529
tel.: +43 1 2600 22417
email: sales.publications@iaea.org
www.iaea.org/publications

Printed by the IAEA in Austria
November 2023
STI/DOC/010/489

IAEA Library Cataloguing in Publication Data

Names: International Atomic Energy Agency.
Title: Status of molten salt reactor technology / International Atomic Energy Agency.
Description: Vienna : International Atomic Energy Agency, 2023. | Series: Technical reports series (International Atomic Energy Agency), ISSN 0074–1914 ; no. 489 | Includes bibliographical references.
Identifiers: IAEAL 23-01611 | ISBN 978–92–0–140522–7 (paperback : alk. paper) | ISBN 978–92–0–140622–4 (pdf) | ISBN 978–92–0–140722–1 (epub)
Subjects: LCSH: Molten salt reactors. | Molten salt reactors — Technological innovations. | Molten salt reactors — Research. | Nuclear reactors.
Classification: UDC 621.039.5 | STI/DOC/010/489

FOREWORD

The IAEA supports Member States in the development of advanced reactor technology by serving as a major focal point for information exchange and collaborative research programmes. The activities of the IAEA in this field are mainly carried out within the framework of several work areas, typically supported by technical working groups that assist in the implementation of corresponding IAEA support and ensure that all technical activities are in line with the expressed needs of the Member States.

In recent years, there has been renewed global interest in molten salt reactors — advanced reactors that are fuelled and/or cooled by molten salt — and the number of activities related to the design and technology of these reactors is growing. The molten salt reactor is one of the six reactor technologies selected by the Generation IV International Forum for further research and development. The technology is appropriate for small modular reactors, and molten salt reactors are expected to have advantages over light water reactors in terms of safety, environment, economics and non-proliferation. High operating temperatures leading to increased efficiencies in electricity generation, passive decay heat removal and flexible fuel cycles are some of the additional benefits of this reactor technology.

This publication summarizes current knowledge on the status of research, technological developments, reactor designs and experiments in the area of molten salt reactors. It presents a balanced view of the status and potential advantages of the technology and identifies challenges and areas in which research and development are required before deployment is achievable.

The IAEA technical officers responsible for this publication were G. Martinez-Guridi, L. Peguero and F. Reitsma of the Division of Nuclear Power of the Department of Nuclear Energy.

EDITORIAL NOTE

CONTENTS

1. INTRODUCTION

1.1. BACKGROUND

The IAEA fosters the international exchange of information on advances in nuclear reactor technology, and supports Member States by providing objective and reliable information and knowledge of various reactor technologies. This includes the promotion of international collaborative research and development (R&D) in the area of advanced reactor technologies that are needed to meet increasing energy demands, such as molten salt reactors (MSRs), small and medium sized reactors and non-electric applications of nuclear power.

Member States are showing a growing interest in MSR technology, and an increasing number of developments and deployment activities are being reported for the near term. In this regard, the IAEA aims to enhance the prospects for the demonstration and implementation of MSRs in the future.

Advanced reactor technologies currently supported in this way include advanced light water reactors, fast reactors and gas cooled reactors. Subprogrammes are also dedicated to small modular reactors (SMRs) and non-electric applications. No specific programme currently exists for MSRs, but activities are undertaken in accordance with resolution GC(60)/RES/12 of the 60th regular session of the General Conference of the IAEA [1], which:

> "Encourages the Secretariat to explore, in consultation with interested Member States, the need for closer collaboration in technology development for advanced reactor lines by hosting a workshop with the aim of considering launching a new project on molten salt and molten salt cooled advanced reactors".

Resolutions from the 61st and 62nd regular sessions of the General Conference also reflected the interest of Member States in MSRs, in particular, resolution GC(61)/RES/11 [2], which includes the following paragraph:

> "Welcoming the increased participation at the meeting, organized in November 2015, to 'present and share important information on the interest and status of technology developments in the area of molten-salt and molten-salt cooled advanced reactors' and welcoming the meeting that took place in November 2016, ..."

In resolution GC(62)/RES/9 [3], "the increased interest in technology developments in the area of molten salt and molten-salt cooled advanced reactors, …" was noted.

In both sessions, the General Conference stated that it [2, 3]:

> "Recommends that the Secretariat continue to explore, in consultation with interested Member States, activities in the areas of innovative nuclear technologies, such as … Generation IV nuclear energy systems including … molten salt nuclear reactors, with a view to strengthening infrastructure, safety and security, fostering science, technology, engineering and capacity building via the utilization of existing and planned experimental facilities and material test reactors, and with a view to strengthening the efforts aimed at creating an adequate and harmonized regulatory framework so as to facilitate the licensing, construction and operation of these innovative reactors".

Over the past few years, the IAEA has organized several meetings on MSR technology. It was concluded at a consultancy meeting, held on 18–20 November 2015, that there were no known fundamental technical hurdles that would prevent MSR technology from being a safe and feasible energy solution. It was also concluded that, with the necessary funding, this technology could be commercially demonstrated within the next two decades. However, many engineering challenges will need to be solved and the economic competitiveness studied further. Subsequently, and to fulfil the above resolution of the 60th regular session of the General Conference of the IAEA, a Technical Meeting on the Status of Molten Salt Reactor Technology was held from 31 October to 3 November 2016. The need to prepare a publication documenting MSR technology was recognized at this meeting. A series of consultancy meetings were held between September 2017 and July 2020 to develop this publication.

1.2. OBJECTIVE

This publication summarizes the current knowledge on the status of research activities, technological developments, reactor designs and experiments in the area of advanced reactors that are related to MSRs. In this publication, an MSR is defined as any reactor in which a molten salt has a prominent role in the reactor core (i.e. as fuel, coolant and/or moderator).

The publication is targeted at government officials from Member States with technical backgrounds, research institutes and university students working on related topics, commercial organizations working or planning to design and build MSRs, and IAEA staff working in this field. The publication will specifically benefit newcomer Member States that wish to understand the technology and R&D needs of MSRs for electricity and cogeneration applications.

The publication aims not only to present a balanced view of the status and potential advantages of the MSR technology, but also to identify challenges and technology areas in which R&D is required before deployment is achievable. The work undertaken in the preparation of this publication provided Member States with an opportunity to share information on their MSR programmes and projects. The global development of advanced reactor technology is changing and advancing rapidly. Efforts were made to ensure that the technical information presented in this publication is as up to date as practicable.

Guidance and recommendations provided here in relation to identified good practices represent expert opinion but are not made on the basis of a consensus of all Member States.

1.3. SCOPE

This publication is intended to document the current status of MSR technology in Member States. It also includes a short history of the development of the technology and a classification system for MSRs. This technology is still evolving, and it continues to change as new developments and concepts appear. In some cases, proprietary and other restrictions have limited the scope of the information provided in this publication.

1.4. STRUCTURE

Six sections follow this introductory section. Section 2 gives a brief history of MSR technology. Section 3 discusses the advantages and technological challenges of MSRs. Section 4 offers a classification (taxonomy) of MSRs by class and family and includes a description of each. Section 5 elaborates on current R&D activities for MSR technology. Section 6 discusses current challenges for deploying MSRs and Section 7 provides a summary and conclusions.

This publication also contains five appendices. Appendices I–IV provide details about the history of MSR technology in Poland and Switzerland, China, France and the Russian Federation, respectively. Appendix V describes MSR concepts, including those that may still be at a conceptual level;[1] it also briefly

---

[1] For a more detailed description of MSR concepts, see the Advanced Reactor Information System (ARIS), an on-line database designed and maintained by the IAEA: https://aris.iaea.org.

demonstrates how each concept fits into one of the families presented in Section 4 by introducing the main characteristics of the concept.

2. HISTORY OF MSR TECHNOLOGY

2.1. DEVELOPMENT EFFORTS IN THE UNITED STATES OF AMERICA

The technology for MSRs has originated from multiple sources. Much of the salt handling and measurement technology was originally developed for the chemical processing and aluminium smelting industries in the early part of the twentieth century. The corrosion resistant alloy technology employed for the piping and containers derives from the alloys originally developed for high temperature gas turbines and jet engines. The concept of employing a liquid–slurry mixture of fuel and moderator dates from shortly after the discovery of fission, when H. Halban and L. Kowarski performed experiments with a slurry of uranium oxide (U_3O_8) in heavy water at the Cavendish Laboratory in the United Kingdom (UK) [4]. Heterogeneous reactors[2] using solid fuel were, however, selected as the primary path for the Manhattan Project because neither enriched uranium nor deuterium were available. Interest in reactors with a fluid fuel grew in 1943 when larger quantities of heavy water became available. This was achieved through the group led by H.C. Urey at Columbia University, which was investigating slurry reactors using U_3O_8 and D_2O. Heterogeneous configurations of slurry reactors then became the focus of attention. According to the theory of reactor physics, providing a reactor region without fuel will slow down the neutrons while avoiding resonance capture. As a consequence, the neutron multiplication factor is maximized. Various slurry designs were pursued during the Second World War as backups to the nuclear reactors located at the Hanford site for producing plutonium. Once the Hanford reactors became operational, interest in alternative plutonium production reactors diminished, and nearly all developmental research had been discontinued by the end of 1944.

The scientific investigation of reactors using a fluid fuel continued, however, at both Los Alamos National Laboratory (LANL) and Oak Ridge National Laboratory (ORNL). The objectives were to provide power for remote locations and to produce radioisotopes. The physicists were also interested in this type of reactor as a research facility to produce high neutron fluxes. The possibility of

[2] See Section 4 for a brief description of homogeneous and heterogeneous reactors.

the ^{232}Th–^{233}U breeding cycle in a homogeneous reactor was described in 1944 by L.W. Nordheim of ORNL [5]. Work on a Th–U aqueous homogeneous reactor with a three year doubling time continued in 1945. However, a number of issues led to a temporary cessation of design and development activities of breeder reactors with fluid fuel. These included technical difficulties with the radiolytic decomposition of water, which resulted in bubbling reactor instability; a lack of container materials with high strength and low absorption of neutrons to enable high pressure operation that will reduce bubbling; corrosion; solution stability; and the explosive potential of the hydrogen bubbles. Experiments to establish the feasibility of molten salt fuels began in 1947 on the initiative of V.P. Calkins, K. Anderson and E.S. Bettis [6].

In early 1949, A.M. Weinberg, research director of ORNL, recommended reconsidering reactors with fluid fuel in the light of the technical developments that had been achieved since the end of 1945. By July 1949, a development effort on homogeneous reactors was under way at ORNL. In September 1949, ORNL was designated by the Atomic Energy Commission of the United States of America (USA) as the lead for conducting research on reactors for the aircraft nuclear propulsion (ANP) programme, which divided the development effort for reactors with fluid fuel into an aqueous branch at low temperature and a salt branch at high temperature. Caustic soda (NaOH) was initially a leading candidate carrier salt for high temperature as fluoride salts are not sufficiently self-moderating to enable a homogeneous configuration. However, problems with corrosion, the limited solubility of uranium and the almost complete lack of thorium solubility in NaOH resulted in a focus on fluoride salts by mid-1950 [7].

R.C. Briant of ORNL suggested the use of the molten mixture of uranium tetrafluoride (UF_4) and thorium tetrafluoride (ThF_4), together with alkali metal fluorides, as the fluid fuel [8]. Fast reactors using chloride salt were also considered in the early stages [9, 10], but the relatively high neutron capture cross-section of chlorine-35 (^{35}Cl) (and the lack of available technology for chlorine isotope separation at large scale) led to the focus on fluoride salts [11]. Additionally, reactors with fast spectra would require power densities that were very high, necessitating unproven heat transfer technologies to avoid excessive fissile inventories. Nevertheless, the fused salt fast breeder reactor based on fluoride salts and the Th–U cycle [12] was proposed by students at Oak Ridge School of Reactor Technology in 1953. It was a predecessor of the Molten Salt Fast Reactor (MSFR) design (see Section 4.7.1.1). Since it was based on LiF–BeF_2 carrier salt, the performance, especially the doubling time, was worse than for a reactor moderated by graphite.

The ANP programme grew rapidly in the early 1950s, and many of the technologies of current MSR designs have their origins in this period. The first MSR, the Aircraft Reactor Experiment, was built and operated at ORNL in 1954.

A basic understanding of molten salt production [13] as well as the container materials [14] and components [15] had to be established prior to operating the reactor. Research on salt phase equilibria [16, 17], salt purification methods [18] and corrosion chemistry [19] was carried out during this period. As the Aircraft Reactor Experiment was not intended to operate for an extended period, much of the remainder of the ANP programme was devoted to developing technologies to extend the period of operation. Development of a Ni–Mo alloy container material having the property of low corrosion (INOR-8[3], now UNS 10003 or Hastelloy N) was among the most significant MSR technical developments of the latter half of the 1950s [20]. These technical accomplishments were summarized in the book Fluid Fuel Reactors [4] produced by the Atomic Energy Commission for the Second United Nations International Conference on the Peaceful Uses of Atomic Energy, better known as the 'Atoms for Peace' conference, in 1958.

The ANP programme was wound down at the end of the 1950s. The emphasis placed on the fluid fuelled reactor programme transitioned to generating power for the civilian grid in the USA in the late 1950s. In 1956, H.G. MacPherson, a research scientist and future ORNL deputy director, formed a group to study the performance and characteristics of converter and breeder MSRs [11]. In the mid-1950s, the USA was pursuing three different reactor classes using liquid fuel (i.e. aqueous homogeneous, liquid salt fuel and liquid metal fuel). As the Atomic Energy Commission lacked the resources to pursue all three reactor classes, a task force for reactors using fluid fuel was commissioned to evaluate which (if any) to pursue [21]. The report of the task force began with the statement "The molten salt reactor has the highest probability of achieving technical feasibility." The report also noted that maintenance is the most important factor influencing the practicability of MSRs. Substantial efforts were made in the late 1950s to demonstrate that remote maintenance of highly radioactive components was possible using long handled tooling operated from overhead cranes. By the late 1950s, adequate progress had been made in all areas of MSR technology to proceed with the design and construction of an experimental reactor to "demonstrate the safety, dependability, and serviceability of a molten-salt reactor and to obtain additional information about graphite in an operating power reactor" [22].

The design of the Molten Salt Reactor Experiment (MSRE) began in 1960, construction was initiated in 1962 and initial criticality was achieved in 1965 [23]. The MSRE reactor fuel mixture nominally consisted of 65 lithium fluoride (^{7}LiF), 29.1 beryllium fluoride (BeF_2), 5 zirconium tetrafluoride (ZrF_4) and 0.9 UF_4 (mol%). Unclad fine-grained graphite served as the moderator. All the other salt wetted components were fabricated from Hastelloy N. The MSRE

[3] At ORNL, Hastelloy N was formerly called INOR-8.

reached full power (7.34 MW) in 1966. Operations with ^{235}U (~32% enrichment) continued until 1968, when this uranium isotope was removed from the salt and replaced with ^{233}U. Operations with the latter isotope (~91.5% [24]) continued until December 1969. The last few refuelling additions were performed using plutonium trifluoride ($^{239}PuF_3$). Overall, operation of the MSRE was highly successful and essentially no problems were encountered with the primary system during operation.

However, four significant technology issues arose during the late 1960s that impacted the design of future MSRs:

(1) Neutron embrittlement of nickel based alloys at high temperature;
(2) Radiation damage and dimensional changes to graphite at high fast neutron fluences;
(3) Need for liquid–liquid chemical extraction processes for removing protactinium and uranium from molten salt fuel;
(4) Rapid tritium permeation of alloys that are tolerant to salt at high temperature.

In 1972, following the successful operation of the MSRE and prior to initiating a larger technology demonstration programme, the Atomic Energy Commission performed an evaluation of MSR technology, documented in the WASH-1222 report [25]. In addition to the technology issues listed above, the report also indicated that the engineering development of large components, a better understanding of the behaviour of fission products, and adequate remote inspection and maintenance techniques would be necessary before MSRs would be suitable for development. The report also cautioned that independent of technology capabilities, MSRs were not high enough on the development priority ranking of the US Government to have reasonable assurance of the required sustained resource allocation. It stated:

> "When significant evidence is available that demonstrates realistic solutions are practical, a further assessment could then be made as to the advisability of advancing into the detailed design and engineering phase of the development process including that of industrial involvement. Proceeding with this next step would also be contingent upon obtaining a firm demonstration of interest and commitment to the concept by the power industry and the utilities and reasonable assurances that large scale government and industrial resources can be made available on a continuing basis to this program in light of other commitments to the commercial nuclear power program and higher priority energy development efforts."

Much of the work on the development of MSRs by the USA during the 1970s focused on addressing the identified technology issues. Two tracks were pursued to alleviate the embrittlement vulnerability of containment alloy: shielding and improved alloy design. The shielding approach added an interior graphite lining of approximately half a metre to the interior of the reactor vessel to prevent a significant neutron flux from reaching the vessel wall. The improved alloy design approach was based on creating large numbers of helium traps (finely dispersed carbides) within the microstructure of Hastelloy N to prevent generated helium from migrating to the grain boundaries. Niobium modified Hastelloy N, which exhibited improved resistance to neutron embrittlement up to 650°C, was a key MSR technology advancement of the 1970s [26]. Accommodating the radiation damage characteristics of graphite was also approached both through reactor design and graphite technology development. Designs for MSRs lowered the power density in the core and hence the rate of radiation damage to graphite components. The MSR design with two fluids was also abandoned, in large part because of the requirement for graphite plumbing to be in a region of high flux. The development of graphite with acceptable tolerance to radiation damage was a key focus of the gas reactor programme during the 1960–70s. The use of moderator materials for high temperatures with an increased tolerance to radiation damage would enable a higher power density core in MSRs. For this, adequate nuclear grade graphite (largely derived from the high temperature gas cooled reactor programme) was developed in the 1960–70s.

A key technology requirement of the MSR with a thermal spectrum and a single fluid was to demonstrate the chemical steps in a liquid–liquid extraction process for removing protactinium and uranium from molten fluoride salts [27, 28]. Calculations indicated that the process could be carried out rapidly and continuously and that the process equipment would be relatively small [6]. Engineering development studies on the processing of fuel salt continued through the mid-1970s [29]. Tritium can be a radiation hazard, and the problem of its escape into the environment was addressed by employing a loop with coolant salt that chemically trapped the tritium before it could reach the steam cycle. Technology for tritium trapping was demonstrated at engineering scale in the mid-1970s [30]. The MSR programme in the USA also began the process of large scale molten salt hydraulic components. Thermal and hydraulic design studies for a nuclear qualified steam generator for a large MSR were completed by an industrial designer in the mid-1970s [31].

The issue of proliferation vulnerability of nuclear fuel cycles also arose during the 1970s. Designs for MSRs that existed in the USA in 1976 were not focused on making the diversion of fissile material difficult or easily detectable [32]. President Carter's nuclear power policy statement of 7 April 1977 [33] announced that there would be "direct funding of US nuclear research and development programs to accelerate our research into alternative nuclear fuel cycles which do not involve direct access to materials that can be used for nuclear weapons."

A conceptual design for an MSR that avoided direct access to materials that could be used for nuclear weapons was subsequently developed in the late 1970s [34]. Overall, the MSR programme in the USA largely overcame the technical issues identified in the WASH-1222 report, which were considered necessary to resolve prior to beginning engineering development, and developed designs that were compliant with the policy directive for proliferation resistance. However, the programme never became of sufficiently high importance to obtain the required resources. Large scale activities in the USA related to MSRs had ended by 1977.

2.2. DEVELOPMENT EFFORTS OUTSIDE THE UNITED STATES OF AMERICA

Several Member States began evaluating MSR technology following the reporting by the USA of its activities at the second United Nations 'Atoms for Peace' conference in 1958.

2.2.1. Development efforts in Poland and Switzerland

Research in Switzerland on MSR technologies started in the late 1960s when M. Taube joined the Swiss Federal Institute for Reactor Research (Eidgenössisches Institut für Reaktorforschung, EIR), the predecessor of the Paul Scherrer Institute. One of the pioneers in chloride salt reactor research, Taube published his first paper related to a chloride fast MSR [35] at the Department of Radiochemistry of the Institute of Nuclear Research in Warsaw in 1961, and in 1967 he proposed the concept of cooling by boiling aluminium trichloride ($AlCl_3$), which was in direct contact with the fuel salt [36]. After Taube joined EIR, the boiling $AlCl_3$ was assessed as a coolant of a fast reactor with solid fuel [37]. Later research activities in Switzerland were mainly oriented towards heterogeneous chloride fast MSRs [38], where the blanket salt was often used as a coolant for the fuel salt. At the time of suspension of the ORNL MSR project, three EIR labs were partly involved in neutronics, chemistry and materials research for MSRs. These research activities at EIR were not stopped; however, from the mid-1970s, the focus moved to homogeneous fast chloride MSRs, including combined breeding in Th–U and U–Pu fuel cycles [39]. At the end of the 1970s, Taube and his team proposed the Salt reactor On site reprocessing Fast converter Task (SOFT) reactor [40] as the final concept before MSR research activities were suspended at EIR. The thesis of E. Ottewitte [41], who was involved with EIR in the 1970s, represents to some extent a continuation of this work. Appendix I provides further information on the history of MSR development in Poland and Switzerland.

2.2.2. Development efforts in Germany

The molten salt epithermal reactor concept, known as MOSEL, was studied in the Federal Republic of Germany in the mid-1960s [42]. In this breeder concept, thorium dissolved in a salt mixture is used as the primary coolant fluid. The fuel is contained in tubes or plates within the core connected to inlet and outlet manifolds. This fuel mixture is circulated primarily for the purpose of fuel mixing, addition and removal [43].

2.2.3. Development efforts in the United Kingdom

Efforts in the UK to develop nuclear graphite that is tolerant to radiation damage for gas cooled reactors contributed significantly to the development of MSRs. The work concentrated on reactors using a chloride salt, a fast neutron spectrum and the ^{238}U–Pu fuel cycle. Work on the chemistry of chloride fuel salts began in 1965 and was summarized in 1969 [44]. The UK Atomic Energy Authority examined three variants of a 2500 MW(e) molten chloride fast reactor, or MCFR: (i) with in-core direct contact with lead, (ii) with in-core cooling by passing a blanket salt through molybdenum tubes and (iii) with out-of-core cooling [45].

2.2.4. Development efforts in the Netherlands

The Netherlands began researching MSR technology in 1963 as a joint undertaking between the Joint Research Centre (JRC) of the European Commission (Petten site) and Delft University of Technology (TU Delft). Staff from the laboratory of thermal power engineering at TU Delft were recruited to investigate molten salt steam generators with bayonet tubes. These efforts were superseded by the Delft Molten Salt Project, which lasted from 1969 until 1978. This project mainly focused on component and engineering developments of primary fluoride and secondary nitrate molten salt loops for the scale of a pilot plant [46–48]. Valves, pumps, heat exchangers, steam generators and flowmeters were investigated, along with simulation software capable of predicting operating limits imposed by steam generators of various geometries [46].

2.2.5. Development efforts in China

In February 1970, Premier Zhou Enlai stated that Shanghai should develop a nuclear power plant to solve the disparity between Shanghai's economic development and its energy shortage. An MSR for high temperature was initially selected for the project, which would generate 25 000 KW(th) [49]. In the 1970s, a critical experiment device was established to research the physics characteristics of an MSR in the Shanghai Institute of Applied Physics (SINAP) (then Shanghai Institute of Nuclear Research), and a series of zero power experiments were conducted and related results were obtained. The device was later transformed into a uranium–water lattice experiment device owing to a change in research priorities. A major advantage of this device is the flexibility of the arrangement of the fuel elements and moderator elements in the core allowing different moderating ratios. The purpose of the device was to verify experimentally the results from theoretical calculations; determine important characteristics, such as the static and dynamic features, the unit inventory gram reactivity and the temperature effect of the molten salt used; and obtain the relevant design data such as the control rod calibration and its temperature effect, and the fertile fuel conversion ratio [50–53]. Appendix II offers additional information on the history of MSR technology in China.

2.2.6. Development efforts in Japan

Activities relating to MSR technology in Japan began in the late 1970s [54, 55] and focused primarily on small FUJI designs with a self-sustaining capability for power generation, owing to the adoption of thorium as fertile material and ^{233}U as fissile material. In addition, an off-line reprocessing facility and an accelerator driven system for producing fissile material were proposed.

2.2.7. Development efforts in France

In France, the French Alternative Energies and Atomic Energy Commission (Commissariat à l'Énergie Atomique, CEA) and Électricité de France (EDF) pursued MSRs from 1973 to 1983 [56]. This work involved four experimental loops, including the development of a pump and a heat exchanger that were made from graphite. Direct contact molten lead cooling was evaluated to minimize fissile inventory while avoiding metallic materials in the core.

After the shutdown of the Superphénix sodium fast breeder reactor in December 1998, and in the framework of the 1991 French law on wastes [57], innovative solutions were studied in order to burn the plutonium and minor actinides (americium, curium, neptunium) coming from French pressurized water reactors. Among the different reactors, MSRs were identified as one of the most promising types in many papers [58–61]. While those studies considered subcritical systems driven by a particle accelerator, EDF proposed a critical burner for plutonium and minor actinides called the Actinide Molten Salt Transmuter (AMSTER) to simplify the transmutation.

French R&D on MSRs had two phases. The first phase from 1998 to 2005 was an in-depth optimization and study of a reactor of the type molten salt breeder reactor (MSBR). AMSTER was defined by EDF and the thorium molten salt reactor (TMSR) was defined by the French National Center for Scientific Research (Centre national de la recherche scientifique, CNRS). The first phase concluded that no design of an MSBR-like reactor moderated by graphite achieved all the design goals (i.e. iso-breeding, negative reactivity feedback coefficients and reasonable graphite lifetime). The decision was taken to avoid using graphite and to develop non-moderated MSR concepts.

Accordingly, the second phase of French R&D, from 2004 to 2021, involved the above concepts. The CNRS proposed the non-moderated TMSR [62–64] in 2004, which uses a fluoride salt and is an iso-breeder. This concept was renamed MSFR and is described in Section 5.7 on French R&D activities. Electricité de France also proposed a non-moderated iso-breeder MSR in 2004, the REBUS-3700 [65, 66] with a classical depleted U–Pu cycle. This reactor uses a chloride salt ($38UCl_3$–$7TRUCl_3$–55NaCl mol%) because a preliminary study showed that ^{238}U–^{239}Pu breeding is difficult to reach with a fluoride salt, since the neutron moderation by fluoride is too high. Both non-moderated reactors show largely negative reactivity feedback coefficients and are iso-breeders with limited fuel reprocessing (around 40 L of salt reprocessed per day).

Appendix III provides detailed technical information about the French MSR designs developed from 1998 to 2008.

2.2.8. Development efforts in the Russian Federation

The Soviet MSR programme began in 1976 [67, 68] and included molten salt cooled reactors [69], thermal spectrum MSRs and eventually fast spectrum systems. It also included an in-core, natural circulation molten salt fuelled loop [70].

The Kurchatov Institute was the main organization under which various specialized institutions collaborated. Following the nuclear power plant accident that took place in Chornobyl in 1986, activity declined and the nuclear industry stagnated. However, by the end of the 1980s, the number of conceptual studies increased as interest in inherently safe, new generation reactors grew. These studies of MSR technology were mainly directed at developing Th–U concepts.

The MSR programme addressed the following topics:

- — Reactor physics and reactor safety;
- — Container materials for fuel and coolant salts;
- — Physical and chemical properties of molten salt mixtures;
- — Heat transfer and hydraulics of fuel and coolant salts;
- — Handling and circulation of fuel and coolant salts;
- — Process and radiochemical bench tests of model installations;
- — Radiochemistry of molten fluoride fuel salt.

The first topic only included theoretical studies; the rest involved both theoretical and experimental studies. Some of these topics are described in Appendix IV.

3. ADVANTAGES AND TECHNICAL CHALLENGES OF MSR TECHNOLOGY

3.1. ADVANTAGES OF MSR TECHNOLOGY

Significant research on MSR technology is under way in various Member States. Early development work, conducted as part of the Molten Salt Reactor Programme at ORNL beginning in the late 1940s, is the origin of most of the work currently being carried out on a variety of MSR concepts. As most of these concepts use fuel in liquid form (i.e. molten salt contains the fuel and is the coolant) or solid fuel with cooling by molten salt, a very different set of features characterizes this class of nuclear reactor when compared with traditional water cooled reactors.

An overview of the significant advantages of MSRs that use molten salt to contain the fuel and as the coolant over conventional water cooled reactors is provided below. Some of the advantages discussed are also applicable to MSRs with solid fuel and cooled by molten salt. Almost all the advantages listed below can be divided into two overarching categories, namely safety and economics; these are described in Sections 3.1.1 and 3.1.2, respectively. Generally, MSRs also have potential advantages in terms of the environment and the non-proliferation of nuclear material, which are also provided for simplicity in Section 3.1.1. Some features of MSR technology are beneficial to both safety and economics; for example, operating the primary system at near atmospheric pressure benefits both. Section 3.2 discusses the technical challenges to deploying MSRs and the R&D needed to address those challenges.

3.1.1. Safety advantages

Features of MSRs that provide potential safety advantages include the following:

— *Near atmospheric operating pressure.* The possibility to operate the reactor at near atmospheric pressure brings several advantages in terms of operation, safety and economics. Owing to a reduction of mechanical stresses to the structural materials, the risk due to potential accident scenarios is lower than for a typical water cooled reactor, as are costs. In addition, a salt circuit and other confinement barriers at ambient pressure reduce the possibility of radioactive nuclide dispersion.

— *Radioisotope retention.* Several benefits are gained from the chemical and thermophysical properties of molten salts. Molten salts are exceptional solvents and strongly bond many of the fission products, even at high temperatures. This is a major, intrinsic safety feature, as the salt (both in liquid and in solid fuel systems) becomes an additional barrier to the release of radioisotopes to the atmosphere. Conversely, water cooled reactors generate a much wider range of unretained radioisotopes. These isotopes accumulate in the fuel pins and increase the internal pressure within the pins. The volatile compounds in the fuel pins are highly mobile and are the main source of radioactive contamination in the event of an accident.

— *Chemically inert and not flammable.* Molten salts have a low chemical reactivity. Not reacting exothermically with nearby materials, air or water is a significant advantage, as there is no potential for hydrogen explosions or sodium fires. Overall, the containment of radioisotopes in MSRs is easier than in nuclear reactors using high pressure fluids (such as water or helium) or reactive fluids such as sodium [71].

— *Removal of gaseous fission products.* Since gaseous fission products can be removed by separating them from the fuel salt during reactor operation, pressure buildup within the core can be avoided. This mitigates the safety risks of the reactor compartment and the consequences of accident scenarios involving the core.

— *No irradiation damage or mechanical failure of fuel.* Molten salt is an ionic liquid, and irradiation damage or mechanical failure of the fuel does not occur.

— *Strong negative reactivity temperature feedback.* This feedback is predominantly caused by the Doppler effects, fuel salt expansion, moderator temperature or a combination of these effects. This results in a strong negative reactivity temperature coefficient that characterizes most MSR concepts using liquid fuel. This characteristic provides an intrinsic stability for the core and enables passive shutdown for any temperature rise. It enhances nuclear safety and other benefits such as load following. Some early designs were moderated by graphite. The ORNL's MSBR in the 1970s was an example. It featured only a weak negative overall temperature reactivity coefficient because the positive density and graphite terms are compensated by a strong negative Doppler term. These issues are being overcome in modern efforts. The density term can be made negative by utilizing a larger fraction of graphite, resulting in a higher carbon to fissile ratio. The positive graphite term of the MSBR was due to the use of ^{233}U; however, ^{235}U gives a beneficial negative graphite temperature reactivity term for MSRs on burner cycles using low enriched uranium. Thus, MSR designs with fast and thermal spectra can achieve strong total negative reactivity feedback with negative density, graphite temperature and Doppler terms.

— *Molten salt as a heat sink.* The molten salt will absorb some of the decay heat after reactor shutdown. The total heat sink of the core itself can be very high in the case of moderated designs because of the large mass of moderator, typically graphite. In fast and thermal MSRs with liquid fuel, thermal inertia tends to be higher than in their respective counterparts with solid fuel. For most MSR designs, the liquid form of the fuel provides a simpler way to remove heat by circulating the fuel in a heat exchanger. Moreover, heat removal has to occur in the reactor core as for water cooled reactors, eliminating the risk of circulating coolant fluid in the core.

— *High boiling point.* Molten salts are usually characterized by a high boiling point. This physical property, in combination with the heat capacity of the salts, allows a large margin (up to 700°C) between the operational and the boiling temperatures to accommodate temperature transients in case of potential accident scenarios without losing the cooling capability of the salts.

— *Excellent neutron economy.* The technology of MSRs allows for the better utilization of fresh fuel loaded in the core than some of the other reactor technologies. This is mainly because of the absence of metallic structure, burnable poisons or other parasitic neutron capture elements within the core. The addition also of semi-continuous fission product removal processes can further improve neutron economy to allow break even or breeding operation.

— *Flexible fuel cycle.* A reactor core with liquid fuel can host all actinide elements and enable their continuous recycling. Since the fuel manufacturing stage can be omitted, even elements producing intensive decay heat can be hosted and recycled with the liquid fuel. Solid fuel, on the contrary, cannot be chemically adjusted following manufacturing. Solid fuel is also vulnerable to mechanical and radiation damage, and therefore needs to be checked and replaced periodically. The liquid fuel cycle flexibility supports both the breeding of fissile isotopes by using ^{232}Th or ^{238}U (i.e. enhancing resource utilization), as well as the recycling of transuranic (TRU) stockpiles (i.e. burning existing waste). This is a significant advantage from an environmental and a non-proliferation perspective. Furthermore, if recycling is performed on-site, the risks related to spent fuel transportation are reduced.

3.1.2. Economic advantages

The intrinsic safety features of MSRs described in Section 3.1.1 contribute to a simple and safe nuclear power plant. Naturally, this has a potentially positive impact on the economics of MSRs, as less engineering would be required to ensure safe operations. Features of MSRs that provide potential economic advantages include the following:

— *High temperature.* MSRs have the possibility to operate at high temperatures (>600°C) with the resulting advantage of very high thermodynamic efficiency (up to 50%). This is possible with either a conventional and standard steam Rankine cycle, or with the advancing field of gas Brayton technology, thus improving the economics of the operation of a nuclear power plant based on an MSR. In particular, the supercritical CO_2 Brayton cycles offer the potential for a dramatically reduced turbine size in the future. The high operating temperatures also support non-electrical applications that require high temperatures. Whereas water cooled technology is limited to the 300°C range, a large potential market for industrial heat exists for output temperatures in the 550°C range that MSRs can provide. The petrochemical industry, steam electrolysis at high temperature, ammonia and liquid fuel production along with desalination facilities are a few of the potential applications.

— *Effective load following.* Owing to the liquid fuel, some MSR concepts offer very promising characteristics in terms of operation flexibility. As mentioned above, many MSR concepts using liquid fuel have a negative reactivity temperature coefficient. The negative reactivity feedback acts very rapidly when the heat is produced directly in the coolant, that is, when the fuel salt itself is used as the coolant. Some MSRs are well suited to load follow the off-site electric grid. This is because of the ability to rapidly adjust the power generated to the power extracted while keeping the salt temperature variations small to limit temperature fluctuations [72]. Indeed, as soon as the fuel salt temperature varies because the power extracted has changed, the quasi-instantaneous variation of the salt density modifies the power generated. Thus, the salt and reactor structure temperature excursions are limited. This property fits well with an electricity grid whose energy mix gives a larger share to intermittent sources than a conventional grid. Moreover, an MSR may adjust to the hardly foreseeable needs of the grid without requiring a control rod system. In commercial use, limitations on load following will be due not to reactor physics, but either to limitations in ramp rates of turbine technologies or to limit temperature fluctuations of the reactor. Furthermore, the use of liquid fuel, which allows the passive release of the fission product xenon, results in a substantial reduction of the xenon poisoning that severely limits load following capabilities in reactors with solid fuel and a thermal spectrum, such as water cooled reactors.

— *High resource utilization.* In many MSR concepts, the fuel cycle can be closed also for minor actinides [73], unlike in most solid fuel reactors. Accordingly, only reprocessing losses would become a waste, while all actinides from the spent fuel of existing reactors can be utilized by MSR breeders or burners in the closed cycle. Even in the open cycle, the utilization of resources in MSRs is higher than in solid fuel reactors. Excellent neutron economy, radiation stability of the carrier salt and the possibility to continuously remove fission products and adjust the fuel composition allow for higher burnup and lower initial fissile fuel enrichment.

— *Fuel qualification.* Whereas fuel qualification of any new solid fuel design requires years of testing to ensure the integrity of the solid fuel and its cladding, this fuel qualification process can be far simpler with liquid fuel. While the fuel composition and thermophysical changes across all stages of an MSR lifetime need to be tested, this work can be done without the need for extensive and time consuming irradiation programmes, as the fluoride salts (and likely chlorides as well) are not damaged or undergo radiolytic decomposition.

— *Compact form factor.* In liquid fuel designs, the liquid fuel allows for an optimal utilization of the reactor core volume. For thermal spectrum reactors, the use of liquid fuel opens the possibility of modularity and decentralized nuclear power production. The modularity and form factor, combined with the load following capabilities, make MSRs an attractive technology for the energy mix of modern power grids.

3.2. TECHNICAL CHALLENGES OF MSR TECHNOLOGY

This subsection discusses the technological challenges for deploying MSRs. Current R&D efforts for this deployment mainly focus on the areas presented here.

While MSR technology operability was demonstrated in the 1960s during the MSRE programme at ORNL, a few technical and developmental challenges lie ahead of modern MSR commercial deployment. These can be divided into three macro areas: (i) reactor physics, (ii) salt chemistry and materials science and (iii) engineering. While R&D efforts in these areas are often intertwined, the following discussion treats them separately for the sake of clarity. Challenges that are of a less technical nature, such as those related to the supply chain, licensing and regulatory aspects, are better characterized as non-technical challenges and are addressed in Section 6. However, since most challenges are ultimately related to technical issues, there is some overlap between the challenges discussed in this subsection and those in Section 6.

3.2.1. Challenges associated with reactor physics

Performance and fuel cycle requirements have been demonstrated for several MSR designs from a reactor physics standpoint. In general, the physics of MSRs is well understood and there are no fundamental scientific obstacles to the technology. However, some significant hurdles remain. One is the lack of MSR multiphysics simulation software that may be acceptable by a regulatory body. To build confidence in the software, an experimental verification is required, but regulatory approval is needed for experimental set-ups to collect data. Such licensing loopholes might be overcome with the joint efforts of MSR developers.

The low energy thermal scattering kernels for various salts are currently not available, nor are displacement per atom data for candidates of structural materials that are corrosion resistant. These gaps can be filled only via significant modelling and irradiation experiments. Furthermore, high fidelity simulations are needed to examine a potentially variable source term involving delayed neutron precursors, which pose increased requirements on reactor control mechanisms.

3.2.2. Challenges associated with salt chemistry and materials science

The most critical development challenges for MSRs revolve around salt chemistry and materials. Generally, the identification and characterization of suitable salt and material combinations for use in MSRs is an absolute priority. Albeit lengthy and resource intensive, the qualification of new structural materials for advanced reactor applications might become necessary. In this sense, the acceleration of materials discovery and testing via multiscale simulations and machine learning approaches will play a central role [74]. A better definition of the interface between materials and salt is sought to drive the selection of the appropriate material and salt candidates.

A critical issue concerning the salt phase is the accurate definition of thermophysical and thermochemical properties of fuel and coolant salts. The quality of the existing data is a concern due to the lack of quality assurance protocols and standardized experimental methods to generate these data. In general, any MSR vendor would be required to generate data specific for its salt choice, which constitutes a significant effort. A validated salt properties database providing data of fuel and coolant salts as a function of salt mixture composition, temperature and pressure can be used as a practical and efficient tool for reactor design and modelling of MSR performance. The prediction of these properties via, for example, molecular dynamics simulations, is a second independent tool that can be useful to inform engineering choices during the reactor design phase but will not suffice alone during validation by regulatory bodies.

Thermochemistry data and models that describe the speciation of the fuel, the salt, activation products and fission products will also need development, as they are necessary for performance and safety evaluations.

To mitigate the degradation of structural materials in contact with a molten salt, it is paramount to maintain a strict control of the salt composition, from fuel loading throughout the core lifetime. The compelling challenges here are the definition of manufacturing and purification protocols for the salt compositions, which are needed for a reliable supply chain for the MSR fleet. The purity of the salts needs to be high to prevent radioactive activation and excessive corrosion. The exact composition of the salts upon loading in the reactor needs to be ensured, and hence, storage and transportation protocols also need to be developed in collaboration with fuel suppliers. As an example, the infrastructure and capacity for separation of lithium and chloride isotopes are currently missing.

Some core components experience exceptionally fast degradation. In moderated (epithermal) MSR concepts, the choice of graphite as moderator poses certain limitations on the core lifetime or imposes relatively fast replacement. Graphite degradation occurs because of volume change and structural contraction followed by swelling. The process is dependent on the neutron flux, but more importantly is dependent on temperature, which is distinctively high in MSRs. This determines a relatively short lifetime for the moderator material and periodic substitution. A significant challenge here lies in the ability to extend the lifetime of graphite materials for moderation, or to implement the use of water without losing many of the technological advantages.

During operation, the salt composition has to be maintained within carefully selected limits. A key aspect to mitigate corrosion is the monitoring and control of the salt redox potential, which has been demonstrated in experiments at laboratory scale, but whose successful implementation in MSR prototypes presents significant challenges. Electrochemical methods to monitor and alter the oxidative power of the salt anions are still to be demonstrated in prototype set-ups. Electrochemical set-ups withstanding highly corrosive and irradiated environments ought to be developed and their functioning over suitable time frames validated.

In general, instrumentation to sample the salts' physicochemical and dynamic properties has already been developed for other applications, but needs to be adapted for use in MSRs, as the harsh conditions place exceptional stress on instrumentation components. Most of the instrumentation in MSRs will not be in contact with the liquid salt, implying that indirect measuring principles are to be used that are radiation tolerant. Otherwise, salt compatibility needs to be demonstrated for any instrumentation. In some cases, such as MSRs with fast spectra, the higher power densities will make in-core instrumentation impractical, increasing the burden on the development of monitoring techniques.

Furthermore, some MSR concepts (e.g. those using the thorium fuel cycle) include an on-line fuel salt reprocessing system, which can increase the fuel circuit complexity significantly. Setting aside the security concerns related to fuel inventory and proliferation, the functioning of a chemical process unit needs to be demonstrated to perform sufficiently well in the treatment of highly radioactive material, separating nuclear waste from purified fuel. While it can be said with confidence that the processing of a liquid fuel form is greatly simplified compared with the use of solid fuels, the significant development and potential operational costs of on-site processing systems for fuel salt that are proliferation resistant can be in no way trivialized.

3.2.3. Challenges associated with engineering design

Several challenges also concern the process and system engineering of MSRs. In particular, an environment at high temperature and the associated issues of thermal stresses on vessels, heat exchangers and mechanical components are a significant challenge. Fluoride salts are also fluxing agents such that any protective oxide coating of metallic surfaces is absent and makes items such as valves a significant challenge because of self-welding concerns. While early development at ORNL showed promise (e.g. the MSRE in the 1960s), much work was still required. Furthermore, methods had to be developed to eliminate the need for mechanical valves. Accurate thermohydraulic studies need to be carried out to generate data on fluid mechanics and heat transients. The transport of both heat and mass across the core is set to vary during the fuel cycle, as fission products are generated and travel across the fuel circuit. Loop test facilities utilizing full scale components should be developed to test molten salt circulation and purification at time and size scales relevant to commercialization.

For MSRs, the design of a reliable off-gas system to separate gaseous fission products from the fuel represents a key challenge, although some similarities exist with current water cooled reactors. Degassing undeniably represents a convenient feature of liquid fuels; in the past it was also considered in the form of vented pins for some solid fuel reactors. Even though not all MSR concepts rely on it, degassing benefits the neutron economy by removal of the neutron poison isotope ^{135}Xe, easily separated from the melt because of its very low solubility, and increases the plant safety as no pressure buildup occurs since this isotope does not accumulate within the core (during normal operations). On the other hand, studies from the MSRE revealed that non-negligible amounts of salt aerosols, solid particulates and mechanically volatilized metal fluoride compounds are directed outside the primary circuit by degassing operations [75, 76]. Together with other naturally volatile radionuclides (e.g. tritium, and molecular and noble gases), they form a complex feed for any purification system, which poses greater process design challenges. Established technology, such as filtering, has the obvious drawback of clogging and has limited efficacy; hence developers are devoting resources to find more comprehensive separation and purification methodologies [77, 78]. Finally, depending on the operation philosophy, degassing systems may significantly increase the void fraction present in the primary circuit, causing complex consequences for reactivity control [79].

The liquid fuel form and the removal of some fission products through the degassing or filtering both have consequences for the fuel cycle strategy applicable to MSRs. In liquid fuel, the share of fission products does not need to correspond to the fissioned mass of actinides. From the neutron economy perspective, the achievable burnup in MSRs with degassing can thus be higher than in solid fuel reactors. At the same time, the homogeneous nature of the liquid fuel means that there is no difference between average and discharge burnup. In solid fuel reactors, the burnup of discharged assemblies is up to twice as high as that of the average burnup of the remaining assemblies in the active core. From this perspective, the homogeneous fuel represents a disadvantage.

The refuelling operation in solid fuel reactors has two major fuel cycle functions: (i) to replace the fissioned mass of actinides and (ii) to reduce the amount of fission products in the active core. Whereas the actinides can be refilled to the homogeneous MSR fuel easily, the removal of the remaining fission products is more complicated. The simplest option is to discharge the fuel salt with average burnup and declare it as waste in the respective open cycle strategy. A more complex option is to separate the fission products in a reprocessing unit and recycle the actinides and carrier salt back to the reactor in a closed fuel cycle. The frequency of reprocessing can range from small daily removed volumes to entire salt discharge at once after a few years of irradiation. The advantage of frequent reprocessing is the minimal actinides load needed for operation and minimal excess reactivity and salt composition oscillations. The disadvantage is that the irradiated and reprocessed salt have the same average burnup. The advantage of the second extreme option, entire salt discharge at once, is that the discharged burnup is twice as high as the average burnup in the core over time. However, the disadvantage of the latter option is that almost double the actinide load is necessary to operate the reactor and the share of fission products in the reactor oscillates. These oscillations can be smaller and therefore more acceptable for fast spectrum MSRs.

In between the two options discussed above is a full spectrum of reprocessing frequencies. Furthermore, there are options to swiftly recycle fissile actinides from the discharged salt and to postpone the separation of fission products and remaining actinides by several years. In general, the salt reprocessing is an additional challenge and the major conceptual decision in this respect is the location of the reprocessing unit. It can be either located at each plant or centralized at a place outside the MSR plant(s). In the second case, the lower reprocessing frequency will be preferable to minimize the count of radioactive material transport and to prolong the time span for the reprocessing. Nonetheless, the truly on-line methods of degassing and filtration should be utilized whenever possible to remove part of the fission products. A quasi-batch system, which is intermediate between purely off-line (perhaps on-site) and on-line processing, might be the most appropriate for reprocessing salt fuel.

Some MSR concepts, typically foreseen as thermal breeders in the closed Th–U cycle, may require ^{233}Pa separation from the salt to improve the neutron economy (see Section 4.6). This separation needs to be done locally and frequently, and it could be an issue from a proliferation and a safeguards perspective.

As for all solid fuel reactors, the high level waste produced by MSR operation strongly depends on the fuel cycle strategy. Actinides can be declared as waste when discharged from a reactor operated in an open fuel cycle. Alternatively, some of them become waste as the reprocessing losses in a closed fuel cycle. Hence, the quantity of actinides in the waste stream strongly differs depending on the fuel cycle used; however, the difference between solid and liquid fuelled reactors is minimal. Another challenge is the chemical form of the waste. A complex treatment may be needed before the discharged salt can be temporarily or permanently stored. At the same time, there is a possibility that processes using a single step, such as vitrification, can be applied to the waste in salt form. Furthermore, owing to the degassing and filtration processes, part of the waste stream may require other treatment or storage strategies. In the case of gases, many small, separated storage volumes may be required, so that a single failure does not result in a release of the entire inventory. It is difficult to assess the final volume of waste per produced unit of energy because the molar share of actinides and the method of waste treatment vary greatly between various MSR concepts. The immobilization of waste also still needs to be assessed. Similarly, the share of fissile actinides in the waste stream will have a broad range because MSRs can be fuelled by enriched uranium, recycled plutonium or recycled TRU elements and an MSR can be developed to be operated as a breeder, converter or burner.

Depending on the concept, MSRs can also produce a substantial volume of intermediate to high level waste. There are three major sources of radioactive waste in water cooled reactors: (i) the production of fission products, (ii) the transmutation of primordial actinides and (iii) the activation of irradiated materials. There is one more major source in MSRs: the contamination of materials by actinides or fission products. Whereas in water cooled reactors the water coolant can contain traces of actinides and fission products released from leaking fuel pins, the liquid fuel in MSRs contains a very large quantity of actinides and fission products. Accordingly, solid structural materials in direct contact with the salt or the graphite moderator (for a thermal MSR using this kind of moderator) can be strongly contaminated. Since they will be irradiated at the same time, their limited lifespan can result in a substantial stream of intermediate to high level waste.

The design and manufacturing of certain components, for which little prior experience is available, are among the key challenges that MSRs pose. In particular, great efforts will be required to produce reliable and long lasting components (such as circulation pumps, heat exchangers, valves and flanges) that are in contact with the harsh salt system. Suitable instrumentation to monitor the status of these components during their lifetimes is yet to be developed and demonstrated. Remote maintenance is likely to be required because of the harsh environment, with the consequent rise in the complexity of the operations and cost. Certain MSRs will require components to be replaced periodically, which adds design requirements to allow accessibility to the core for the substitution of degraded or spent components, increasing the overall costs.

Several accident scenarios need to be evaluated before commercialization, independently of the MSR design. To develop accurate source term analysis to gauge performance and safety, developers need to describe the chemical speciation and transport behaviour of fission and activation products in the fuel circuit and within the containment area. This analysis has to be tailored for non-equilibrium conditions. However, its complexity increases because radionuclide transport across the plant systems within the containment zone changes according to several scenarios. Normal operation with and without an off-gas system, maintenance, abnormal events and accident scenarios have a different complexity [80]. To promote public confidence in the technology, developers can furthermore focus on producing accurate data about the thermodynamics and chemical speciation of fuel salt mixing with the external atmosphere, in the event of loss of coolant accidents. The claims of exceptional solvent ability and inertness of fuel salt should be verified for a variety of spillage scenarios, some general and some that are specific to an MSR design. Spillage of the fuel salt may also occur as part of monitoring, maintenance or refuelling operations; effective protocols to remediate it have to be demonstrated, although the solidification of the salt matrix is expected to simplify the cleaning of contaminated areas. Nuclide inventory considerations are to be addressed too in the event of spillage; in this regard, defining the barriers for preventing release of radioactive material for liquid fuels is also necessary. Spilling fuel salt during any operation is not especially technically challenging if a catch basin and drain tank (with adequate cooling) are provided. All known designs incorporate this feature. Conceptually, fuel salt flows like water down to the drain tank where it is safely cooled. Some uncertainty remains, however, as to which radionuclides evolve out of the fuel salt during both normal operations and spills.

Finally, the constitution of standards for various aspects of the materials qualification, such as metals and ceramics fabrication, joining and weld inspection, is also a significant challenge to deployment.

To summarize, the MSRE showed that MSRs can be safely designed and operated. The present challenge is to design and manufacture reactors with a sufficient operating lifetime to enable them to be economically viable.

For thermal spectrum MSRs, both the fine-grained graphite and Hastelloy N (nickel–molybdenum–chromium) based nickel alloys necessary to enable construction are currently available, albeit with a limited supply chain. Although these alloys are not included in major high temperature nuclear structural design codes such as Section III, Division 5 of the Boiler and Pressure Vessel Code of the American Society of Mechanical Engineers [81] or the AFCEN mechanical components code RCC MRx [82], MSRs are low pressure systems, which means that construction to other codes and standards may be possible. Furthermore, as the reactor vessel is exposed to relatively low neutron fluences, it is also possible to construct the reactor vessel using a conventional high temperature nuclear material such as stainless steel 316 (SS316) that is coated or clad to provide the required corrosion resistance. The use of coated or cladded materials is more challenging for reactor internals and piping, and hence it is likely that unless new advanced materials are used, these components will have to be routinely replaced since their operational lifetimes will be shorter than that of the reactor vessel.

The neutron fluence in fast spectrum MSRs can be significantly larger and unless sufficient secondary salt shielding is inherent in the design, neutron absorbing materials may have to be also included in the coating or cladding of the reactor vessels. For the same reason, unless novel, more radiation resistant materials can be developed, the operational lifetimes of reactor internals in fast spectrum MSRs will be short, necessitating frequent replacement and maintenance.

4. CLASSIFICATION OF MSR FAMILIES

4.1. INTRODUCTION

Molten salt reactors are a broad and diverse category of reactors, in which a molten salt plays one or more significant functions in the reactor core, such as fuel chemical form, liquid fuel carrier, coolant or moderator. All historical and recent concepts rely either on fluoride or chloride salts. Other halides or other salts (e.g. the industrially utilized nitrate salt) are not considered because one or more of their properties, such as radiolysis of the salt without recombination, high neutron absorption or activation, low solubility of fuel salt or unfavourable melting points, disqualify them for use in the reactor core. Also, the homogeneous mixture of fluorides and chlorides is not considered. It was evaluated briefly early in the ANP programme in the USA and is mentioned in Ref. [83]. Such a mixture would have high neutron absorption in moderated reactors and a softer neutron spectrum than standalone chlorides in fast reactors. Furthermore, it would strongly increase the complexity of the chemistry of the fuel salt.

The development of MSRs spans approximately seven decades, that is, since the late 1940s. Some of the conceived MSR concepts are based on a relatively simple idea, typically related to the fuel cycle arrangement, and rely only on neutronic simulations; other concepts were supported by enormous R&D efforts and their technology readiness level [84] is very high. A prominent example of a design class that is well developed is the graphite based MSR family.

Several efforts have been undertaken in the past to classify MSRs (see, for example, Refs [45, 85]). The contribution described in Ref. [85] to the International Nuclear Fuel Cycle Evaluation Project, which was supported by the IAEA, is especially relevant because it includes detailed discussion of an MSR classification. It uses the following classification parameters: method of cooling, flux intensity, number of core zones, type of fuel cycle, neutron spectrum and salt type. The method of cooling was selected as the foremost criterion and defines three major MSR groups based on the location of the heat exchange between the fuel and dedicated coolant and on whether the coolant and liquid fuel are in direct contact or not. The three groups defined were (i) internally indirectly cooled MSRs, (ii) externally indirectly cooled MSRs and (iii) internally directly cooled MSRs. For internally cooled reactors, the heat exchange between fuel and coolant takes place in the active core; for externally cooled, it is outside of the core. This classification is, however, outdated because (i) the directly cooled MSRs, where liquid fuel and coolant are in direct contact, are no longer viable designs, (ii) several classification parameters from the extensive list of possible parameters for classifying MSRs[4] were not considered at that time and (iii) some recent MSR concepts introduced non-graphite moderating materials, which are different to those used in the past ANP [86] and MSBR [8, 11] research programmes at ORNL. Unlike graphite, these moderators are not compatible with the salt and require coating or cladding material to avoid direct contact. Accordingly, they have different technological and safety requirements than those for the MSRs that are moderated by graphite. From this perspective, these concepts are similar to internally indirectly cooled MSRs, where structural material is needed in the core to separate the fuel salt from the dedicated coolant.

4.2. NEUTRONIC CHARACTERISTICS OF MAJOR CONSIDERED NUCLIDES

The taxonomy described in this section divides MSRs into three major classes according to the type of materials present in the core. The fact that graphite is the sole known moderator directly compatible with the salt plays a dominant role in the taxonomy. To provide some insight into the neutronics performance of different MSR concepts, the neutronics characteristics of major nuclides considered for MSRs are discussed here. The selected nuclides are either part of the salt or part of the moderator. Structural materials are not included. The discussion is sorted by nucleon number and does not follow the actual frequency of appearance of the isotopes in the concepts. Data for the total microscopic cross-section and the neutron capture probability of selected nuclides presented in Figs 1 and 2, respectively, were taken from the Evaluated Nuclear Data Files ENDF/B-VII.0[5] cross-section library.

---

[4] Parameters that have typically been considered for MSR classification are technology readiness level (low, medium, high); fuel state (liquid, solid); neutron spectrum (thermal, epithermal, fast, hybrid, variable); fuel cooling (in-core, ex-core); contact with coolant (direct, indirect); moderator (graphite, hydrogen based, deuterium based, beryllium based, other); number of actinides containing salts (single-fluid, two-fluid, multifluid); salt type (fluorides, chlorides, mixture); salt purpose (fissile, fertile, coolant, moderator); core criticality (critical, subcritical); primary circuit layout (loop, pool); core structure (homogeneous, heterogeneous); use of neutron leakage (reflector, blanket, multizone core); neutronics design (burner, converter, breeder, breed-and-burn); make-up fuel (enriched uranium, existing spent fuel, none); fuel cycle (open, partially closed, closed); breeding (Th–U cycle, U–Pu cycle, combination); fuel reprocessing location (in situ, ex situ); fuel reprocessing pace (as fast as possible, with years of delay, combination, no reprocessing); fuel refilling/removing (continuously, batchwise, at once, none); and power level (micro, small, large).

[5] IAEA database. US Evaluated Nuclear Data Library, issued 2006.

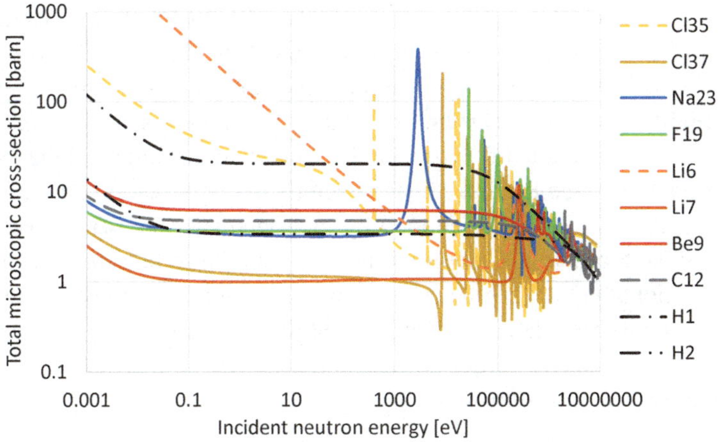

FIG. 1. Total microscopic cross-section of selected nuclides (courtesy of J. Křepel, Paul Scherrer Institute).

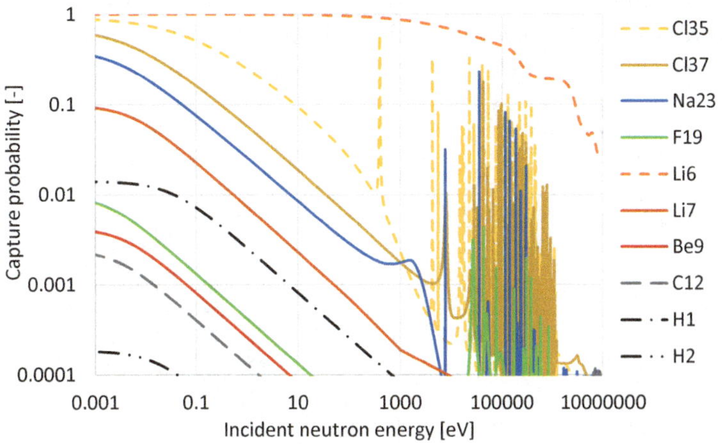

FIG. 2. Neutron capture probability of selected nuclides (courtesy of J. Křepel, Paul Scherrer Institute). [-] — no unit.

4.2.1. Hydrogen and deuterium

Two isotopes of hydrogen, ^{1}H (referred to simply as hydrogen in this publication for brevity) and ^{2}H (deuterium), offer the fastest moderation in nature. Only 14 or 20 collisions, respectively, are needed to slow down a neutron from 2 MeV to 1 eV (see Fig. 3). The total microscopic cross-section of hydrogen is much higher than for deuterium and both are without resonance (see Fig. 1). The higher cross-section of hydrogen is related to its much higher neutron capture probability (see Fig. 2). This characteristic might deteriorate the neutron economy in thermal MSRs moderated by hydrogen (see Fig. 4). However, any one of these isotopes would require structural material to be separated from the fuel salt. Optionally, they can be embedded in a compound diluted in the salt. In this case, some issues related to chemical stability may arise.

4.2.2. Lithium-6 and lithium-7

Lithium is considered as a possible cation for both fluoride and chloride salts. It has two isotopes, ^{6}Li and ^{7}Li, with natural abundances of 7.6% and 92.4%, respectively. Since they are light nuclides, only around 50 collisions are needed to slow down a neutron. In contrast to hydrogen and deuterium, both lithium isotopes have a strong resonance at around 260 keV (see Fig. 1). There is a huge difference between ^{6}Li and ^{7}Li in the neutron capture probability (see Fig. 2). Accordingly, lithium cannot be practically applied without isotopic enrichment and removal of ^{6}Li. The need is much stronger than in the case of the ^{35}Cl and ^{37}Cl isotopes, discussed below. The resonance at 260 keV is strongly exposed

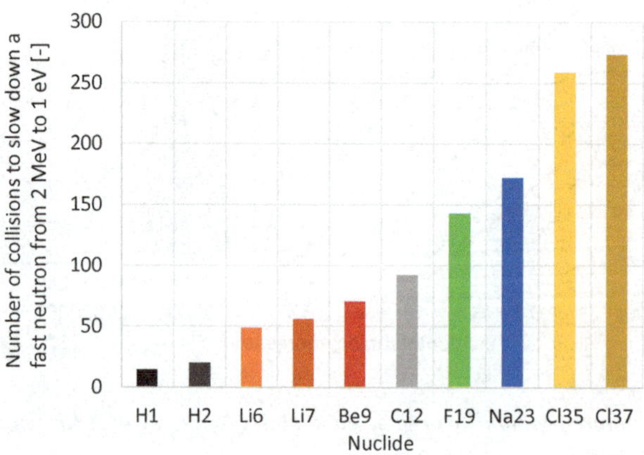

FIG. 3. Number of collisions to slow down a fast neutron from 2 MeV to 1 eV for selected nuclides (courtesy of J. Křepel, Paul Scherrer Institute). [-] — no unit.

by fast neutrons, and hence it shifts the neutron spectrum towards lower energy areas and the reactors relying on ^{7}Li have quite a low share of fast neutrons in the spectrum. Outside of this resonance, ^{7}Li has a very low total microscopic cross-section and ^{6}Li has a very high neutron capture probability. Accordingly, lithium is usually not considered as a moderator.

4.2.3. Beryllium-9 and carbon-12

Unlike lithium, ^{9}Be and ^{12}C (graphite) are considered as neutron moderators. They both have reasonably high total microscopic cross-sections (see Fig. 1). The resonances are small and in a very high energy range. Nuclides such as ^{9}Be and ^{12}C need 70 or 90 interactions with a neutron, respectively, to slow them down from 2 MeV to 1 eV. A very low neutron capture probability is therefore important (see Fig. 2). The performance of ^{12}C is very similar to ^{9}Be. However, as Fig. 4 shows, ^{9}Be outperforms ^{12}C in moderation capability. Nonetheless,

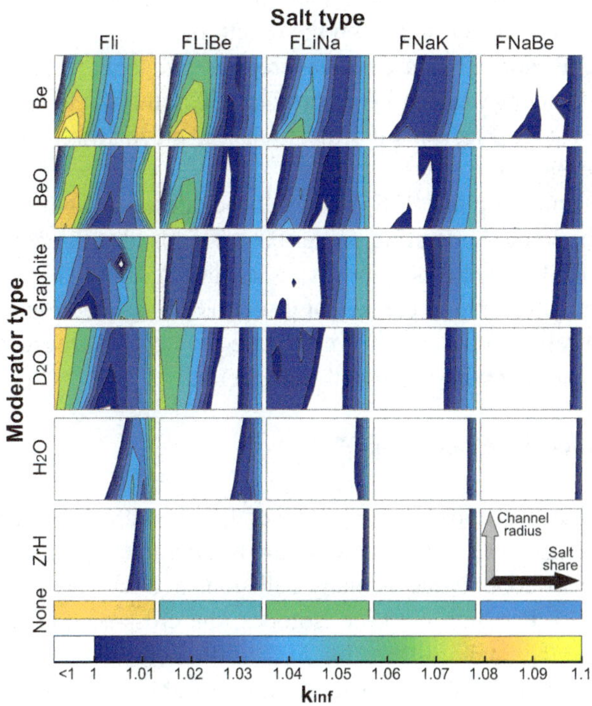

FIG. 4. Equilibrium k_{inf} (neutron multiplication factor) as a function of carrier salt type, moderator type, salt share in the core and salt channel radius (reproduced from Ref. [87] with permission).

since it is not directly compatible with the salt, ^{9}Be needs cladding or coating, which will deteriorate the neutron economy of the reactor. Accordingly, graphite is considered in more MSR concepts. Its features are therefore discussed in a dedicated subsection below (see Section 4.2.6).

4.2.4. Fluorine-19 and sodium-23

Fluorine and sodium have only one stable isotope, ^{19}F and ^{23}Na, respectively. Fluorine-19 is considered as the major component of many coolants and fuel salts. It has a similar total microscopic cross-section to ^{23}Na, which is considered as a coolant for fast reactors with solid fuel and as a component of the salt in several MSRs. The cross-section of these two nuclides has several resonances. There is one dominating broad resonance for both nuclides: ^{23}Na has a broad resonance with substantial amplitude at 3 keV, whereas ^{19}F has a broad resonance at around 100 keV but with a smaller amplitude (see Fig. 1). Similarly, as for ^{7}Li, these resonances shift the neutron spectrum to lower energy areas. The occurrence of the ^{19}F resonance at higher energies is exposed to higher neutron flux and thus has a stronger impact on the spectrum.

Neither ^{19}F nor ^{23}Na are considered as moderators. For both these nuclides, an interacting neutron needs around 150 collisions to slow down from 2 MeV to 1 eV (see Fig. 3). Fluorine-19 has a much lower neutron capture probability (see Fig. 2).

The frequently applied $^{7}LiF–BeF_2$ salt includes a combination of ^{19}F, ^{7}Li and ^{9}Be and has some moderating power [11]. This is also the reason that early homogeneous MSFRs were not always called fast reactors [12]. The ^{7}LiF salt without ^{9}Be is not moderating per se, but it softens the fast neutron spectrum. The MSFR concept relying on this salt is considered to be a fast reactor; however, it has the softest fast spectrum among all fast reactors.

4.2.5. Chlorine-35 and chlorine-37

Chlorine has two stable isotopes, ^{35}Cl and ^{37}Cl , with natural abundances of 76% and 24%, respectively. Both isotopes are heavier than ^{19}F and the interacting neutron thus needs about twice as many collisions to slow down from 2 MeV to 1 eV (see Fig. 3). For both chlorine isotopes, the cross-section resonances are quite narrow and the total cross-section in the fast energy range is lower than for ^{19}F and ^{23}Na (see Fig. 1). The neutron capture probability is higher for ^{35}Cl than for ^{37}Cl. However, nuclear data libraries for this nuclide show some discrepancies and this probability may therefore be revised. The difference in neutron capture probability between these two isotopes is nonetheless much smaller than the difference between ^{6}Li and ^{7}Li.

Reactors that use chlorine isotopes have a very hard spectrum. This is because of the low total microscopic cross-section of chlorine and the high number of collisions needed to slow down the neutrons. The higher capture probability in the thermal area (see Fig. 2) practically disqualifies chlorides for application in the thermal spectrum. Still, they may be considered for cases with a relatively relaxed neutron economy, such as in waste burners that can profit from the chemical features of chlorine.

4.2.6. Graphite moderator

Graphite is the only solid moderator directly compatible with the fluoride salts considered for thermal MSRs. Since it is the most frequently appearing moderator in MSR concepts, it defines one class of the taxonomy and its features are described here in more detail:

- In power reactors, graphite usually has a higher temperature (700°C) than moderators based on ^{1}H or ^{2}H. Hence, the Maxwellian spectrum peak is shifted towards higher neutron energies than in reactors using the latter two moderators. This feature, in combination with a fission resonance of ^{233}U slightly above this peak, can result in a positive temperature reactivity effect of the graphite moderator. Nonetheless, it depends on the fuel composition and core size, because the spectral shift is competing with increased neutron leakage, which is caused by the high neutron energy of the Maxwellian spectrum peak.
- The volumetric share of salt in the graphite moderated MSR for optimal neutronics performance is typically around 13%. In the Th–U closed fuel cycle, the reactor often acts as undermoderated. Fuel salt removal from the core of undermoderated reactors results in positive reactivity. This positive reactivity effect is inevitably connected with the fuel salt temperature increase and thus with stronger negative reactivity insertion due to the Doppler effect. The total effect is therefore negative.
- The $^{7}LiF–BeF_2$ salt has some moderating power, but it also has higher neutron absorption than graphite. Therefore, the graphite moderated MSR could outperform the homogeneous reactor with $^{7}LiF–BeF_2$ salt [8, 11], from the perspectives of conversion ratio and doubling time.
- To slow down a neutron from 2 MeV to 1 eV with a graphite moderator, 92 collisions are necessary. At the same time, it has a 10 times smaller scattering cross-section and 200 times smaller capture cross-section than water. Accordingly, the migration area for neutrons in a graphite moderated MSR can be twice as large as for a homogeneous fluoride fast MSR. Hence, the treatment of neutron leakage in graphite moderated MSRs is of high

importance. To achieve breeding, often features such as a blanket, reflector, layout with two zones or a bulky core, or a combination of these features, need to be used.

— The specific neutron moderating process in graphite is sensitive to neutron parasitic captures. This has two consequences. First, the parasitic capture by the graphite becomes significant before the well moderated state is reached. Graphite moderated MSRs always act as somewhat undermoderated. Second, the graphite for reactor use needs to be extremely clean, because impurities (e.g. boron) may strongly deteriorate its performance.
— The fissile to fertile isotope ratio in an MSR moderated by graphite with a Th–U closed cycle is around 0.015. Together with the optimal 13% salt share in the core, this ratio results in a substantially smaller mass of fissile material load than in a homogeneous fluoride fast MSR (ratio of 0.15 and 100% salt share in the core). The initial fissile load is thus lower for graphite moderated MSRs. At the same time, a reactor with very low fissile content in the core is more sensitive to the parasitic absorption of neutrons by fission products.

4.3. TAXONOMY OF MSRs

The development of an MSR taxonomy can consider many parameters and options. Similar to animal taxonomy, some of the parameters might or might not be important for its structure. The taxonomy developed in this publication to classify major MSR concepts is partly inspired by a previous classification provided in Ref. [85], but the technological similarities play a more important role in this taxonomy. The taxonomy covers most of the concepts that have been proposed to date, and has three layers.

The first layer of the taxonomy is organized according to three major classes of MSRs:

— Class I: Graphite based MSRs (see Section 4.4.1);
— Class II: Homogeneous MSRs (see Section 4.4.2);
— Class III: Heterogeneous MSRs (see Section 4.4.3).

Since some MSRs, usually specific or smaller power concepts, can be on the border between two classes or are not fully congruent with a class definition, a fourth class is introduced to address them:

— Class IV: Other MSRs (see Sections 4.4.4 and 4.11).

Each class has its own specific features:

- *Class I: Graphite based MSRs.* These are characterized by the presence of graphite in the core and employ exclusively fluoride salt and a thermal spectrum. Since the graphite is directly compatible with the salt, no other structural material is needed in the core. This class has the highest technology readiness level because of the successful operation of the MSRE, and for some concepts because of their similarity to high temperature reactors.
- *Class II: Homogeneous MSRs.* These are characterized by the absence of structural materials in the core that separate the fuel salt from a coolant or a moderator. The core is thus filled solely by the fuel salt. Optionally, structural materials (e.g. baffles for flow direction) can be present in the core; however, they do not have the separation function. The technology readiness level of this class depends mainly on the fuel salt, and for some of the MSRs using similar salts to those of the MSRE or the MSBR, the technology readiness level can be higher than for others.
- *Class III: Heterogeneous MSRs.* These are characterized by the presence of structural materials in the core that separates the fuel salt from a dedicated coolant or from a moderator not directly compatible with the fuel salt. The technology readiness level of this class depends also on the separation material, and it can be lower than for the previous two classes because these materials are typically not yet qualified.
- *Class IV: Other MSRs.* These do not fit into the three major classes and generally have the lowest technology readiness level. Class IV: 'Other MSRs' is discussed as a single entity in Sections 4.4.4 and 4.11.

The second layer of the taxonomy is defined by the six major MSR families:

- Family I.1: Fluoride salt cooled reactors (see Section 4.5);
- Family I.2: Graphite moderated MSRs (see Section 4.6);
- Family II.3: Homogeneous fluoride fast MSRs (see Section 4.7);
- Family II.4: Homogeneous chloride fast MSRs (see Section 4.8);
- Family III.5: Non-graphite moderated MSRs (see Section 4.9);
- Family III.6: Heterogeneous chloride fast MSRs (see Section 4.10).

The six major families defined for the second layer of the taxonomy often share many features across the classes of the first layer; however, they usually differ within one class. Table 1 offers a comparison of the significant parameters for the six families. Appendix V provides a brief description of several current MSR concepts, organized by the six major families.

TABLE 1. COMPARISON OF PARAMETERS FOR THE SIX MAJOR MSR FAMILIES

Class	I. Graphite based MSRs		II. Homogeneous MSRs		III. Heterogeneous MSRs	
Family	I.1. Fluoride salt cooled reactors	I.2. Graphite moderated MSRs	II.3. Homog. fluoride fast MSRs	II.4. Homog. chloride fast MSRs	III.5. Non-graphite moderated MSRs	III.6. Heterog. chloride fast MSRs
Fuel state	Solid	Liquid	Liquid	Liquid	Liquid	Liquid
Spectrum	Thermal	Thermal	Fast	Fast	Thermal	Fast
Salt type	Fluorides	Fluorides	Fluorides	Chlorides	Fluorides	Chlorides
Neutronics performance	Burner, converter	Burner, converter, breeder	Burner, converter, breeder	Burner, converter, breeder, breed-and-burn	Burner, converter, breeder	Burner, converter, breeder, breed-and-burn
Actinides	Enriched U, TRU, Th as semi-inert matrix	Enriched U, TRU, closed Th–U cycle	Enriched U, TRU, closed Th–U cycle and U–Pu cycle	Enriched U, TRU, closed Th–U cycle and U–Pu cycle	Enriched U, TRU, closed Th–U cycle	Enriched U, TRU, closed Th–U cycle and U–Pu cycle
Irradiation induced issues	Limited burnup of fuel in graphite matrix	Limited graphite moderator lifespan	Limited vessel lifespan	Limited vessel lifespan	Limited vessel and structural material lifespan	Limited vessel and structural material lifespan
Fuel extensive pumping	No	Yes	Yes	Yes	Yes (no if cooled by moderator)	No

TABLE 1. COMPARISON OF PARAMETERS FOR THE SIX MAJOR MSR FAMILIES (cont.)

Class	I. Graphite based MSRs		II. Homogeneous MSRs		III. Heterogeneous MSRs	
Family	I.1. Fluoride salt cooled reactors	I.2. Graphite moderated MSRs	II.3. Homog. fluoride fast MSRs	II.4. Homog. chloride fast MSRs	III.5. Non-graphite moderated MSRs	III.6. Heterog. chloride fast MSRs
Heat transport medium	Fluoride coolant salt	Fluoride fuel salt	Fluoride fuel salt	Chloride fuel salt	Fluoride fuel salt (or moderator)	Molten salt or lead coolant
Primary heat exchange	In-core	Ex-core	Ex-core	Ex-core	Ex-core (in-core)	In-core

The third layer of the taxonomy is the classification of each family into different types — see Sections 4.5–4.10. Each subsection introduces a family and the major reactor types.

Figure 5 illustrates the three layers of the MSR taxonomy as (a) a block diagram and (b) a circular chart transposed from the block diagram.

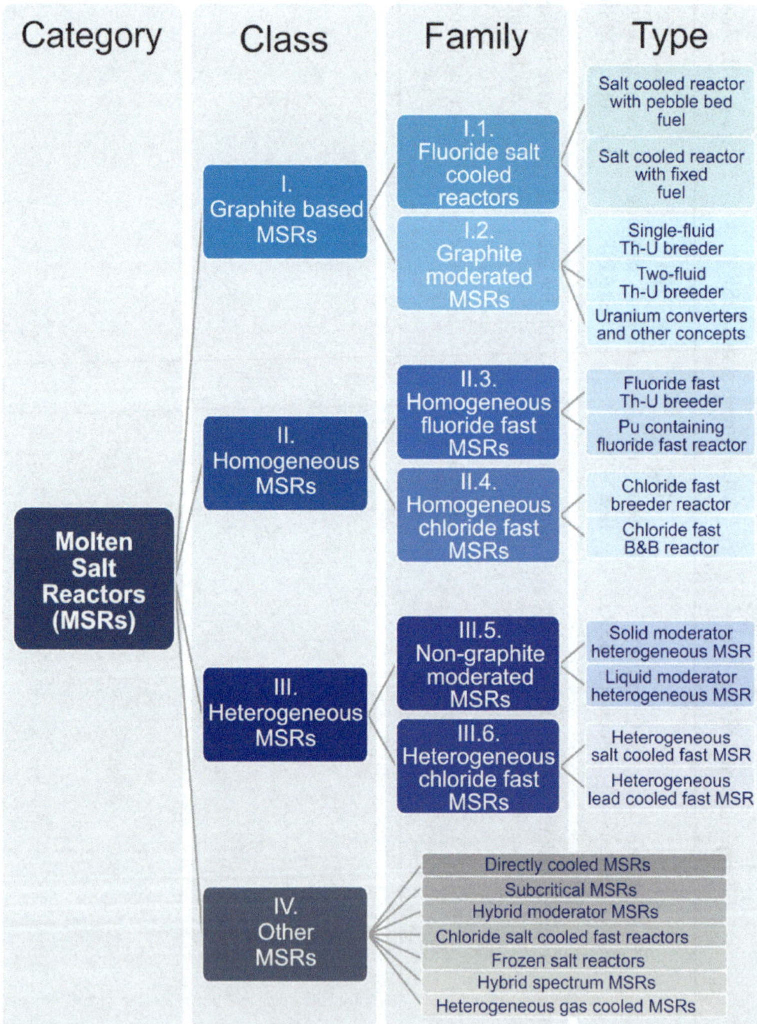

FIG. 5. (a) The three layers of the MSR taxonomy as a block diagram (courtesy of J. Křepel, Paul Scherrer Institute).

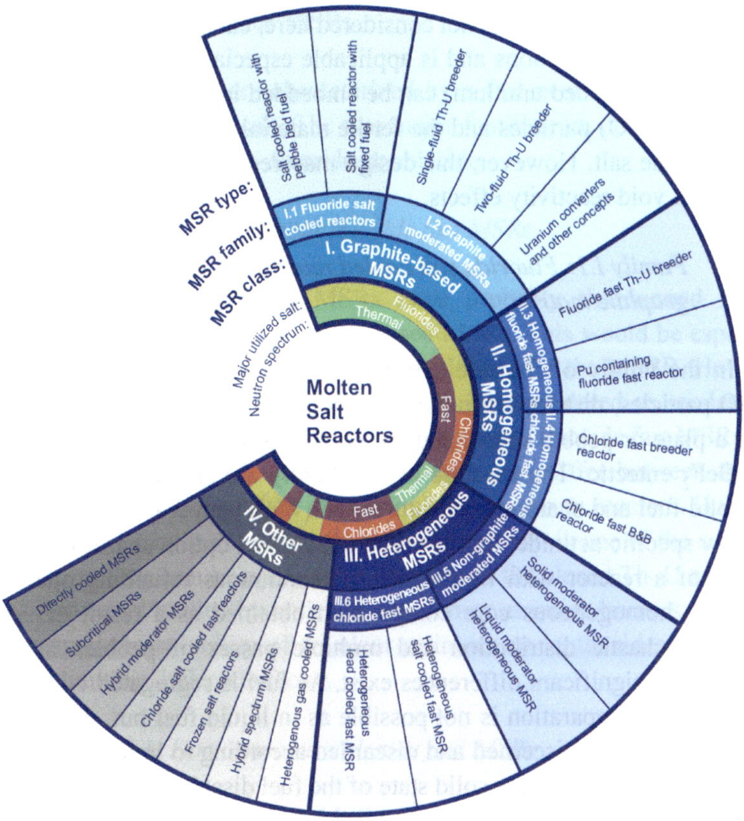

FIG. 5. (b) The three layers of the MSR taxonomy as a circular chart.

4.4. CLASSES AND FAMILIES

4.4.1. Class I: Graphite based MSRs

Graphite based MSRs are moderated by graphite and can be further subdivided into two families according to the location of the fuel. In both graphite based families, graphite has a limited lifespan due to irradiation. In reactors cooled by fluoride salt, graphite acts as a fuel matrix and is naturally exchanged when the fuel is depleted. The volume of discharged spent fuel is typically much higher than for existing light water reactors, but its form is suitable for final repository. In a graphite moderated MSR, the graphite has to be regularly replaced solely because of its limited lifespan. It generates voluminous activated waste.

class are defined by the material that needs to be separated — the moderator or the dedicated coolant.

4.4.3.1. Family III.5: Non-graphite moderated MSRs

Non-graphite moderated MSRs utilize fuel in the form of fluoride salts diluted in fluoride carrier salt and are characterized by a thermal spectrum and the presence of structural materials in the core. There are some moderators (e.g. based on beryllium or deuterium) that can ensure better neutronics performance than graphite (see Fig. 4). Such concepts can be potentially better breeders or waste burners than graphite moderated MSRs. Nonetheless, the performance will strictly depend on the characteristics of the separating structural material. It appears that only silicon carbide (SiC) can preserve the excellent neutronic performance features. However, the lifespan of some of these moderators and structural materials can be significantly longer than graphite, hence producing much less waste. The lifespan of these structural materials determines the lifespan of the reactor. Since the technology readiness level of this family depends also on the separation material, it can be lower than for other MSRs. However, the only operational experience of an MSR besides the MSRE is the Aircraft Reactor Experiment, which belongs to this class and is operated with a beryllium oxide moderator and fluoride fuel salt, separated by Inconel[6] tubes.

Heterogeneous fast reactors using fluoride fuel salt and dedicated coolant have inferior neutronic performance than both heterogeneous chloride fast MSRs and homogeneous fluoride fast MSRs and are not considered here.

4.4.3.2. Family III.6: Heterogeneous chloride fast MSRs

Heterogeneous chloride fast MSRs have their chloride based fuel separated from the dedicated coolant by the structural material. The coolant can be another salt or liquid metal, typically lead. So far, none of the concepts consider sodium as the coolant. In the salt case, the coolant is usually chloride salt, which does not carry actinides. In some historical concepts, a blanket salt was used to cool the core. Owing to the structural material and dedicated coolant presence in the core, these reactors have higher parasitic neutron absorption in the core than homogeneous chloride fast MSRs. At the same time, the fuel salt does not serve as the primary heat transport medium and is therefore not extensively pumped out of the core. Owing to the structural materials, which are often not yet qualified, the technology readiness level of these reactors is lower.

---

[6] Nickel–chromium based high performance alloys that are oxidation and corrosion resistant.

4.4.4. Class IV: Other MSRs

The class 'Other MSRs' has to be introduced because not all MSR designs or concepts documented in the literature can be covered by the three major classes defined above. Many of these concepts are historical or discontinued, have the lowest technology readiness level and have adopted a non-standard design approach. Some of the types included in this class are represented by more than one concept; for instance, directly cooled MSRs. They were proposed to avoid extensive fuel salt pumping through the high surface heat exchangers or the structural material presence in the active core. The high heat exchange surface is needed because the salts have low thermal conductivity, which is comparable to water. Since the direct contact leads to strong contamination of the direct coolant, all concepts of this type have been discontinued. Subcritical MSRs are another example of a type of MSR that is represented by more than one concept. These were proposed during the 1990s as a liquid fuel option of the accelerator driven system. Since the only major function of the costly external neutron source is criticality safety, it can be avoided in inherently safe MSRs. The remaining types are typically based on single, albeit rather hypothetical, concepts that do not fit one of the six defined families or which are sometimes not meant for large power production. Gas cooled MSRs or frozen salt MSRs are examples.

4.5. FAMILY I.1: FLUORIDE SALT COOLED REACTORS

Fluoride salt cooled reactors feature solid fuel, mostly based on coated fuel particles embedded in or surrounded by a graphite moderator, cooled by fluoride salt. Similarly, as in the case of high temperature reactors that are gas cooled, the fuel particles can be dispersed in a graphite pebble bed or concentrated in sealed plates or compacts and surrounded by graphite blocks. The technology of coated fuel particles was already suggested in 1957 [88] for high temperature reactors, at around the same time as the graphite moderated MSR became a major option for ORNL. Early designs for the fluoride salt cooled high temperature reactor (FHR) were based on prismatic blocks and TRISO-bearing compacts replicating the prismatic block geometry of high temperature gas cooled reactors, or HTGRs. However, such an option was quickly discarded as the average density of the fuel blocks would be lower than that of the salt, causing the blocks to float and making some operations, such as refuelling, very challenging. Designs featuring plate, pebble or stringer fuel have instead gained popularity.

Fluoride salt cooled reactors combine two previously existing technologies: (i) reactor coolants based on molten fluoride salts operating at high temperature and low pressure, developed in the ANP programme of the 1950s in the USA, and

(ii) the subsequent breeder reactor programme that was based on molten salt and solid fuel consisting of coated particles (TRISO) embedded in a graphite matrix originally developed for reactors operating at high temperature and cooled by gas in Member States such as Germany and the USA. As fuel is based on TRISO particles embedded in or surrounded by a graphite moderator, fluoride salt cooled reactors are characterized by a soft neutron spectrum. This feature restricts the choice for coolant to neutronically favourable fluoride salts. Among those, the $^{7}LiF–BeF_2$ (2:1) eutectic remains the preferred option based on neutronics (if enriched in ^{7}Li at 99.995% or a higher percentage) and thermophysical properties. Although graphite is the main moderator, the salt's light components also contribute to the slowdown of the neutrons. Moreover, to obtain negative coolant temperature or coolant void reactivity feedback, it is necessary to have sufficient moderation by the salt. Therefore, the coolant scattering cross-section should have a stronger impact on the neutronics than the capture cross-section. When the salt temperature increases (causing the salt's density to decrease) or when a void appears in the salt, two competing reactivity changes occur: (i) a positive change due to reduced absorption of neutrons in the salt and (ii) a negative effect due to the spectrum hardening from reduced moderation in the salt. For the latter effect to be dominant and the overall feedback to be negative, neutron moderation in the salt has to be sizeable and neutron capture minimal [89]. Therefore, (i) only fluoride salts are suitable for the coolant, (ii) ^{6}Li is excluded for its high capture cross-section and (iii) ^{9}Be is included for its high scattering cross-section.

As mentioned in Section 4.4.1.1, the low specific fuel density and the solid state of the fuel disable sustainable breeding in these reactors. Accordingly, they rely on enriched uranium (typically >8% ^{235}U); optionally they can act as TRU burners. Using fluoride salt as the coolant enables higher specific power than in a high temperature reactor cooled by helium. On the other hand, it results in higher decay heat that is removed by coolant natural circulation rather than passive radiation as in the case of a high temperature reactor.

4.5.1. Major reactor types

The major reactor types in this family are described in this subsection and the major concepts are introduced in Table 2.

4.5.1.1. Salt cooled reactor with pebble bed fuel

A pebble bed reactor concept cooled by salt is first mentioned in a paper from 1983 [69], although no development followed. Currently, the most common pebble bed variant is the FHR initially proposed in the early 2000s [93].

TABLE 2. FAMILY I.1. FLUORIDE SALT COOLED REACTORS

Class	Family	Type	Conceptual design	Designer	Appeared in year	Ref.
I.	1.	Salt cooled reactor with pebble bed fuel	FCSR	Kurchatov Institute	1983	[69]
			Pebble bed fluoride salt cooled high temperature reactor (PB-FHR), Kairos Power Fluoride Salt Cooled High Temperature Reactor (KP-FHR)	University of California, Berkeley, Kairos Power	2016	[90]
		Salt cooled reactor with fixed fuel	Advanced high temperature reactor (AHTR), Small modular advanced high temperature reactor (SmAHTR)	ORNL	2010	[91]
			Advanced gas cooled reactor–fluoride salt cooled high temperature reactor (AGR-FHR)	Forsberg	2019	[92]

The design based on a pebble bed was developed at the University of California, Berkeley, as part of a multiuniversity project (together with Massachusetts Institute of Technology and the University of Wisconsin Madison). The concept, a pebble bed FHR named Mk1 PB-FHR [90], uses annular pebbles that are 3 cm in diameter and contain fuel (TRISO particles) in a thin layer at the outer region of the pebble to limit the operating temperature of the particles. Pebbles are slightly buoyant in the salt and therefore they are inserted from the bottom of the core and extracted at the top, which is opposite to the direction in pebble beds cooled by gas. Typical of a system based on a pebble bed, the Mk1 PB-FHR features on-line refuelling that enables low reactivity excess. Commercial deployment of a PB-FHR is being pursued by Kairos Power and its proprietary design is known as Kairos Power Fluoride Salt Cooled High Temperature Reactor, or KP-FHR.

Owing to the low heavy metal content of fuels based on particles, FHRs require 10–20% uranium enrichment. The better heat transfer capabilities of the salt allow these reactors to achieve a power density that is double that of reactors cooled by gas (about 20 W/cm^3).

4.5.1.2. *Salt cooled reactor with fixed fuel*

Multiple MSR designs have been proposed that rely on stationary fuel. At ORNL, a study on the use of plate fuel embedding TRISO particles in the development of the small modular advanced high temperature reactor, or SmAHTR, cooled by a fluoride salt [91] was conducted. As this was mostly a feasibility study, no further development has occurred in recent years. An early study evaluated the use of conventional fuel pellets or stringer fuel and found such an option viable, although more in-depth studies were not pursued [94]. A recent design variant based on the advanced gas cooled reactor (often abbreviated to AGR), which was developed in the UK, employs cylindrical fuel stringers made of graphite within which are channels for coolant and channels containing alternative fuel [92]. The fuel stringers contain TRISO bearing compacts and are housed in a graphite core. The advanced gas cooled reactor–fluoride salt cooled high temperature reactor, or AGR-FHR, aims to benefit from the long operational experience of the advanced gas cooled reactor and to adapt its technology of refuelling at high temperature.

4.6. FAMILY I.2: GRAPHITE MODERATED MSRs

The reactor concepts belonging to this family use graphite as a moderator, hence the neutron spectrum is thermal. The fuel is liquid in the form of actinide fluorides dissolved in a fluoride carrier salt or their mixture, for instance, BeF_2, ^7LiF or NaF. The fuel simultaneously has a coolant function, so it not only conveys the fission heat out of the core, but also cools the graphite moderator. From the perspective of the fuel cycle, this reactor can be operated as a breeder in a closed Th–U fuel cycle, as a converter fuelled by enriched uranium or as a TRU burner in an open or closed fuel cycle. This reactor type has a prominent position from a historical perspective and the highest experimental and theoretical efforts so far have been invested in this technology.

The fuel and the moderator are in direct contact. The capability of graphite to withstand this contact was recognized in the early pioneering time of the nuclear era. During the ANP programme at ORNL between 1947 and 1961 [95], reactors moderated by graphite were excluded as an option for nuclear powered jet engines because of their size. The associated shielding weight penalty was too big [86]. Nonetheless, the benefits of graphite features were recognized, and graphite was used or proposed to be used as outer shielding of the reactor moderated by beryllium [96, 97]; as graphite packed seals [98]; as graphite powder for extinguishing material [99]; as vessels lined with graphite for fuel salt preparation [99]; as a graphite anode for electrochemistry in chlorides [100]; or in the compound UF_6–C_7F_{16} [101]. Tests were performed during the ANP programme on the compatibility of graphite of high density with other materials [102] and changes in a chemical system's behaviour resulting from the presence of graphite [103]. Many other studies were conducted later [104], including carburization of Inconel and INOR-8, irradiation time and radiation damage of graphite and graphite permeability for molten salts.

The three major requirements for a graphite moderator, which are still valid today, are recapitulated in Ref. [105]:

(1) It needs to withstand neutron irradiation for a sufficiently long time.
(2) It needs to have pores small enough that capillary forces exclude fuel salt.
(3) It needs to have a low enough permeability to gases to keep down the absorption of xenon.

In around 1957, ORNL decided to develop a civilian MSR breeder. In Refs [8, 11], the Th–U cycle is recognized as an alternative to the U–Pu cycle, and a fluoride salt mixture consisting of BeF_2, ^7LiF (or NaF), UF_4 and ThF_4 was proposed for both a homogeneous fluoride fast MSR and a graphite moderated MSR. The performance of both reactor concepts was comparable; the moderated reactor based on a ^7LiF–BeF_2 salt even slightly outperformed the fast reactor [106, 107] from the perspectives of conversion ratio and doubling time. At that time, other moderators were excluded because they were not compatible with the fuel salt, and the parasitic neutron absorption by their respective claddings would disable sustainable breeding in the Th–U cycle [11, 106]. The Th–U cycle generates the intermediate isotope ^{233}Pa, which decays with 27 days half-life to ^{233}U and causes the parasitic capture of neutrons. The parasitic capture of neutrons is especially pronounced in the thermal spectrum. Hence, it is desirable to remove this isotope or to minimize this capture in a graphite moderated MSR. To achieve breeding in the Th–U closed cycle with a graphite moderated MSR, several design solutions are needed. These include leakage minimization with a blanket, reflector, undermoderated outer core zone, bulky core, or a combination of these features; intensive salt purification from fission products; and minimization of the ^{233}Pa parasitic capture of neutrons. Other considerations related to graphite that are relevant for MSR application are the following:

— Potentially positive effect on reactivity due to temperature;

— Large thermal inertia;
— Large cores;
— Slightly undermoderated cores;
— Limited lifespan of graphite due to irradiation;
— Penetration by salt and gases;
— Shrinkage and growth under irradiation;
— Radioactive graphite waste;
— Low fissile inventory;
— Need for intensive salt cleaning and ^{233}Pa removal in breeder reactor concepts;
— Complexity of reactor concepts using two fluids;
— Reduced neutron fluence of the reactor vessel due to the presence of graphite reactor;
— Neutronic compatibility of graphite with moderating $^{7}LiF–BeF_2$ salt;
— Lower salt melting temperature due to BeF_2.

4.6.1. Major reactor types

The major reactor types in this family are described in this subsection and the major concepts are introduced in Table 3.

TABLE 3. FAMILY I.2. GRAPHITE MODERATED MSRs

Class	Family	Type	Conceptual design	Designer	Appeared in year	Ref.
I.	2.	Single-fluid Th–U breeder	Molten salt breeder reactor (MSBR)	ORNL	1971	[108]
			FUJI	Furukawa	1992	[109]
			Actinide Molten Salt Transmuter (AMSTER)	EDF	2000	[110]
			Thorium molten salt reactor (TMSR)	SINAP	2013	[111]
		Two-fluid Th–U breeder	Two-fluid molten salt breeder reactor (MSBR2f)	ORNL	1970	[112]

TABLE 3. FAMILY I.2. GRAPHITE MODERATED MSRs (cont.)

Class	Family	Type	Conceptual design	Designer	Appeared in year	Ref.
I. (cont.)	2. (cont.)		Liquid fluoride thorium reactor (LFTR)	Flibe Energy	2011	[113]
			Stable Salt Reactor-Thorium (SSR-Th)[a]	Moltex Energy	2018	[114]
		Uranium converters and other concepts	Molten Salt Reactor Experiment (MSRE)	ORNL	1965	[22]
			Molten Salt Breeder Experiment (MSBE)	ORNL	1967	[115]
			Molten salt demonstration reactor (MSDR)	ORNL	1971	[116]
			Denatured molten salt reactor (DMSR)	ORNL	1980	[34]
			FUJI	Furukawa	1987	[55]
			Integral Molten Salt Reactor (IMSR)	Terrestrial Energy	2013	[117]
			ThorCon	ThorCon	2015	[118]
			Stable Salt Reactor-Uranium (SSR-U)[a]	Moltex Energy	2018	[114]

[a] The SSR-Th and SSR-U as reactors moderated by graphite belong to the respective reactor types. However, they also use metallic structural materials to separate the fissile and fertile salts or the fuel and coolant salts, respectively. As such, they have some common features with the heterogeneous MSRs.

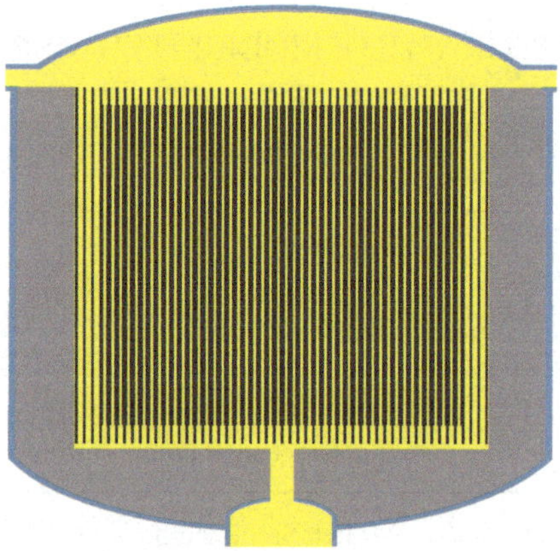

FIG. 6. Simplified figure of the graphite moderated two zone MSBR design proposed at Oak Ridge National Laboratory in 1971 (courtesy of J. Křepel, Paul Scherrer Institute).

4.6.1.1. Single-fluid thorium–uranium breeder

Since it is desirable to remove ^{233}Pa and fission products because of their parasitic neutron capture, a design with a single fluid was proposed when new techniques were developed in 1971 for the extraction of ^{233}Pa using liquid bismuth and of rare earth fission products from a salt with mixed fertile and fissile isotopes [119]. The resulting MSBR design (see Fig. 6) with a single fluid [108] had two zones with a salt to graphite ratio of 13% and 37% in the central fissile and in the peripheral fertile zones, respectively. This design became an inspiration for an entire family of MSBR concepts, such as Japan's FUJI [109, 120], France's AMSTER [110] and China's TMSR [111]. The complexity of designs with two fluids can be avoided by using a single fluid. The major drawbacks, typical for an MSR moderated by graphite, cannot be avoided and include the limited lifespan of graphite, the requirement for ^{233}Pa extraction and the need for the intensive removal of fission products. Since the salt share in the peripheral fertile zones was around 37% in the ORNL MSBR design, the graphite pebbles were considered as an alternative in one of the concepts [121]. Nonetheless, pebbles are not suitable for the core region, which requires a much lower salt share to obtain an optimal fuel to moderator ratio.

4.6.1.2. Two-fluid thorium–uranium breeder

Historically, the first MSR concepts using graphite were based on a layout with two fluids. The separation of the fertile salt containing thorium from the fissile salt has some neutronic and reprocessing advantages. Nonetheless, for breeding to occur, this separation can only be implemented with material with low absorption of neutrons (e.g. graphite). The MSBR of 2225 MW(th) was a reference Th–U breeder concept that was initially proposed with two fluids until 1968 at ORNL when design efforts transitioned to a single-fluid design [112, 122, 123]. The fuel salt was separated from the blanket salt by graphite tubes. A similar solution is nowadays adopted by the liquid fluoride thorium reactor, or LFTR, concept [113]. The presence of two fluids in the core results in a complex design. This complexity was underlined when the limited graphite lifespan due to irradiation was recognized. Accordingly, regular and remote graphite replacement is required to ensure the tightness of the barrier between the two fluids [115]. The Stable Salt Reactor with thorium fluoride based coolant, or SSR-Th, concept in Ref. [114] is also included within this reactor type, even though it uses structural material to separate the fissile and fertile salts.

4.6.1.3. Uranium converters and other concepts

The MSRE was designed as a burner of highly fissile fuel [22] with a very high ratio of fissile actinides. Even though it was developed in the framework of the MSBR programme, thorium was never used in this reactor. In Ref. [115], the Molten Salt Breeder Experiment, or MSBE, was proposed as a reactor concept with thorium and two fluids. Later, a molten salt demonstration reactor (also known as MSDR), with 300 MW(e) was discussed with an industrial consortium and presented in Ref. [116]. The last major concept from this category proposed by ORNL was the denatured molten salt reactor, or DMSR. The addition of ^{238}U addressed the proliferation risk of ^{233}U. The ^{233}Pa parasitic neutron capture was minimized by low specific power, without the need for removing this isotope. The low specific power also enabled 30 years of operation without exchanging the graphite [34]. As a trade-off, the active core diameter was 830 cm and 1000 cm without and with reflector, respectively. The enormous salt volume in the large denatured MSR core provided a fuel reserve for 30 years of operation without extensive removal of fission products.

Nowadays, there are several converter concepts moderated by graphite, such as FUJI [55], the Integral Molten Salt Reactor (IMSR) [117] and ThorCon [118]. Without the self-sustaining breeding in a closed cycle, a design margin is provided to optimize other parameters. Accordingly, the IMSR burns enriched uranium in a compact integral layout. The lifespan of graphite, of around seven years, matches the lifespan of the vessel material. The fuel cycle for ThorCon includes both enriched uranium and thorium. Hence the same denaturation as in the denatured MSR can be used; however, the core would probably be more compact. The FUJI MSR designed in Japan had several versions, and one was dedicated to a high burnup cycle based on uranium. The SSR-U reactor in Ref. [114] is also included within this reactor type, even though it uses structural material to separate the fuel and coolant salts.

4.7. FAMILY II.3: HOMOGENEOUS FLUORIDE FAST MSRs

The reactor concepts belonging to this family do not use any moderator, making the neutron spectrum fast. The fuel is liquid in the form of actinide fluoride and the carrier salt is usually ^7LiF even if other options with BeF_2, NaF and KF are possible. The fuel mixture acts also as a coolant and thus the homogeneous fluoride fast MSR belongs to the externally indirectly cooled MSRs. The main options for the fuel cycle are based on thorium (Th–U and Th–TRU), allowing the reactors in this category to be suitable for breeder and burner applications. Utilization of enriched uranium or of the U–Pu cycle is also possible but may face an issue with trivalent fluorides (e.g. the PuF_3 solubility limit).

Homogeneous fluoride fast MSRs have one of the softest fast spectra, making the U–Pu cycle less appropriate. The Th–U cycle is less dependent on the spectrum and thus acts as a reference fuel cycle for these reactors. The combination of liquid fuel and fast spectrum allows for both waste incineration of the actinides that are long lived and breeding for resource sustainability. This possibility brings different benefits in the reactor design, safety and operation. First, the absence of graphite or any other moderator material leads to a simplification of the layout of the fuel circuit, typically characterized by the absence of internal structures, providing advantages in terms of neutronics and thermohydraulics. However, this absence also reduces thermal inertia of the core and results in a higher actinides load. Second, because of the reduced capture cross-section of the fission products and the increased fissile fuel share in the fast spectrum, these reactors are less sensitive to fission products and frequent salt treatment is not needed.

From a safety point of view, when compared with thermal reactors, these fast MSRs are characterized by (i) a highly negative reactivity coefficient, both for the salt void (thermal expansion) and the temperature Doppler effect, (ii) a small reactivity swing, (iii) a lower tritium production, (iv) a much shorter lifetime of prompt neutrons and (v) a lower thermal inertia. On the other hand, a fast reactor has a hard neutron spectrum because many of the neutrons have high energies, and this spectrum causes higher irradiation damage in terms of displacement per atom to those structural materials that are in direct contact with the salt and not protected and shielded by a graphite reflector.

The fuel salt carrier employed by this MSR family is characterized by mixtures based on fluoride (typical cations are alkali or alkaline earth ions) owing to the favourable features in terms of melting and boiling points (compatible with reactor design and structural material limitations), low capture cross-section and high solubility of actinide and lanthanide fluorides. The latter property is particularly important in this family since the fast spectrum requires a much larger fissile and fertile inventory than thermal MSRs. Fluoride salts have good thermophysical properties with respect to other candidates (general candidates or salt candidates) in terms of density, heat capacity and thermal conductivity. However, they are not as good as liquid metals, which are an excellent medium for heat exchange, especially in passive decay heat removal. With respect to other candidate salts for a fast MSR (mainly chloride salts), experience with fluoride salts is relatively extensive given their use in MSRs moderated by graphite operated in the past, such as the MSRE, and in other experimental tests, even if their behaviour in an environment with fast neutrons is not well established. (Examples of such tests are the SALIENT irradiation tests at the Nuclear Research and Consultancy Group (NRG, Netherlands), the thermohydraulic tests such as the liquid salt test loop at ORNL; the forced fluoride flow for experimental research (FFFER) and the salt at wall: thermal exchanges (SWATH) projects at the Laboratory of Subatomic Physics and Cosmology (LPSC) in Grenoble; the ADETTE technological loop programme in the Czech Republic; and the Compact Integral Effects Test facility at the University of California, Berkeley.) In addition, as compared with reactor concepts based on chloride salts, reactor concepts based on fluoride salts show a less hard spectrum, a proven compatibility with structural materials (e.g. Hastelloy N), fewer chemical issues (e.g. chlorine isotope separation is required to decrease the production of ^{36}Cl but it is not ready at industrial scale) and no sulphur production. On the other hand, using a fluoride salt bearing lithium or beryllium can be a concern due to tritium generation, and lithium also needs isotope separation to decrease tritium production and increase the neutron economy.

Different options for fuel cycles are available in this family, according to the reactor operation mode, that is, whether the reactor works as a TRU burner, converter or breeder. Since the solubility limit of trivalent fluorides often defines the applicable carrier salt, homogeneous fluoride fast MSRs can be generally divided into Th–U breeders and reactors containing plutonium. A reactor containing plutonium can be designated as a TRU burner or breeder in a closed U–Pu cycle. At the same time, ^{233}U is not available in sufficient amounts, and Th–U breeders will need to be started with enriched uranium, reactor grade plutonium or a mixture of the two. The reactor design for such fuel cycle transition will be classified according to the applied fuel type.

4.7.1. Major reactor types

The major reactor types in this family are described in this subsection and the major concepts are introduced in Table 4.

TABLE 4. FAMILY II.3. HOMOGENEOUS FLUORIDE FAST MSRs

Class	Family	Type	Conceptual design	Designer	Appeared in year	Ref.
II.	3.	Fluoride fast Th–U breeder	Fused salt (fast) breeder reactor (FSBR)	ORSORT	1953	[12]
			Two region homogeneous MSR	ORNL	1958	[124]
			Molten Salt Actinide Recycler and Transformer (MOSART, breeding option)	Kurchatov Institute	2001	[125]
			Molten Salt Fast Reactor (MSFR)	CNRS	2005	[62]
			Indian Molten Salt Breeder Reactor (IMSBR)	Bhabha Atomic Research Centre	2015	[126]
		Pu containing fluoride fast reactor	Molten Salt Actinide Recycler and Transformer (MOSART, burner option)	Kurchatov Institute	2001	[127]
			Fast Molten Salt Reactor (FMSR)	VNIINM[a]	2015	[128]
			Molten Salt Fast Breeder Reactor (MSFBR)	Hirose	2016	[129]

[a] A.A. Bochvar High-Technology Research Institute of Inorganic Materials.

4.7.1.1. Fluoride fast thorium–uranium breeder

As mentioned in Section 2, the fast breeder reactor based on fluoride salts and the Th–U cycle was proposed by a group of students at Oak Ridge School of Reactor Technology in 1953 [12]. It was a predecessor of the MSFR design. However, it was based on LiF–BeF_2 carrier salt and reflected by graphite. Accordingly, the average neutron spectrum was epithermal to fast. The performance of the fast breeder reactor based on fluoride salts, especially the doubling time, was worse than the performance of graphite moderated MSRs. The two-fluid layout of the homogeneous fluoride fast MSR concept is presented in an ORNL report from 1958 [124]. It is a very representative concept, showing the typical core dimensions (e.g. 2 m diameter) and fuel blanket layout used by many later concepts (see Fig. 7). However, this concept was published around the time that the graphite moderated MSR was selected as a reference concept at ORNL and thus abandoned.

The most advanced concept of this type is the MSFR, developed in the framework of several European projects such as the project on the evaluation and viability of liquid fuel fast reactor system (the EVOL project), the project on the safety assessment of the MSFR (the SAMOFAR project) and the project on severe accident modelling and safety assessment for fluid fuel energy reactors (the SAMOSAFER project), which builds on the results of the SAMOFAR project. It is a 3000 MW(th) reactor, employing LiF as the salt carrier with

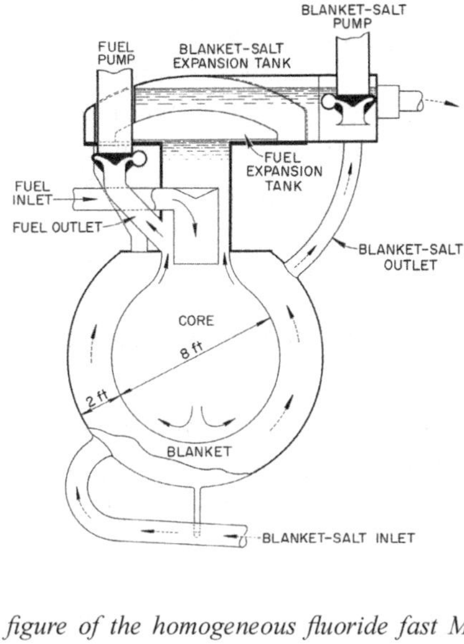

FIG. 7. Representative figure of the homogeneous fluoride fast MSR concept proposed by ORNL in 1958 (reproduced from Ref. [124] with permission). 1 ft = 0.3048 m.

ThF_4–$^{233}UF_4$ or ThF_4–UF_4–$(TRU)F_3$ as fuel options. The Bhabha Atomic Research Centre (India) has developed the Indian Molten Salt Breeder Reactor, or IMSBR. This name identifies reactor concepts having a loop or a pool. These molten salt breeder reactors are 850 MW(e) reactors, employing LiF as the salt carrier and ThF_4–$^{233}UF_4$ as the fuel.

4.7.1.2. Plutonium containing fluoride fast reactor

Since the solubility of PuF_3 is limited in LiF–ThF_4 and LiF–UF_4 salts, usually other carrier salts are envisioned for plutonium containing reactors. The same can be valid for TRU burners, which are included in this type. From the possible cations ^{7}Li, sodium and potassium, lithium has the lowest capture cross-section. At the same time, ^{7}Li has significant resonance of the scattering cross-section at 260 keV and it is the lightest element from the comparison. Therefore, it softens the neutron spectrum. It can be the best option for the closed Th–U cycle, which is more sensitive to parasitic capture than to spectrum softening. In the U–Pu cycle case it is the opposite. Accordingly, when lithium is avoided and the reactor spectrum becomes harder, the neutron balance of the U–Pu cycle improves. However, the increased parasitic neutron absorption of the salt slightly deteriorates this improvement.

The closed U–Pu cycle was utilized by the fast molten salt reactor, or FMSR, developed at A.A. Bochvar High-Technology Research Institute of Inorganic Materials (VNIINM, Russian Federation). It is a 3200 MW(th) reactor concept, employing FLiNaK as the salt carrier and UF_4–PuF_3 as the fuel [128]. A similar concept with a closed U–Pu cycle and FNaK carrier salt was analysed in Ref. [129]. The major concept of a plutonium containing fluoride fast reactor is the Molten Salt Actinide Recycler and Transformer (MOSART) system [127], developed at the Kurchatov Institute (Russian Federation). It is a 2400 MW(th) reactor, employing ^{7}LiF–BeF_2 or NaF–^{7}LiF–BeF_2 as the salt carrier and $(TRU)F_3$ as the fuel.

4.8. FAMILY II.4: HOMOGENEOUS CHLORIDE FAST MSRs

Homogeneous chloride fast MSRs are very similar to homogeneous fluoride fast MSRs. The obvious major difference is the use of chloride salts. Different options in terms of carrier salts are available [35], ranging from the alkali (sodium, potassium) and alkaline earth (calcium, magnesium) compounds to copper chloride (CuCl) and zinc chloride ($ZnCl_2$). This family is characterized by a hard fast spectrum given the absence of light nuclei in the carrier salt, such as lithium and beryllium, usually employed in concepts based on fluoride. This feature is exploited in the fuel cycle where the most selected option is the breeder U–Pu cycle, or less frequently the Th–U cycle. Burner and converter applications are also possible, as well as the self-sustaining breed-and-burn U–Pu cycle, which avoids the need for fuel enrichment, reprocessing and separation of actinides.

The use of chloride salt was one of the first options considered during the first phase of the development of MSR concepts. The use of salts based on fluoride was more favourable than chloride salts. The problem with chloride salts is the activation of ^{35}Cl into ^{36}Cl, which is an isotope that is long lived and a 0.7 MeV beta emitter. A solution to this radiotoxicity issue is the chlorine isotopic separation with enrichment in ^{37}Cl, but this technology was not considered mature at that time. As a result, R&D efforts concentrated on fluoride carriers, making experimental data on chloride very limited. In the 1950s, different concepts that involved the use of chloride in fast MSRs were proposed. This includes the homogeneous option and a U–Pu cycle [10, 130]. During the 1960s, chloride salts were considered also for pyro-reprocessing [44] and studied in several countries [35, 38, 131, 132].

In homogeneous chloride fast MSRs, the active core is composed of fuel salt with no structural material that may affect the neutronics feature. With the power directly released in the fuel salt, the heat exchange with a secondary salt takes place outside the active region, leading to a loss in delayed neutron precursor and an overall increase of the required fissile inventory due to the salt volume outside the core active region. For this purpose, the higher solubility for the actinides granted by the chloride salts allows for an increased load capacity of actinides compared with MSRs with fluoride salts. From a neutronics point of view, the chloride salts are characterized by the presence of two stable isotopes, ^{35}Cl (75.76%) and ^{37}Cl (24.24%). In addition to the production of ^{36}Cl, ^{35}Cl has a higher capture cross-section than ^{37}Cl, leading to the need for isotopic separation in a chloride fast MSR. The low scattering properties of ^{37}Cl, in combination with the low neutron absorption, result in a hard spectrum in this kind of MSR. On the other hand, this results also in a large neutron mean free path, requiring a reflector or a blanket for counterbalancing the neutron transparency. Issues related to chemistry, such as corrosion, compatibility with structural material and stability of chloride salts, require further investigation. From a thermohydraulic perspective, the chloride salts share similar characteristics with fluoride salts, having good heat transfer properties and showing small differences in terms of heat capacity, density and viscosity. In terms of melting point, the chloride salts show a slightly lower value than fluoride salts.

4.8.1. Major reactor types

The major reactor types in this family are described in this subsection, and the major concepts are introduced in Table 5.

4.8.1.1. Chloride fast breeder reactor

Reactors of this type rely on fuel recycling in a closed U–Pu fuel cycle. A Th–U fuel cycle would also be possible. However, chloride's transparency for neutrons requires a high neutron multiplication factor to keep the core size reasonable. Since the multiplication factor of the Th–U cycle is lower, the operation is possible, but the core size would be much larger.

The possibility to close the U–Pu cycle in homogeneous chloride fast MSRs was assessed as early as in 1952 [130]. It was followed by several studies (see Table 5). All these concepts were relying on ^{37}Cl enrichment to improve the neutron economy. However, a number of concepts were designed for natural chlorine [40]. Similarly, one concept was designed for the combined Th–U and U–Pu breeding cycle [39]. Another non-breeder reactor concept dedicated to waste burning [134] is included here. The concept is proposed to be coupled to a U–Pu cycle breeder. The most recent concept was proposed in 2004 by EDF and is labelled REBUS [65, 66]. All the concepts mentioned rely on NaCl carrier salt.

4.8.1.2. Chloride fast breed-and-burn reactor

Homogeneous chloride fast MSRs have the capability to operate in an open breed-and-burn fuel cycle. In this specific cycle, fuel is not recycled. The reactor is operated on natural or depleted uranium, because the excess of fissile fuel bred in the core is equal to or higher than the amount of fissile fuel discharged from the core. Hence, fuel recycling is not needed. The principle of the breed-and-burn cycle has been known for solid fuel reactors for decades. Nonetheless, the first publications related to the breed-and-burn cycle in MSRs are from 2015 [135, 136]. Since the breed-and-burn mode requires a much higher neutron economy than the breeder in a closed fuel cycle, the size of the breed-and-burn reactor should be much larger for comparable blanket or reflector conditions. The neutrons saved by the decreased leakage are utilized for the breeding of the fuel amount, which is finally discharged with the fuel. A list of breed-and-burn reactors is included in Table 5.

TABLE 5. FAMILY II.4. HOMOGENEOUS CHLORIDE FAST MSRs

Class	Family	Type	Conceptual design	Designer	Appeared in year	Ref.
II.	4.	Chloride fast breeder reactor	Fast converter	Goodman et al.	1952	[130]
			Fused salt fast breeder	Bulmer et al.	1956	[10]
			Homogeneous chloride fuelled fast reactor	Nelson et al.	1967	[133]
			Molten chloride salt fast reactor (MCSFR)	Smith et al.	1974	[45]
			Thorium–uranium fast/thermal breeder	Taube	1974	[39]
			High flux fast molten salt reactor	Taube	1975	[134]
			Salt reactor On site reprocessing Fast converter Task (SOFT)	EIR	1980	[40]
			Thorium–uranium molten chloride fast reactor (MCFR)	Ottewitte	1982	[41]
			REBUS	EDF	2004	[65]
		Chloride fast breed-and-burn reactor	Breed-and-burn molten chloride fast reactor (B&B MCFR)	Hombourger	2015	[135]
			Molten chloride fast reactor (MCFR)	TerraPower	2015	[136]

TABLE 5. FAMILY II.4. HOMOGENEOUS CHLORIDE FAST MSRs (cont.)

Class	Family	Type	Conceptual design	Designer	Appeared in year	Ref.
II. (cont.)	4. (cont.)		Molten chloride salt fast reactor (MCSFR)	Elysium Industries	2017	[137]
			Breed-and-burn molten chloride fast reactor (B&B MCFR) in multizone	Raffuzzi and Krepel	2020	[138]
			Breed-and-burn molten chloride fast reactor (B&B MCFR) with baffles for flow direction	De Oliveira and Hombourger	2020	[139]

4.9. FAMILY III.5: NON-GRAPHITE MODERATED MSRs

The reactor concepts belonging to this family are a cohort of moderated MSR designs, where the neutron spectrum is thermal, and characterized by using non-graphite moderators. From the perspective of the fuel cycle, this type of reactor can be operated as a breeder in a closed Th–U fuel cycle, as a converter fuelled by enriched uranium or as a TRU burner in an open or closed fuel cycle. These designs separate the fuel salt and moderator because of chemical incompatibility, and in some cases thermally insulate the moderator from the fuel salt. This large MSR family used diverse moderator types, but many of these MSR designs were not actively developed after the termination in 1961 of the ANP programme at ORNL.

Broadly, this family can be divided into three main types:

- *Solid moderator heterogeneous MSRs.* The moderator is in the form of metals, metal oxides, metal hydrides or metal carbides. This group excludes MSRs moderated by graphite, which is considered a separate family, as described in Section 4.6.
- *Liquid moderator heterogeneous MSRs.* The moderator is in the form of circulating or stagnant liquid, for example, light water, heavy water, metal hydroxides or metal deuteroxides. This group includes reactors cooled by circulating the fuel salt or the moderator.

— *Hybrid moderator heterogeneous MSRs*. These employ both solid and liquid moderators. Since this type is discontinued, it is included in Class IV: 'Other MSRs' (see Section 4.11).

Homogeneous mixtures of fuel salt and carrier salt, referred to simply as 'fuel salt', are not considered moderating enough to be part of this family, but they are, instead, considered as fast reactors (see Sections 4.7 and 4.8 on homogeneous fluoride and chloride fast MSRs, respectively). However, fluoride fuel salts, especially those containing beryllium difluoride (BeF_2), might be characterized as epithermal rather than fast and contribute significant self-moderation in otherwise unmoderated designs.

Concepts with a homogeneous mixture of solid fuel and moderator in the form of a molten salt (e.g. U_3O_8 particles suspended in molten hydroxide), that is, concepts based on slurry, are characterized as reactors cooled by molten salt. Similarly, concepts with heterogeneous solid fuel and the same type of moderator (e.g. solid fuel assemblies or pebble beds cooled and moderated with molten hydroxide) are also characterized as reactors cooled by molten salts. Accordingly, these concepts are not part of this family.

Since the ANP programme focused on jet engines powered by nuclear energy, reactors with high outlet temperature, compact dimensions, reasonable weight and sufficient control stability were required. However, the requirement for compact dimensions conflicts with the reactor breeding capability for reactors without blankets, and the fertile isotope ^{232}Th was excluded and ^{238}U minimized.

The different moderator types are listed in Ref. [140], including the direct mixture of liquid moderator with a compound bearing fuel. They are evaluated from a breeder perspective and have either low solubility for actinides or include nuclides that absorb neutrons. These characteristics deteriorate the breeding performance in a thermal spectrum.

When the ANP programme was terminated, MSR research at ORNL shifted towards developing reactors that produce power, focusing on thermal breeder reactors. At that time, other moderators were excluded because they were not compatible with the fuel salt and the parasitic neutron absorption of the respective cladding would disable sustainable breeding in the Th–U cycle [11, 106].

Recently, the performance of selected moderators and fuel salts was evaluated in the closed Th–U fuel cycle [87]. When the cladding material is neglected, moderators based on beryllium and deuterium, namely beryllium metal, beryllium oxide ceramic and heavy water, neutronically outperform the graphite (see Fig. 8). The study did not include lithium deuteroxide (^7LiOD), which is likely to have a similar neutronic performance to heavy water. Moderators based on hydrogen, even with the neglected cladding, are not applicable for sustainable breeding; nor are moderators based on beryllium and deuterium when metallic cladding containing Hastelloy N or SS316 is used. The only applicable cladding material that allows sustainable breeding is SiC. However, both beryllium and heavy water with SiC cladding neutronically outperform the breeding performance of graphite moderators, making them the highest neutronically performing thermal spectrum MSRs. For a power reactor, beryllium is likely to require internal cooling while heavy water can be circulated for cooling.

The limited life of graphite under radiation, the issues related to ingress of fuel salt and gaseous species, and a possibly improved neutron economy and lower fissile inventory have led to the liquid moderator being the focus of recent MSR developments. Liquid moderators theoretically have an infinite lifetime, similar to fuel salts, resulting in the cladding material lifetime setting the overall possible reactor lifetime (e.g. SiC cladding has a significantly larger resistance to reactor irradiation than fuel salt compatible metal alloys).

4.9.1. Major reactor types

The major reactor types in this family are described in this subsection and the major concepts are introduced in Table 6.

TABLE 6. FAMILY III.5. HETEROGENEOUS MSRs WITH A NON-GRAPHITE MODERATOR

Class	Family	Type	Conceptual design	Designer	Appeared in year	Ref.
III.	5.	Solid moderator heterogeneous MSRs	Aircraft Reactor Experiment (ARE)	ORNL	1954	[86]
			Aircraft Reactor Test (ART) concept	ORNL	1954	[86]
			Aircraft Reactor Test (ART) concept variation	ORNL	1954	[86]

TABLE 6. FAMILY III.5. HETEROGENEOUS MSRs WITH A NON-GRAPHITE MODERATOR (cont.)

Class	Family	Type	Conceptual design	Designer	Appeared in year	Ref.
III. (cont.)	5. (cont.)		Transatomic Power (TAP) MSR design	Transatomic Power	2013	[141]
		Liquid moderator heterogeneous MSRs	Copenhagen Atomics Waste Burner	Copenhagen Atomics	2015	[142]
			Compact Molten Salt Reactor (CMSR)	Seaborg Technologies	2018	[143]
			Heavy water MSR (HW-MSR)	SINAP	2019	[144]

4.9.1.1. Solid moderator heterogeneous MSRs

The ANP programme studied and considered many non-graphite solid moderators together with molten salt fuels, including beryllium and beryllium oxide (BeO) [83], magnesium oxide (MgO) [145], several hydrides and hydrates [146] and beryllium carbide (Be_2C) [147]. Thermal stress considerations limited BeO moderator application for reactors with power density below 0.5 kW/cm^3 [86]. Metal hydrides were investigated and found to be stable under irradiation, but none was found to have satisfactory physical properties during the early stages of the ANP programme [86]. An unpublished concept for a zirconium hydride (ZrH_x) moderated MSR with the fuel salt separated from direct contact with the moderator via tubes was suggested by Pratt and Whitney [145]. The compatibility of graphite and molten salt was identified early on, but a reactor that was reflected and moderated by graphite required bigger cores than other concepts, and the penalty caused by the shielding weight made them especially undesirable for nuclear powered aircraft [86]. However, several studies of civil reactors that were purely moderated by graphite were conducted as early as 1953 [145], and these reactors were being researched substantially by the late 1950s [147].

In the framework of the ANP programme, the Aircraft Reactor Experiment was operated successfully for nine days in 1954. It was a small reactor with 2.5 MW(th) power. The core outlet temperature at steady state reached up to about 860°C. The fuel was in the form of a mixture of molten salt fluorides (NaF, ZrF_4 and UF_4) and the reactor was moderated by BeO. The thermal stresses in the moderator limited the reactor power [86].

One of the major concepts proposed by the ANP programme was a compact MSR core, also referred to as 'Fireball' [145], with a beryllium reflector and a beryllium central block, both cooled by a liquid NaK alloy. It was known as the Circulation Fuel Reactor Experiment, or CFRE, and later as the Aircraft Reactor Test [101], which was not constructed. The moderator was separated from the fuel salt by cladding in the Aircraft Reactor Experiment and in the Aircraft Reactor Test.

The concept of the now defunct company Transatomic Power also belongs to this type. It was moderated by ZrH_x (where x can range from 1 to 4) in direct contact with the fuel salt and possibly chemically stabilized with lithium hydride (LiH) and yttrium hydride (YH) [141].

4.9.1.2. Liquid moderator heterogeneous MSRs

Moderators containing hydrogen or deuterium, based on liquid organic compounds, were also suggested during the Nuclear Energy for the Propulsion of Aircraft, or NEPA, project, which preceded the ANP programme. However, they were found to have insufficient radiation stability, especially at the high power densities required for nuclear propulsion, lasting for only 20 min of operation at 1 kW/cm^3 [86].

The ANP programme also identified hydrogen gas as too diffuse for use as a moderator and the cooling of a liquid hydrogen moderator as unfeasible in reactors operating at high temperature [86].

After the work on suspensions of uranium particles in molten hydroxides was dropped in 1950, work continued on fluoride salt fuels moderated by molten (^{7}Li, sodium, potassium, beryllium and rubidium) hydroxides, deuteroxides and hydroxide–fluoride mixtures. These moderators were separated from the fuel salts by metal or ceramic cladding [148].

Much of the work in the early 1950s focused on determining the corrosiveness of fluorides and hydroxides, with the latter found to be the more corrosive of the two. At the time, the nature of the corrosiveness of hydroxides, correctly called oxoacidity, was poorly understood. Experiments were conducted to see whether corrosion caused by hydroxides could be reduced by additives. Trace amounts of H_2O were shown to be beneficial, while larger amounts appeared to increase corrosion [16]. Adding sodium carbonate to sodium hydroxide seemed to have little effect [140]. Adding sodium metal to sodium hydroxide seemed to increase corrosion [140, 149]. Studies of hydroxides continued throughout the ANP programme. The tests hinted that corrosion mitigation was possible. However, the lack of a systematic approach hindered successful corrosion control, and by 1955 the hydroxides were deemed to be too harsh on structural materials to permit use in high temperature reactors [7].

A few years later, the redox chemistry of non-aqueous systems was systematically studied by B. Trémillon and co-workers, who demonstrated that the corrosiveness of alkaline hydroxides can be controlled thermodynamically [149, 150], rehabilitating the use of this interesting family of liquid moderators. In 2018, the use of hydroxides for the moderation of thermal and epithermal reactors was patented [143].

Light water and heavy water were given high consideration during the ANP programme as liquid moderators for concepts with circulating salt fuel, although most research went into concepts based on light water because of the smaller core sizes achievable. Concepts were considered using unpressurized and pressurized light water as a moderator, which was thermally insulated from the fuel salt, as well as concepts using supercritical water as a coolant and a moderator [86].

An unpressurized and thermally insulated reactor that is moderated and reflected by light water was estimated to transfer 8–15% of reactor power to the water, and the low temperature difference to ambient air would entail prohibitively large heat exchangers for nuclear powered flight [86, 148]. Pressurized water concepts also entailed pressurizing the salt, resulting in increased challenges to pump shaft seals and heavy reactor pressure vessels [86]. Several hybrid MSR designs, mostly beryllium reflected and with a solid or liquid moderator from the end of the ANP programme, are illustrated in Fig. 8.

The simple reactor core with two regions, cooled by liquid metal and beryllium reflected, has the disadvantages of needing a higher fissile inventory and having a higher probability of stagnant fuel in the larger void than the core with three regions, which is reflected and moderated by beryllium. The core with three regions also has a more uniform power distribution, but there were technical difficulties with cooling the beryllium island in the core centre by liquid metal [86]. The problem of cooling the beryllium reflector and moderator is alleviated by an intermediate cooling layer with liquid metal in the core with five regions. The heat dissipation problem favours designs with a liquid moderator, where the moderator can be circulated and cooled as seen in the designs that are hydroxide moderated and beryllium reflected [86]. Core designs that are beryllium reflected and graphite moderated were also considered, where the graphite with fuel channels or sheets is located at the centre of the core and the graphite is cooled by direct contact with the salt. These designs showed a slight

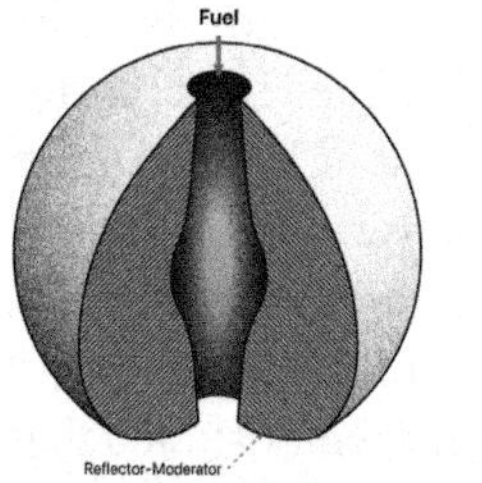

(a) Simple two-region reactor core with a thick, spherical shell of moderator surrounding a spherical chamber containing liquid fuel.

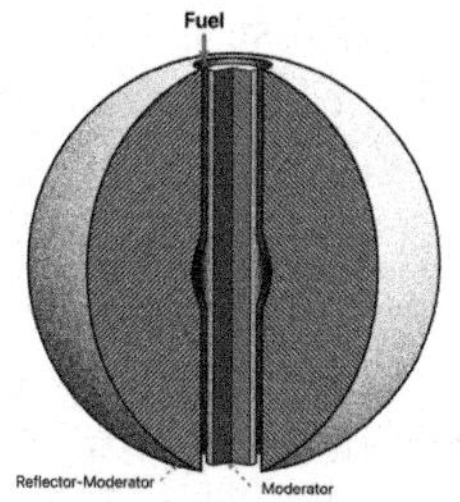

(b) Three-region reactor core with central island of moderating materials.

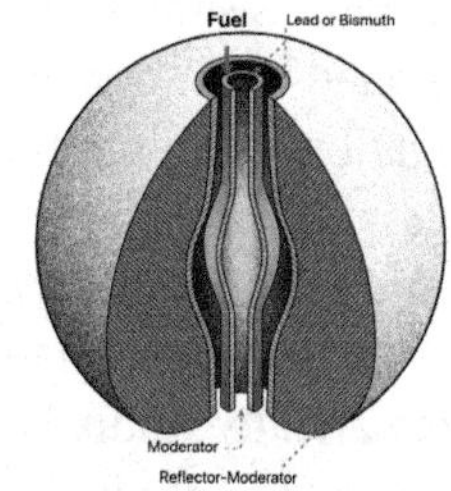

(c) Five-region reactor core with provision for cooling reflector moderator regions.

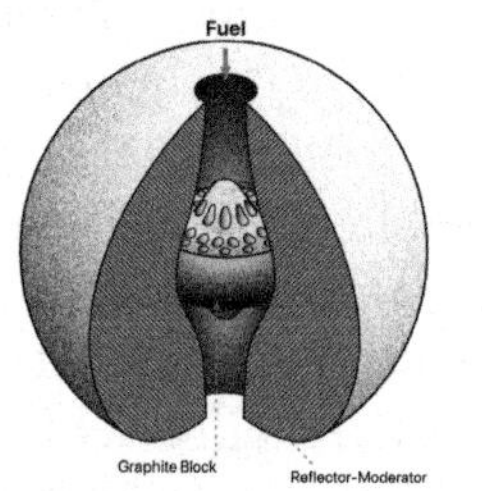

(d) Reactor core with fuel channels in a graphite block.

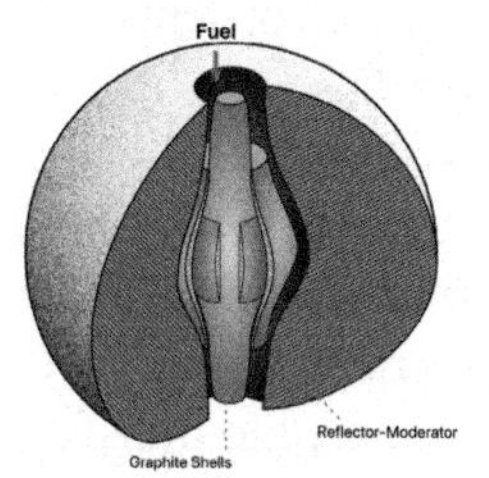

(e) Reactor core with graphite shells in a fuel channel.

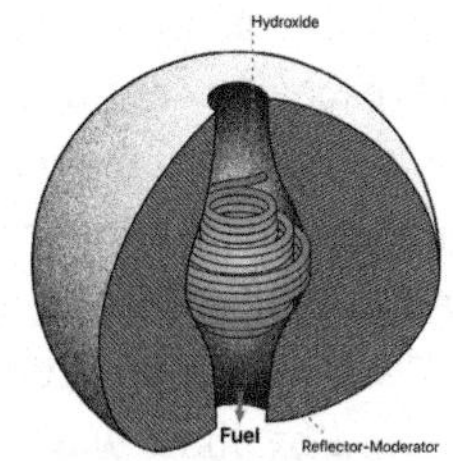

(f) Fluid moderated reactor core with coiled tubes for circulating the moderator.

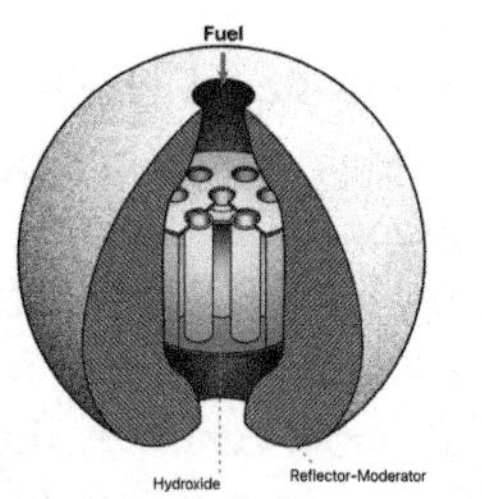

(g) Fluid moderated reactor core with straight-tube fuel passages and provision for circulating the moderator around fuel tubes.

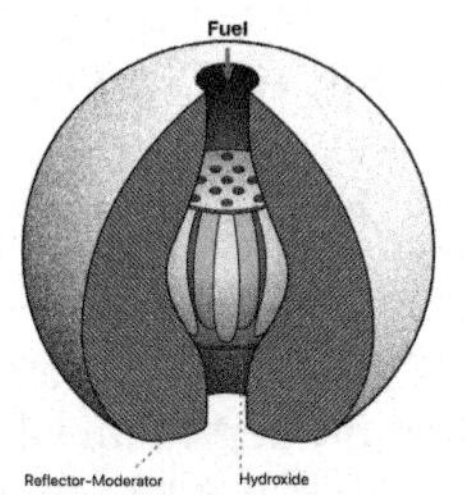

(h) Fluid moderated reactor core with spheroidized fuel passages and provision for circulating the moderator around fuel passages.

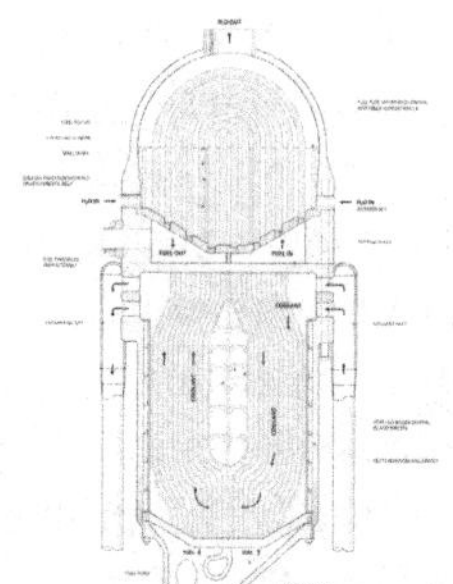

(i) Reactor design for use with liquid moderator (water or hydroxide) and circulating fluoride fuel.

FIG. 8. Core concepts from the Aircraft Reactor Experiment (reproduced from Ref. [86] with permission).

improvement from a nuclear perspective over the design purely reflected by beryllium [86].

4.10. FAMILY III.6: HETEROGENEOUS CHLORIDE FAST MSRs

The reactor concepts belonging to this family do not use any moderator and the neutron spectrum is fast. The fuel has the form of liquid actinide chlorides with or without chloride carrier salts. The carrier salts can be NaCl, $MgCl_2$, KCl, $CaCl_2$ or a mixture of them. Other options for the carrier salts are possible but seldom reported. As a common practice for this family, the fuel salt does not perform the coolant function and is physically separated from the dedicated coolant. In some older concepts, the blanket salt was used as the fuel coolant [151, 152]. The options for the fuel cycle can be U–Pu burning or breeding, Th–U burning or breeding or TRU element burning, thus allowing the reactors in this family to be suitable for different applications. Owing to the presence of structural materials and dedicated coolant, the neutron economy of heterogeneous chloride fast MSRs is tighter than for homogeneous chloride fast MSRs. Nevertheless, the once through breed-and-burn U–Pu cycle is possible [153, 154].

Reactors in this family share many characteristics with other fast spectrum MSRs. They allow the possibilities of waste incineration of the actinides that are long lived and breeding for resource sustainability. The installed core power depends on the heat exchange surface of the separating material. At the same time, liquid fuel can withstand higher temperatures than the solid fuel and the dedicated coolant circulation can be fast. Hence, compact reactor core designs with a high core power density are possible. The absence of a moderator also eliminates one of the constraints on the reactor core lifetime set by moderator materials such as graphite. Nonetheless, the lifetime of fuel pins is also limited. Owing to the reduced capture cross-section of the fission products in the fast spectrum, there is no need for frequent salt treatment to remove neutron poisons. However, fast reactors have hard neutron spectra causing more severe radiation damage than thermal reactors in terms of displacements per atom to those structural materials that are in direct contact with the salt and not shielded by a moderator or reflector.

Unlike homogeneous MSRs, the reactors in this family usually use fuel pins or assemblies in the reactor core to contain fuel salts. The approach reduces the volume of the fuel salt. Difficulty in modelling the reactor core can be reduced since the core arrangement is similar to the arrangements commonly seen in reactors cooled by water or in sodium fast reactors. For reactors that do not circulate the fuel salt, the contamination of the components by the fuel salt can be minimized. However, the introduction of structural materials into the reactor core will impact the neutron economy and require the qualification of the structures such as the fuel pins or assemblies. The fuel of heterogeneous chloride fast MSRs usually utilizes enriched ^{37}Cl as the salt base.

The fuel salts employed by this family are usually mixtures of chlorides because of their favourable features in terms of melting and boiling points, low capture cross-sections and high solubility of actinide and lanthanide chlorides. The high solubility of actinides such as plutonium is particularly important for this family, since fast reactors require a much larger fissile and fertile inventory than their thermal counterparts. Chloride fuel is expected to have a good stability at high burnup and at high temperatures. Chloride fuel also has good thermophysical properties compared with other types of nuclear fuel in terms of density, heat capacity and thermal conductivity.

With respect to other candidate salts for fast MSRs (mainly fluoride salts), chloride salts have a superior solubility of actinides. Reactors based on chloride fuel tend to have harder neutron spectra, which may result in better burning rates if designed as waste burners. Historically, more effort was devoted to developing MSRs using fluoride salts than to those using chloride salts. The issues of using $2LiF–BeF_2$ (FLiBe) salt (which was the salt used in the MSRE by ORNL) caused by the ^{6}Li isotope and the beryllium health hazard are widely recognized. However, ^{36}Cl that is produced mainly from the activation of ^{35}Cl may be a challenge for the disposal of spent nuclear fuel from the reactors using the chloride fuel salt because ^{36}Cl has an extremely long half-life and a high solubility in water.

Different options for fuel cycles are available in this family. For TRU burning, the reactors are usually fuelled by spent nuclear fuel from reactors cooled by water. For U–Pu burning or breeding, the reactors are usually fuelled by uranium and reactor grade plutonium. For Th–U burning or breeding, the reactors are usually fuelled by thorium, reactor grade plutonium or uranium.

4.10.1. Major reactor types

Historically, the non-circulating salt fuel was considered in the ANP programme at ORNL but rejected for that particular application. In the 1960s, a gas cooled reactor concept with molten chloride fuel was proposed at ORNL [151]. Nonetheless, major research focused on MSRs moderated by graphite and not on heterogeneous chloride fast MSRs. The major reactor types in this family are described in this subsection and the major concepts are introduced in Table 7.

TABLE 7. FAMILY III.6. HETEROGENEOUS CHLORIDE FAST MSRs

Class	Family	Type	Conceptual design	Designer	Appeared in year	Ref.
III.	6.	Heterogeneous salt cooled fast MSRs	Stable Salt Reactor-Wasteburner (SSR-W300)	Moltex Energy	2014	[155, 156]
			SSR-B&B	Kasam and Shwageraus	2016–2020	[153, 154]
			Molten chloride fast breeder reactor (MCFBR)	EIR	1972–1974	[38, 152]
		Heterogeneous lead cooled fast MSRs	Dual Fluid Reactor (DFR)	IFK Berlin	2015	[157]
			Hard Spectrum Reactor (HSR)	Aristos Power	2018	[158]

4.10.1.1. Heterogeneous salt cooled fast MSRs

This type of MSR relies on molten chloride fuel, which is typically cooled by a dedicated chloride salt coolant. In the 1970s, several concepts of a blanket salt cooled molten chloride fast breeder reactor, or MCFBR, were proposed at EIR [38, 152], where the coolant salt also acted as fertile material. In recent history, a fluoride salt cooled chloride fuel salt system was proposed in Refs [155, 156]. The concept was later modified for chloride salt coolant. The SSR-W300, a stable salt reactor (SSR), was developed by Moltex Energy in Canada and the UK. It is a 300 MW(e) waste burner using a mixture of reactor grade plutonium chlorides and depleted uranium chlorides as the fuel. The fuel salt is contained in the fuel pins and is not forced to circulate; hence the name of the design. The same design was used to assess the breed-and-burn capability of this concept [153, 154]. It is possible to operate the SSR in this fuel cycle type, but the neutron economy can be tight.

4.10.1.2. Heterogeneous lead cooled fast MSRs

Other coolants have been considered for the liquid chloride salt encapsulated in fuel pins. As already mentioned, a gas cooled option was proposed in the

1960s [151]. From the other two typical fast reactor coolants (i.e. liquid metals), only lead has been proposed in the literature so far. The Dual Fluid Reactor, or DFR, series was developed by the Institute for Solid-State Nuclear Physics (Institut für Festkörper-Kernphysik, IFK) in Berlin, Germany. These reactors are so named because the fuel salt and the coolant are circulated via different loops [157]. Their power levels are in the range of 130–1500 MW(e). They use uranium, plutonium or thorium chlorides as the fuel without carrier salts and liquid lead as the coolant. The Hard Spectrum Reactor HSR50 concept [158], developed by Aristos Power in France, currently also exists. This is a fast spectrum reactor using chloride salt fuel and liquid lead as coolant that generates 50 MW(th). So far, none of the concepts have considered sodium as the coolant.

4.11. OTHER MSRs

There are several MSR concepts that do not fit the six major families. Even though these concepts have the lowest technology readiness level and have a non-standard design approach, they are still classified into several types. Each is briefly described in this subsection.

4.11.1. Major reactor types

The other major reactor types are described in this subsection and the major concepts are introduced in Table 8.

TABLE 8. CLASS IV. OTHER MSRs

Class	Type	Conceptual design	Designer	Appeared in year	Ref.
IV.	Directly cooled MSRs	U solution in molten bismuth cooled by KCl–LiCl salt	Burch et al.	1955	[159]
		Chloride fuel salt cooled by lead	Moore and Fawcett	1966	[132]
		Chloride fuel salt cooled by boiling $AlCl_3$	Taube et al.	1967	[36]

TABLE 8. CLASS IV. OTHER MSRs (cont.)

Class	Type	Conceptual design	Designer	Appeared in year	Ref.
IV. (cont.)	Directly cooled MSRs (cont.)	Fluoride fuel salt cooled by lead	Gat	1967	[160]
		Chloride fuel salt cooled by lead	Smith et al.	1974	[45]
	Subcritical MSRs	Accelerator transmutation of waste (ATW)	Bowman	1997	[161, 162]
		Japan Atomic Energy Research Institute (JAERI) OMEGA	Katsuta	1997	[163]
		SPHINX	Hron	2000	[164]
	Hybrid moderator MSRs	Moderator diluted in the salt	Ellis and Thompson	1950	[95]
		Reflector moderated homogeneous MSR	Wehmeyer	1953	[12]
		Aircraft Reactor Test (ART) concept variation	ORNL	1954	[86]
	Chloride salt cooled fast reactors	Carbide fuelled fast breeder reactor	Taube	1969	[37]
		Liquid salt cooled fast flexible conversion ratio reactor (LSFR)	Todreas et al.	2009	[165]
		Molten chloride cooled fast reactor (MCCFR)	Lin et al.	2020	[166]

TABLE 8. CLASS IV. OTHER MSRs (cont.)

Class	Type	Conceptual design	Designer	Appeared in year	Ref.
IV. (cont.)	Frozen salt reactors	Zero power (cold) MSR	SINAP	1971	[167]
	Hybrid spectrum MSRs	Flux trap reactor	Křepel et al.	2014	[168]
		Variable spectrum reactor	Transatomic Power	2015	[169]
	Heterogeneous gas cooled MSRs	Chloride salt fuel cooled by gas	Alexander	1963	[151]

4.11.1.1. Directly cooled MSRs

Directly cooled MSRs have been considered in the past, with five concepts developed between 1955 and 1974. Similar to water, the salts in MSRs have a low thermal conductivity and need a large heat exchange surface area. The aim of directly cooled MSRs is to avoid this large heat exchange surface required for the salts. The first reactor in this family was proposed in 1955 using a uranium solution in bismuth as fuel and KCl–LiCl as salt coolant [159]. The more common approach of chloride fuel salt cooled by direct contact with lead was first proposed in 1966 [132]. A similar solution was also considered in Ref. [45]. The fluoride salt cooled by lead was analysed in Ref. [160], and a rather exotic concept of chloride fuel salt cooled by boiling $AlCl_3$ was proposed in Ref. [36]. The biggest disadvantage of the family of 'directly cooled MSRs' is the contamination of the coolant by actinides and fission products.

4.11.1.2. Subcritical MSRs

Subcritical MSRs were studied in the 1990s, when the accelerator driven system was of great interest. The purpose was solely waste burning. There were concepts moderated by graphite with fluoride salt [161, 164], homogeneous fluoride MSRs moderated by a graphite reflector [162] and chloride fast MSRs [163]. The major disadvantage of subcritical reactors is the cost of the accelerator. The function in the reactor is for the accelerator driven system to serve as a specific form of reactivity control, which compensates for the insufficiency in thermal feedback coefficients of some accelerator driven systems. Since an MSR using liquid fuel can be designed as a waste burner with all thermal reactivity feedback coefficients negative, the accelerator becomes unnecessary in concepts like MOSART.

4.11.1.3. Hybrid moderator MSRs

Many of the reactor concepts considered by the ANP programme were hybrids of solid and liquid moderators. Solid metal beryllium moderator performance was deemed the best. Reflector materials were treated similarly, and NaOH, sodium deuteroxide (NaOD) and ^{7}LiOD were also considered. However, their moderating performance was limited by the predominant forward scattering, and beryllium performed better because it is heavier [86]. Hybrid moderator MSRs utilize a homogeneous mixture of moderator and fuel salt: this could be a mixture of liquid moderator and fuel salt, for example, hydroxide–fluoride mixtures, or principally a mixture of fuel salt and moderator slurry with solid particles suspended in fuel salt. The ANP programme aimed at developing a supersonic aircraft that was nuclear powered, and focused development on light and compact reactor designs that operated at high temperature and shielding that was lightweight. Early concepts included solid fuel reactors, liquid metal alloy reactors and reactors with boiling coolant and solid fuel that were cooled and moderated by sodium hydroxide [146, 170]. By 1950, the ANP programme had narrowed in on the circulating fuel systems, when E. Bettis and R.C. Briant proposed molten fluoride salt systems, or fused salts, as they were referred to initially [171]. Concepts early in 1950 considered NaF–BeF_2–UF_4 as a self-moderating circulating fuel [95] and suspensions of uranium compounds in molten sodium hydroxide, including uranium chloride and fluoride [140]. However, the latter concepts were dropped by mid-1950 because of the lack of solubility to actinide compounds and reactions between hydroxide and fluoride species [7].

Reflector moderated MSRs can also be included under hybrid moderator MSRs. They lie between graphite moderated MSRs and homogeneous fluoride fast MSRs. The fluoride salt includes beryllium and thus has moderation power. The homogeneous core is, furthermore, surrounded by a graphite reflector. The importance of the thermal spectrum decreases with increasing core size. Accordingly, some of the bigger cores surrounded by graphite, such as the first homogeneous fluoride fast MSRs proposed at the Oak Ridge School of Reactor Technology in 1952 [12], have epithermal or fast spectra. An example of the 'Fireball' concept developed within the ANP programme belonging to this category is discussed in more detail in Section 4.9.

4.11.1.4. *Chloride salt cooled fast reactors*

Chloride salt can be used also as a coolant for a fast reactor with solid fuel and a bonding material in the fuel pin. Boiling $AlCl_3$ was considered as a coolant in Ref. [37]. The disadvantages of the usual coolants for fast reactors can be avoided by using chloride salt as a coolant. However, salt coolant introduces additional complexity. A more recent evaluation of molten salt as the coolant of fast reactors in concepts with a liquid salt cooled reactor with flexible conversion ratio, or LSFR, is provided in Ref. [165] and with long refuelling intervals (molten chloride cooled fast reactor, or MCCFR) in Ref. [166].

4.11.1.5. *Frozen salt reactors*

Frozen salt reactors rely on solid fuel in salt form. Accordingly, they belong to heterogeneous MSRs that require dedicated liquid coolant and separating structural material. The frozen state enables low operating temperatures. At the same time, the fuel can be occasionally re-melted and purified. Therefore, this concept has similar advantages and disadvantages to a reactor with solid fuel including the flow of coolant through an orifice and the possible presence of hotspots. A major advantage and disadvantage is the fuel melting temperature. It may be too low to be frozen during nominal operation and too high to be re-melted for purification. In 1971, SINAP built a frozen salt zero power test reactor [167].

4.11.1.6. *Hybrid spectrum MSRs*

Hybrid spectrum MSRs are characterized by the presence of both thermal and fast spectra in the core. This design option is usually applied to minimize the neutron leakage. The core often acts as a flux trap, where the multiplication factor is much higher in the centre than on the periphery [168]. Concepts also exist with variable neutron spectra, where manipulation with the core composition or geometry results in spectral changes [169].

4.11.1.7. *Heterogeneous gas cooled MSRs*

Gas coolant is considered for heterogeneous chloride fast MSRs in one paper by L.G. Alexander from 1963 [151].

5. RESEARCH AND DEVELOPMENT ACTIVITIES

5.1. INTRODUCTION

This section presents the R&D activities in Member States with major programmes for MSR technology in the public sector, private sector or both. The following subsections are ordered alphabetically.

5.2. RESEARCH AND DEVELOPMENT ACTIVITIES IN CANADA

5.2.1. Background

The nuclear sector in Canada is currently focused on SMRs that have the potential to provide clean energy for applications, ranging from use in remote communities to heavy industry or to electricity generation at grid scale [172]. In November 2018, the Canadian Government released the Canadian SMR roadmap, which outlines recommendations for collaboration between federal, provincial and territorial governments and other stakeholders to support SMR development in Canada [173]. In February 2020, the Canadian Government announced that the launch of the Canadian SMR action plan would be in November 2020 [174][7]. At the provincial level, Ontario, Saskatchewan and New Brunswick signed a memorandum of understanding in December 2019, and were joined in August 2020 by Alberta, to work together on the development and deployment of SMRs in Canada, including on aspects such as technological readiness, regulatory frameworks and nuclear waste management.

Many SMR developers have initiated a pre-licensing vendor design review with the Canadian Nuclear Safety Commission [175]. Two small modular MSRs are among a range of SMR designs under the pre-licensing vendor design review: Terrestrial Energy and Moltex Energy. Terrestrial Energy is developing the Integral Molten Salt Reactor, also known as the IMSR, which has fluoride fuel salt, thermal spectrum (graphite moderator) and integral heat exchangers. Moltex Energy is developing the Stable Salt Reactor, which has chloride fuel salt, a fast spectrum and is tailored to burn spent fuel from Canada Deuterium Uranium (CANDU) plants.

In parallel, experienced Canadian operators of nuclear power plants are working with SMR developers to advance engineering and design work and evaluate potential demonstration projects. The IMSR by Terrestrial Energy

[7] The actual launch was in December 2020.

is one of the SMR designs selected by Ontario Power Generation, and the stable salt reactor by Moltex Energy is one of the SMR designs affiliated with New Brunswick Power. As part of these efforts, the Canadian Government announced a strategic innovation fund investment of 20 million Canadian dollars in October 2020 to assist Terrestrial Energy to complete a key pre-licensing milestone through the Canadian Nuclear Safety Commission [176]. The fund's programme is designed to attract and support business investments of high quality across all sectors of the Canadian economy.

5.2.2. Research and development activities

Molten salt technology is a new area of nuclear R&D in Canada. The leading nuclear science and technology organization in Canada is Canadian Nuclear Laboratories (CNL). To support the government's areas of focus, CNL has identified SMRs as one of its key strategic initiatives. As part of this programme, CNL is building its expertise and capabilities to support the development and deployment of different SMR technologies. In particular, several MSR research projects were funded by the Canadian Government through Atomic Energy of Canada Limited and launched in 2016 at CNL's Chalk River Laboratories site. The site is the largest science and technology complex in Canada. Historically, the expertise and research facilities at the site were used to develop the CANDU technology. These capabilities are expanding, and new capabilities are being developed, towards qualification and demonstration of molten salt technology among other SMR advanced technologies. At present, the MSR R&D programmes at CNL address the following areas:

- Development of experimental protocols, integrated with quality assurance plans, for verification and characterization of key transport and thermodynamic properties of fuel and coolant salts to support safety assessment and licensing of MSR technology;
- Development of experimental facilities for synthesis of actinide fluorides and chlorides;
- Investigation of molten salt's fuel behaviour in accident conditions, including the measurement of release of fission products from molten salt fuels;
- Molecular dynamics for predicting thermophysical properties of fuel and coolant salts;
- Corrosion of materials that are in contact with fuel and coolant salts;
- Passive decay heat removal in MSRs;
- Development of simulation tools that couple computational fluid dynamics (CFD) code and neutronic code for MSR transients;

— Development of modelling capabilities for beyond design basis events for MSRs;
— Phenomena Identification and Ranking Table analysis for severe accidents in MSRs;
— Evaluation of waste streams from MSRs considered for deployment in Canada;
— Approaches and techniques for addressing safeguards and non-proliferation issues of MSRs.

Another important initiative is CNL's Canadian Nuclear Research Initiative that was launched in June 2019 to support collaborative advanced reactor research projects with third party proponents in Canada. The goal of the programme is to accelerate the deployment of safe, secure, clean and cost effective advanced reactors in Canada [177]. The following MSR research projects were funded by the initiative:

— Research and engineering of technologies to better separate, analyse and store tritium generated through the operation of Kairos Power FHR design [178];
— Support to Moltex Energy to demonstrate the commercial viability of the waste to stable salts, or WATSS, technology to convert spent CANDU fuel into new fuel that can power SSRs [179];
— Evaluation of safeguards related to the operation of Terrestrial Energy's IMSR [180].

The following universities also perform MSR R&D activities:

— *The Polytechnique Montréal:* A technological university that maintains and develops thermodynamic model databases for salt/oxide systems in the FactSage [181] thermochemical software, which is licensed to universities and companies around the world. Currently, one of the thermodynamic models under development is Kairos Power's thermodynamic/phase equilibrium modelling of molten 2LiF–BeF_2 (FLiBe) salt with corrosion products.
— *Ontario Tech University:* Several research projects related to fluoride salts are under way. Capabilities are being developed to model molten salts with the open source code Thermochimica [182], which is being coupled to various multiphysics codes. In particular, a new code called Yellowjacket

is under development in partnership with Idaho National Laboratory to simulate lower-length scale[8] corrosion by fluoride salts.

— *University of New Brunswick:* The university's Centre for Nuclear Energy Research is a partner in New Brunswick's nuclear research cluster and is focusing on developing experimental capabilities in MSR R&D related to Moltex Energy's SSR technology. Moltex Energy is one of the two developers of SMRs with fast spectrum that committed 5 million Canadian dollars to match funds from New Brunswick's provincial government, through New Brunswick Energy Solutions Corporation to support MSR R&D activities. This corporation is a joint venture formed in May 2017 by New Brunswick's provincial government and New Brunswick Power, operator of the Point Lepreau nuclear power plant [183].

At the international level, Canada has been considering joining the MSR system for international R&D collaboration of the Generation IV International Forum and is working with other forum members to integrate an SMR focus into the forum's programme. The forum's MSR collaboration is currently operating under a memorandum of understanding and is in the process of transitioning to a system arrangement. With support from Natural Resources Canada, Terrestrial Energy became a signatory of the memorandum of understanding in May 2019. Natural Resources Canada supported Moltex Energy's participation in a meeting of the forum's MSR provisional System Steering Committee that was held in September 2019. Canadian Nuclear Laboratories is currently an observer in this steering committee and the European project on severe accident modelling and safety assessment for fluid-fuel energy reactors (the SAMOSAFER project).

5.3. RESEARCH AND DEVELOPMENT ACTIVITIES IN CHINA

SINAP, part of the Chinese Academy of Sciences, restarted a research programme on thorium utilization known as the Thorium Molten Salt Reactor (TMSR) programme in January 2011. This programme focused on the R&D of technologies related to the TMSR, to achieve the safe and economic use of Generation IV MSRs and the efficient use of the thorium resource. The research programme developed a TMSR simulator (mock-up) and test reactor to prove the concept. The TMSR simulator was built in 2019. The construction of a 2 MW(th) TMSR test reactor began in September 2018 and was reportedly completed in August 2021. The research programme plans to develop demonstration reactors

[8] In this context, 'lower-length scale' refers to the direct coupling of phase field calculations of microstructure evolution and computational thermodynamics.

to validate TMSR technologies and finally deploy commercial nuclear power plants. A 168 MW(e) small modular TMSR demonstration reactor, named smTMSR-400, is expected to be deployed before 2030.

5.3.1. Progress of R&D in TMSR technology

The TMSR programme has already achieved several results in key TMSR technologies. Some of the main achievements are presented in the subsections below [184].

5.3.1.1. Molten salts

Fluoride molten salts are important materials in the TMSR. FLiBe is used as a coolant in the primary loop and FLiNaK is used as a coolant in the secondary loop. In the liquid fuel TMSR, thorium and uranium are dissolved in FLiBe. Advanced techniques for the preparation and purification of these salts have been developed as well as the capabilities to test and evaluate these techniques.

The corrosion rates of structural materials in contact with the salts are important for the safe use of the salts in the TMSR. The corrosion rate depends on impurities in the salts, such as oxygen, water and sulphur. Consequently, the impurities need to be reduced to an appropriate level before the salts can be used in the reactor's loops.

The specifications of FLiBe to be used in the primary loop are strictly controlled. The two main requirements for FLiBe are control of the impurities and an abundance of lithium isotope. The isotopic abundance of ^{7}Li for the preparation of the salt has to be over 99.99%. A new centrifugal extraction technique was developed as an alternative to the traditional amalgam method. This new technique is more environmentally friendly than the amalgam method since no mercury is used during salt production.

Beryllium fluoride (BeF_2) is another important raw material for the production of FLiBe. Before purification treatment, beryllium fluoride contains 2000 ppm sulphur and 30 ppm boron. However, after desulphurization at high temperature and recrystallization with ammonium, all the impurities are significantly reduced: the sulphur and phosphorus concentrations are below 20 ppm, and the nitrogen and boron concentrations are below 1.5 ppm. More importantly, these new techniques have been combined to produce nuclear grade FLiBe to be used in both the solid fuel and liquid fuel TMSR designs. Figure 9 shows fluoride salts and their production facilities.

The main requirement for the use of FLiNaK salt in the secondary loop is the control of impurities in the salt. By controlling these impurities, the corrosion rate can be reduced. It was found that with an oxygen concentration of around

FIG. 9. Fluoride salts and production facilities (courtesy of Y. Zou, Shanghai Institute of Applied Physics).

100 ppm, the corrosion rate of the Hastelloy N alloy (UNS N10003) is about 20 μm/a. By using an ammonium hydrogen fluoride method, the oxygen was removed. A tonne of FLiNaK salt was produced using this method and tested in an experimental loop for a year without any problems. To reduce the impurities further, a hydrofluoric acid–hydrogen process (HF–H_2 process) was used. The quality of the salt prepared by this process is dependent on the impurities in the hydrofluoric acid. A new technique was developed to purify commercial hydrofluoric acid. The corrosion rate of the Hastelloy N alloy in FLiNaK salt prepared using this new technique is under 2 μm/a.

5.3.1.2. Alloy based on nickel and graphite

The structural materials of the TMSR will be subjected to extreme environments, that is, high temperature, high neutron fluence and a corrosive coolant. In the case of the liquid fuel design, the dissolved fuel in the fluoride salt will produce a number of radioactive and corrosive substances (e.g. xenon, fluorine, iodine) under neutron irradiation in the core. Accordingly, the structural materials need to have a good strength at high temperature, a good resistance to corrosion and a high tolerance to neutron irradiation. Alloys based on nickel or graphite, the two main structural materials, are to be used in the TMSR.

The alloy based on nickel is considered to be the primary option for metallic structural materials. Such an alloy, named GH3535, was developed and its performance parameters are comparable with those of Hastelloy N produced in the USA. Production of GH3535 at pilot scale was completed and a technique for mass production was established. Furthermore, several processing methods were developed for alloys based on nickel, such as the hot extrusion and rolling process for pipes of large calibre to be used in the reactor loop, as well as tungsten inert gas welding for thick plates that meets the requirements of the American

Society of Mechanical Engineers. These key techniques ensure the successful application of GH3535 in reactor components such as pressure vessels, loops and control rod sleeves.

Owing to high thermal conductivity and low chemical activity at elevated temperatures, graphite is considered an ideal moderator for the TMSR, as well as a structural material. Improvements were made to a fabrication method for isostatic graphite. This new method has been applied to produce an ultrafine grain graphite, NG-CT-50, for the liquid fuel TMSR. This graphite has a high density and other excellent material properties such as high bending and compressive strengths, low porosity and low boron. To assess the effect of molten salt impregnation, a series of compatibility tests of graphite with molten salt were conducted.

5.3.1.3. Corrosion experiment

The corrosion of various structural materials in fluoride salts was studied and the results were used to optimize the composition of the alloys to mitigate the corrosion. Thermal diffusion and bimetallic methods were also tested. The corrosion of metallic samples in fluoride salts of different purity was also studied, and it was found that maintaining the oxygen level below 100 ppm in FLiNaK salt can significantly reduce corrosion. The infiltration of fluoride salts into graphite was also studied, and it was demonstrated that the ultrafine grain graphite to be used in TMSR test reactors has excellent resistance against salt infiltration. A natural circulation loop to study the dynamic corrosion by the fluoride salts was constructed. The dynamic corrosion experiment was completed in 2018.

5.3.1.4. Irradiation of thorium, alloys and graphite

Molten salt reactors are often associated with the Th–U fuel cycle [185]. To obtain irradiation data, thorium, alloys and graphite samples were irradiated. In 2015, a sample containing 9.7 g of thorium oxides was irradiated to conduct basic research on the separation of thorium and uranium. The number of neutrons accumulated on the sample during the irradiation was in the order of 10^{19} per cm^2. In 2018, thorium fluorides with and without other fluoride salt mixtures were irradiated to study the behaviour of the thorium fluoride fuel. Nickel alloy and nuclear graphite samples were also irradiated to 2.5×10^{19} and 5×10^{20} neutrons per cm^2, respectively, at temperatures of 650°C and 700°C. These are the design operating conditions of the test reactors. Following irradiation, the samples were subjected to post-irradiation examinations.

5.3.1.5. *Components and instruments*

Prototypical components and instruments for the test reactors were designed and built. A prototypical fluoride salt pump was manufactured using a nickel alloy. It has a design operating temperature of 700°C, a flow rate of 300 m^3/h and a pump head of 20 metres to meet the requirements of the test reactors. Freezing valves to be used in the test reactors were also developed, and other major components such as salt–salt heat exchangers, salt–air heat exchangers, reactor vessels, in-vessel structures and control rods were designed. Prototypes were produced by manufacturers and have undertaken tests in the design operating conditions.

Instruments for high temperature operation were also developed, such as flowmeters, manometers, thermometers and tank gauges that can be used in the fluoride salt test reactors. In addition, equipment that can be used to calibrate these instruments at temperatures of over 700°C was built, and software that can be used to control the test reactors and to simulate the behaviour of the reactors under different conditions was developed.

5.3.1.6. *Test loops*

Several test loops were constructed that can create an experimental complex to support future MSR R&D. These loops use fluoride salt and other kinds of surrogate fluids. The fluoride salt test loop uses FLiNaK salt and operates with a flow rate of 15 m^3/h at a temperature of 650°C. The loop was mainly used for testing components and obtaining fluoride salt operating experience. A natural circulation experimental loop using nitrates salt as the surrogate fluid is able to conduct experiments of the direct reactor auxiliary cooling system. The loop is capable of removing 20 kW of heat just by natural circulation of the salt and the air, a phenomenon that can be relied upon to passively remove the reactor's decay heat when all power is lost. Another nitrates salt loop with forced circulation was used for thermohydraulic experiments. A specially designed Dowtherm-A oil loop can be used to perform experiments to simulate some fluoride salt experiments when scaling analysis is correctly performed. The advantage of using oil as the surrogate fluid to replace fluoride salts is the significant cost reduction. There are also several water loops for various purposes, such as the hydraulics test of salt pumps, the visualization of the flow field and the performance test of the fluid diode. Figure 10 shows pictures of pumps, heat exchangers and loops. The following list describes some of the test loops used to support MSR R&D.

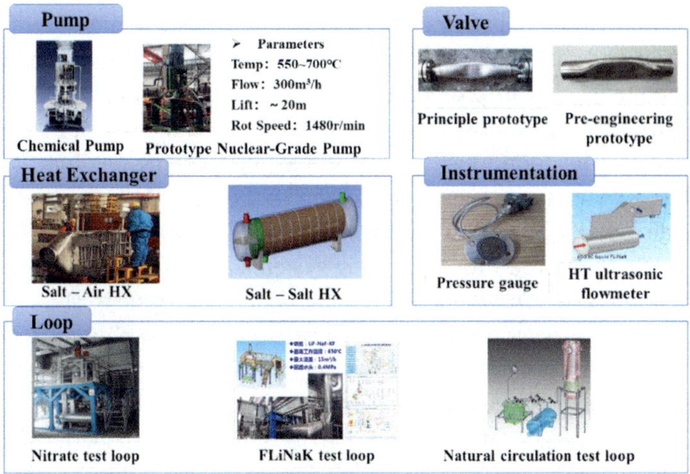

FIG. 10. Key equipment and molten salt loops developed at SINAP (courtesy of Y. Zou, Shanghai Institute of Applied Physics). HT — high temperature. HX — heat exchanger.

— *High temperature molten salt thermohydraulic experiment loop.* This loop is the first molten salt experiment loop at high temperature. The loop system scheme is shown in Fig. 11. It can be divided into five subsystems: (i) molten salt circulation system; (ii) molten salt storage system; (iii) heat conduction test system; (iv) pneumatic system; (v) control and safety interlock system. Various research efforts have been carried out, including (a) research and experimental platforms for key high temperature molten salt equipment; (b) research on the loop's thermohydraulic characteristics to provide a basis for the design, construction and operation of a TMSR loop; (c) research on the molten salt loop control and safety interlock system for the safe and reliable operation of the TMSR; and (d) research on molten salt loop operation technology. Establishing a database for the loop's operating parameters provides a basis for developing the operating procedures for the TMSR loop.

— *FLiNaK molten salt high temperature experiment loop [186].* This loop was designed and constructed to study resistance at high temperature and fluoride salt resistance of this loop and related equipment; to carry out research on molten salt flow and heat transfer characteristics; to provide a research platform for developing and testing key equipment; to perform tests such as a gas simulation test, a fluoride salt deoxygenation purification test and a thermal test; to study the control and safety interlock; and to improve safety. This loop is shown in Fig. 12. It is mainly composed of a molten salt circulation system, molten salt storage and transfer system, gas circuit system, heat tracing and insulation system, measurement system, control and safety interlocking system, pretreatment system and other key components. It can provide design basis and information for the design, construction, operation and safety control of the molten salt loop system of a reactor. This loop uses FLiNaK salt as the working medium. The fluoride salt is a eutectic composite salt composed of LiF–NaF–KF with a melting point of 454°C and a boiling point of 1570°C, which is commonly used to simulate molten salt for MSR loop systems and as a test bench.

— *Nitrate natural circulation experiment loop.* This system, shown in Fig. 13, mainly includes (i) a main circuit system; (ii) an air cooling tower system; (iii) a salt charging and discharging system; (iv) a measurement control system; and (v) a heat tracing insulation system. The main circuit system is the key system of this loop and its function is to perform the natural circulation of molten salt. Its main components are a direct reactor auxiliary cooling system heat exchanger, draft heat exchanger, pipelines, valves, expansion tanks, heating cables, heaters, nitrates and molten salt pools [187]. The cooling tower system is used to enhance the heat dissipation of the draft heat exchanger and is mainly composed of a ventilation tower, air door, heater, fan, lifting mechanism, positioning mechanism, and a temperature and humidity measuring instrument. The salt charging and discharging system mainly includes a gas system and a vacuum system. The measurement control system is used to measure and control the temperature, liquid level, flow and pressure of the loop system and to monitor system failures to ensure the safe and reliable operation of the loop system. It is mainly composed of thermocouples, temperature controllers, molten salt flowmeters, air flowmeters, liquid level gauges, pressure sensors, data acquisition instruments and other equipment. The heat tracing insulation system is used to heat the main circuit before molten salt is injected into the pipeline to prevent the molten salt from solidifying after entering the pipeline and causing blockage. It is mainly composed of heating plates, resistance wires and circuits.

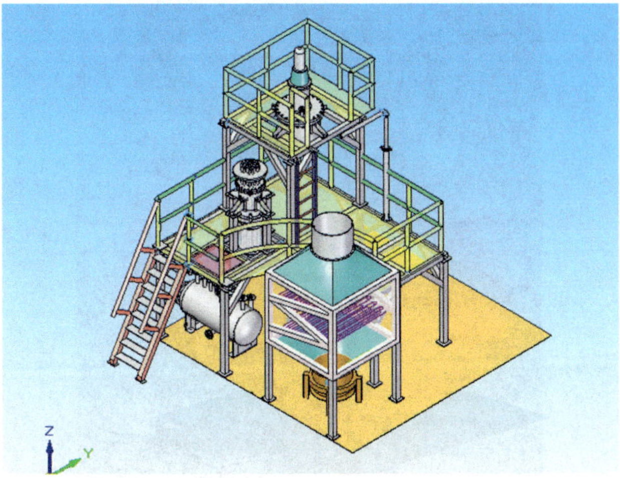

FIG. 11. Overall layout of the high temperature molten salt thermohydraulic experiment loop (courtesy of Y. Zou, Shanghai Institute of Applied Physics).

FIG. 12. Overall layout of the FLiNaK molten salt high temperature experiment loop (courtesy of Y. Zou, Shanghai Institute of Applied Physics).

FIG. 13. Overall layout of the nitrate natural circulation experiment loop (courtesy of Y. Zou, Shanghai Institute of Applied Physics).

5.3.1.7. Radiochemistry

In the thorium fuel cycle designed by the TMSR team, fissile materials including ^{233}U will be separated from the irradiated fuel and reloaded to the reactor. After decontamination, fluoride salt will also be reloaded to the reactor because of the salt's high cost. Research was conducted on the processing of the irradiated fuel. Techniques such as fluorination for uranium recovery, distillation for salt purification and fluorides' electrochemical separation for uranium recovery were developed and tested in 'cold' experiments [188]. The distillation device at laboratory scale can collect fluoride salt at a speed of 6 kg/h, and the decontamination factor (mass fraction ratio of the analysed element in the evaporated sample and condensed salt) is greater than 100 for most neutron poisons. The separation ratio of the electrochemical test set-up for uranium is over 99%. As shown in Fig. 14, hot cells for future 'hot' experiments of the irradiated fuel have been constructed and put into operation in 2018. Advanced instruments were procured and techniques were developed to accurately measure the properties and compositions of the salts. The TMSR laboratories are capable of measuring physical properties such as viscosity,

FIG. 14. Radiochemistry facilities (courtesy of Y. Zou, Shanghai Institute of Applied Physics).

density and thermal conductivity. They also provide an environment for researchers to analyse trace elements such as oxygen, boron, cadmium and rare earth elements in the salts. Examples of radiochemistry facilities are shown in Fig. 14.

5.3.1.8. Tritium management

FLiBe was chosen as the fuel salt for the test reactors. This salt generates tritium when irradiated by neutrons. Research was conducted on stripping tritium from the salt using a bubbling system, tritium separation from noble gases using cryogenics, tritium storage using alloys and continuous tritium monitoring to prevent an uncontrolled release of tritium into the environment [189]. Experiments using surrogate gases such as hydrogen and deuterium were carried out to evaluate the performance of different techniques. Some experiments were carried out on a palladium–silver alloy membrane diffuser, tritium storage alloy bed and oxidation post-treatment bed using hydrogen isotopes to simulate tritium control performance.

5.3.2. Design and construction of the TMSR-0 simulator reactor

The TMSR-0 simulator (mock-up) reactor, which was designed and constructed by SINAP, provides data to validate the thermohydraulic simulation and computer codes for analysing the safety of the TMSR test reactor. It is also a training platform for TMSR construction, operation and maintenance.

The TMSR-0 reactor is a simulator at a scaled size of 1:3 of TMSR-SF1, an experimental 10 MW(e) solid fuelled TMSR. The heat source of TMSR-0 is an electric heater with a maximum power of 400 kW. The reactor consists of a reactor body, a primary loop and a secondary loop (with molten salt to air heat exchanger), a passive residual decay heat removal system for natural gas circulation, and instrumentation and control systems. The layout of TMSR-0 is shown in Fig. 15. Table 9 provides the major technical parameters of TMSR-0.

FLiNaK molten salt is used for the primary and secondary loops of TMSR-0. The nickel based alloy GH3535 developed at SINAP is used as the structural material for the reactor body and both loops of TMSR-0. The inlet and outlet temperatures of the primary loop are 600°C and 650°C, respectively. TMSR-0 can be used to perform verification experiments related to the design, safety and components and instruments of TMSR-LF1, an experimental liquid fuelled TMSR.

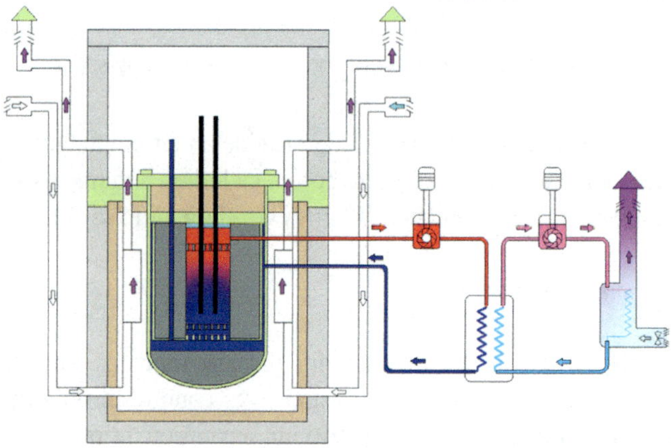

FIG. 15. Schematic layout of the TMSR-0 simulator reactor (courtesy of Y. Zou, Shanghai Institute of Applied Physics).

TABLE 9. MAJOR TECHNICAL PARAMETERS OF TMSR-0

Parameter	Value
Electric heater power (kW)	0–400
Coolant	FLiNaK
Mock-up core inlet/outlet coolant temperature (°C)	600/650
Air heat exchanger inlet/outlet molten salt temperature (°C)	520/536
Air heat exchanger inlet/outlet air temperature (°C)	40/180
Molten salt rated flow of primary circuit (kg/s)	0–10.0
Molten salt rated flow of secondary circuit (kg/s)	0–12.2
Power of the passive residual decay heat removal system (kW)	12.8

For the longer term development of the TMSR programme, technical information acquired from TMSR-0 can be used to support the design of the molten salt experimental reactor and the demonstration reactor, both based on thorium, and accelerate the development of MSRs.

The construction objectives of TMSR-0 include the following:

— To provide an overall engineering verification platform for the key technologies and equipment of the TMSR-SF1 and TMSR-LF1 experimental reactors, and to help gain the necessary experience for the design, construction, operation and maintenance of these reactors.
— To verify the thermohydraulic design and safety of the TMSR-SF1 and TMSR-LF1 experimental reactors, to verify the applicability of the design and analysis programme and to assist with obtaining licences for these reactors.
— To provide an important experimental platform for the future development of reactors cooled by molten salt. For example, this platform supports R&D for procedures for thermohydraulic and safety analyses and the development of key technologies and equipment.

The design and construction of TMSR-0 started in April 2016. The layout is shown in Fig. 16. The project design was completed in July 2016, the arrival

FIG. 16. Layout of the TMSR-0 simulator reactor (courtesy of Y. Zou, Shanghai Institute of Applied Physics).

and acceptance of all equipment in September 2018, and the construction and installation of TMSR-0 in 2019. The reactor started operation in March 2020.

5.3.3. Design and construction of the TMSR-LF1 experimental reactor

5.3.3.1. Introduction

TMSR-LF1 is a molten salt experimental reactor with liquid fuel and a power design of 2 MW(th). It employs an integrated layout where the pump and heat exchanger are located in the reactor vessel. The technical goal is to develop an MSR based on liquid fuel (thorium) and to verify its feasibility. TMSR-LF1 can be used to provide technical support and experience for the design and construction of a small modular molten salt demonstration reactor based on thorium. Table 10 provides the major technical parameters of TMSR-LF1. Figure 17 presents a schematic layout of the TMSR-LF1 reactor vessel and core.

TABLE 10. MAJOR TECHNICAL PARAMETERS OF TMSR-LF1

Parameter	Value
Reactor type	MSR with liquid fuel
Thermal capacity (MW(th))	2
Design life (years)	10
Fuel	LiF–BeF_2–ZrF_4–UF_4– (ThF_4) [a]
Moderator	Graphite
Core inlet/outlet coolant temperature (°C)	630/650
Rated flow of fuel salt circuit (kg/s)	50
U-235 enrichment (wt%)	19.75
Fuel loading/unloading	Initial loading by gas pressure, on-line loading by capsule and unloading by gas pressure
Reactivity control	Two shutdown systems
Control rod absorber material	B_4C

[a] Thorium will be added as experimental material for reactor operation.

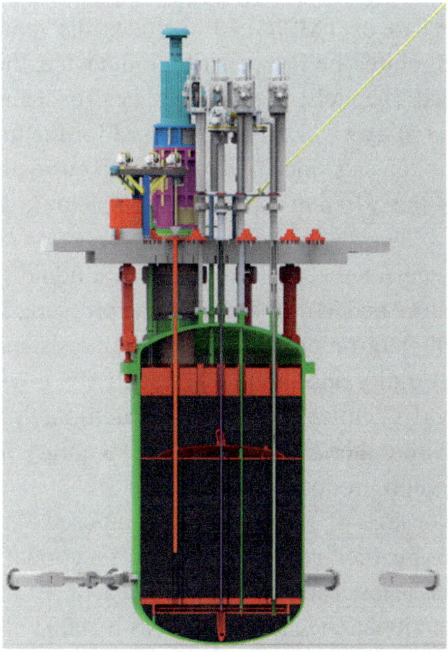

FIG. 17. Schematic layout of the TMSR-LF1 reactor vessel and core (courtesy of Y. Zou, Shanghai Institute of Applied Physics).

Verification activities include (i) the verification of basic principles, such as neutronic characteristics, thermohydraulic characteristics and reactor safety characteristics; and (ii) technical verification, such as verifying equipment and technology related to materials, fuel salt, molten salt, radiochemistry and loop and instrument and control systems.

5.3.3.2. System composition

TMSR-LF1 generates 2 MW(th), has a graphite moderator and an integrated layout with forced circulation. The main systems include an integrated reactor system, a cooling salt circuit system, an instrument and control system and auxiliary systems (such as a gas circuit system, a fuel salt loading/unloading, sampling system, a three-waste treatment system and a heat tracing and insulation system). Heat is generated when the fuel salt flows through the core and is released when it flows through the salt–salt heat exchanger. TMSR-LF1 is an integral reactor with the core, main pump and salt–salt heat exchanger located in the reactor vessel.

The reactor system of TMSR-LF1 includes the main reactor vessel, the core moderated by graphite, the flow distribution device, the control rod system, the relevant functional channels, the salt–salt heat exchanger and the fuel salt circulation pump. The alloy UNS N10003 is used for the metal components that are in contact with molten salt since it is resistant to corrosion. The following list gives a description of the different aspects of the reactor system.

— *Fuel salt.* The nuclear fuel in TMSR-LF1 is a liquid fuel salt that operates at high temperature and at near atmospheric pressure. Liquid fuel salt needs to have excellent neutron properties, radiation resistance, high specific heat capacity, good fluidity and thermal conductivity, a wide range of working temperatures and chemical stability. In addition, it needs to have good compatibility with materials and a certain solubility for fuel (uranium and thorium) and fission products.

— *Reactor core.* The core of TMSR-LF1 is composed of graphite components, on which there are fuel salt passages and functional channels. There are 244 fuel salt passages, each with a diameter of 40 mm. The reactor core contains ten functional channels with UNS N10003 alloy casing, including six control rod channels, one experimental measurement channel, one neutron source channel, one fuel loading/unloading channel and one temperature measurement channel.

— *Systems for reactivity control and shutdown.* The reactivity control system and the shutdown system are used to meet the core reactivity control and shutdown requirements in all possible states. TMSR-LF1 has a total of six control rods, which belong to two independent systems. The reactivity control system includes two compensation rods, one regulating rod and one safety rod. The four control rods of the reactivity control system perform reactivity control functions during normal operation, including power regulation, temperature compensation, xenon poison effect compensation, fuel consumption compensation, delayed neutron flow effect, operation backup and compensation for reactivity changes caused by experimental samples. The shutdown system consists of two safety rods, which have a driving mechanism and rod shape that are different from those of the first shutdown system. All control rod absorbers are made of boron carbide (B_4C). The neutron source is an americium–beryllium source.

— *The main reactor vessel.* This is composed mainly of a cylinder assembly and a roof assembly. The cylinder assembly mainly consists of a flange, a straight cylinder body, a lower head and an inlet and outlet pipe of cooling salt loop. The flange and straight cylinder body are made of UNS N10003 ring rolled pieces. The flange is the supporting part of the whole reactor. The straight cylinder body is composed of a barrel stiffener and a barrel.

The roof assembly mainly consists of the top head and the upper channel's socket. The top head is a butterfly head, welded with the pipe seat of the related channels.

- *Thermohydraulic design.* The basic function of the reactor's thermohydraulic design is to ensure that the reactor's heat transfer capacity is matched with the reactor's heat generation capacity during normal operation. In addition, the design provides a set of reasonable thermal parameters such as loop pressure and temperature for the thermal power transfer system. Furthermore, it can transfer decay heat from the reactor to the ultimate heat sink after shutdown and ensures the integrity of multichannel barriers, which are designed to limit the release of radioactive products during potential accidents.
- *Temperature considerations.* The temperature of UNS N10003 structural material on the fuel salt pressure boundary of the reactor is lower than 700°C under normal operating conditions. The maximum temperature is lower than 800°C under the predicted operating events and design basis accidents. The maximum temperature of fuel salt in the core should be less than 1200°C under design basis accidents. The minimum temperature of fuel salt in the reactor is above 500°C under normal operating conditions and predicted operating event conditions.
- *Fuel management system.* The fuel management system is used for initial fuel loading and fuel salt discharge under normal or accident conditions. The fuel salt in these cases is driven by gas pressure and gravity.
- *On-line fission gas removal system.* Fission gases and some noble metals are removed from the reactor by entrained cover gas. They then flow into the off-gas process system for decay, filtration and finally are released into the atmosphere.

5.3.3.3. Safety features

The following list describes the safety features of TMSR-LF1.

- *Reactivity safety.* Changes in the temperature of a nuclear reactor lead to changes in the reactor's reactivity, which is characterized by the temperature reactivity coefficient. The coefficient for TMSR-LF1 is strongly negative, which supports the control and safety of the reactor at various power levels. As described above, TMSR-LF1 has two different control rod systems, which will automatically operate in accidents. Low excess reactivity allows the reactor to be shut down through the very negative temperature reactivity coefficient. After shutdown, the fuel salt can be drained from the reactor for long term shutdown.

— *Decay heat removal system.* TMSR-LF1 has a passive decay heat removal system. The verification test of this system's capability has been completed on the TMSR-0 simulation reactor, including verifying the reliability of the associated software.
— *Confinement system.* The confinement system performs the safety function for containing the radioactive nuclide after a hypothetical design basis accident. Therefore, the safety vessel of this system contains the fuel salt system and reactor.

5.3.3.4. Plant safety and operational performance

Design principles for TMSR-LF1 are (i) to establish and adopt effective process and design guidelines and (ii) to ensure that basic safety design principles are followed throughout the phases of design and design change.

5.3.3.5. Project progress

The TMSR-LF1 project is undertaken by SINAP, part of the Chinese Academy of Sciences. The site selection, environmental impact assessment, scheme design, preliminary engineering design and construction drawing design have been completed. In December 2018, the Chinese National Nuclear Safety Administration approved Hong Sha Gang Industrial Park in Wuwei, Gansu Province, as the site for TMSR-LF1, and in July 2019, the design was also approved. In August 2019, the technical review of the TMSR-LF1 preliminary safety analysis report was completed. In January 2020, the National Nuclear Safety Administration approved the construction permit for TMSR-LF1.

The construction of TMSR-LF1 officially began in March 2020 and the plant's main building was completed by the end of 2020. The equipment for TMSR-LF1 has been manufactured and its installation started before the end of 2020.

5.4. RESEARCH AND DEVELOPMENT ACTIVITIES IN THE CZECH REPUBLIC

Nuclear reactor systems with liquid molten salt fuel and corresponding pyrochemical technologies have been investigated in the Czech Republic since 1999. The original impetus for this R&D came from the partitioning and transmutation concept, which is based on the combination of pyrochemical partitioning technology and the incineration of TRU elements in a subcritical accelerator driven system with liquid fluoride fuel. After 2005, this idea was gradually abandoned and transformed into the development of the classical MSR technology utilizing the Th–U fuel cycle.

Work in the field of theoretical and experimental development of MSR technology and the Th–U fuel cycle was carried out by a national consortium of institutions and companies managed by the Nuclear Research Institute Řež. Since 2013, these R&D activities have been led by the Research Centre Řež.

This R&D has been managed by the Ministry of Industry and Trade of the Czech Republic, which also financially supported two important domestic projects dedicated to MSR technology. These were the projects 'System SPHINX with liquid fluoride fuel' and 'Fluoride reprocessing of spent fuel from GEN-IV reactors', launched in 2004 and 2006, respectively.

Current activities for developing MSR technology are supported by the Technology Agency of the Czech Republic. These activities cover the following fields:

— *Reactor physics and neutronics.* Physics and neutronics of MSR carrier salt, graphite and comparison of measured data with results from computer codes.
— *Fuel and coolant chemistry and supporting technology.* Liquid fuel processing, pyrochemical separation studies applicable for on-line reprocessing and studies of flowsheeting.
— *Performance of materials.* Development of alloys based on nickel for the environment of a fluoride salt, mechanical and corrosion tests of these alloys and development of impellers and flange and gasket systems.

5.4.1. Reactor physics

In the areas focused on theoretical and experimental reactor physics of MSR systems with liquid fuel and FHRs, efforts were directed mainly at the interconnection of theoretical and experimental studies in the field of thermal neutron spectrum and neutron studies of MSR, as well as studies of FHRs of FLiBe type. Experimental work on the neutronics of pure FLiBe salt and on the neutronics of FLiBE mixture with thorium and uranium fluorides was carried out at the LR-0 experimental reactor of the Research Centre Řež. For these studies, the LR-0 active zone is composed of six pin-type fuel assemblies (VVER-1000 design) with a nominal uranium enrichment of 3.3%, which form the neutron driver, and an empty experimental channel located in their central core, which forms the driven zone. Materials are inserted into this central driven zone and occupy one position in the overall lattice. The FLiBE tests were performed with MSR $LiF–BeF_2$ cooling salt (66–33 mol%) containing the isotope ^7Li (purity 99.994 mol%). This salt was provided by ORNL for the experiments reported here. The tests were aimed at studying the shape of the neutron spectrum and were compared with previous experiments performed with the LiF–NaF salt.

The neutron spectrum after passing through the salt was measured by the recoil proton method in different energy ranges. These spectra were measured independently by a set of hydrogen proportional detectors (in the 0.1–1.3 MeV range) and by organic scintillators (Stilbene scintillator, 10 × 10 mm for the 0.8–10 MeV energy range) with shape discrimination of neutron and gamma pulses. The model used to calculate the criticality and neutron spectra LR-0 was subsequently analysed with the Monte Carlo N-particle 6.1 code using data from various nuclear libraries. Data[9] from ENDF/B VII.0 and ENDF/B VII.1, Joint Evaluated Fission and Fusion (JEFF) 3.1 and 3.2, Chinese Evaluated Nuclear Data Library CENDL 3.1, Japanese Evaluated Nuclear Data Library JENDL-3.3 and JENDL-4 and Russian File of Evaluated Neutron Data RUSFOND-2010 were then compared with data from older versions of ENDF/B VII.0 libraries and JEFF 3.1. This mainly concerned the data from the different nuclear libraries that were used to define the material loaded into the reactor zone. In this case, it was mainly data related to fuel, moderator and structural materials. The reason was to exclude their influence on criticality, as this value was not the subject of the studies described. The thermal neutron scattering in FLiBe, Teflon and stainless steel canisters was described by a free gas model. The described physical model in this case also did not consider photoneutron production [190, 191].

Existing measurements with FLiBE salt were performed at room temperature. However, future neutronic tests will be performed in a special preheated zone put into LR-0. The measurement is planned to be carried out in the temperature range of 500–750°C.

5.4.2. Fuel and coolant chemistry and supporting technology

In the area of chemistry and chemical technology of MSRs, the work covered the verification of processing of liquid fuel. It included the preparation of pure UF_4 and ThF_4 and the final processing of MSR liquid fuel samples (i.e. preparing several mixture types containing FLiBe melts with various contents of UF_4 and

---

[9] The data were obtained from the following nuclear libraries: (i) ENDF/B VII.0: US Evaluated Nuclear Data Library released in December 2006; (ii) ENDF/B VII.1: US Evaluated Nuclear Data Library released in December 2011; (iii) CENDL 3.1: Chinese Evaluated Nuclear Data Library, released in December 2009, containing neutron-induced data up to 20 MeV for 240 materials; (iv) JENDL-3.3: Japanese Evaluated Nuclear Data Library, released in 2002, containing data for 337 materials; (v) JENDL-4: Japanese Evaluated Nuclear Data Library, released in 2010, containing data for 406 materials; (vi) RUSFOND-2010: Russian File of evaluated nuclear data, released in 2020, containing data for 686 materials.

ThF_4). The processing of MSR liquid fuel was verified at high laboratory level, which typically corresponds to several hundred grams per batch [192].

The other area of chemical technology covered the study of electrochemical separation from fluoride molten salt. The effort focused on developing an experimental set-up for molten fluoride salt, including developing a reference electrode, and evaluating the individual redox potentials for uranium, thorium and the most important fission products in selected molten fluoride salts — typically FLiBe, LiF–CaF_2 and LiF–NaF–KF (FLiNaK) mixtures. Another development focused on investigating the technique of fused salt volatilization proposed for the extraction of uranium (in the form of UF_6) [193].

In addition, studies of the technology of the MSR fuel cycle covered the flowsheeting. Research focused on balance simulation calculations and proposals of reprocessing diagrams, safety assessment, and non-proliferation and physical protection aspects of the systems used in the Th–U fuel cycle.

5.4.3. Performance of materials

Research on materials is focused primarily on developing the superalloy MONICR (molybdenum–nickel–chromium), which is suitable as a structural material for MSR technology. Mechanical, corrosion and irradiation tests have been carried out and development is also focused on studying casting, forming and recrystallization. Other studies are focused on the use of MONICR for the manufacture of components for MSR technology, such as impellers, flanges and valves [194]. The ADETTE Molten Fluoride Loop programme is an integral part of the investigation. It is focused mainly on dynamic corrosion tests of structural material and verification of the function of selected devices and methods of measurement in the environment of molten salts. An important part of this programme has been also the design and study of the operation of the so-called freeze safety valve.

5.5. RESEARCH AND DEVELOPMENT ACTIVITIES IN DENMARK

Until only a few years ago, research on nuclear technologies in Denmark had been carried out solely at the Centre for Nuclear Technologies of the Technical University of Denmark (DTU). Project activities for Commercial Danish designs have started. Research and development for MSR technologies in Denmark is mainly carried out at two nuclear startups, Copenhagen Atomics and Seaborg Technologies, as well as at DTU.

Owing in part to the merger of the Risø National Laboratory with DTU in 2007, DTU has a long history of research in nuclear energy. Originally centred around the three research reactors at Risø, in particular the DR3 reactor that was cooled by heavy water and generated 10 MW(th), nuclear research initially focused on reactor physics, nuclear chemistry, dosimetry and health physics and radioecology. For many years, DR3 constituted an important neutron source in Europe and was utilized for a range of applications for materials research, neutron scattering experiments, isotope production and neutron transmutation doping of silicon for high power semiconductors. After the closure of DR3 in 2000, materials science with neutrons continued at DTU and at the University of Copenhagen using other neutron scattering facilities in Europe and elsewhere.

Meanwhile, as Denmark shifted away from including nuclear power in its energy policy, R&D at the Risø National Laboratory expanded to cover a range of renewable energy technologies including wind, solar photovoltaic cells, fuel cells and electrolysis and bioenergy. Research on nuclear technologies focused on radioecology, dosimetry, luminescence physics and neutronics.

Neutronics at DTU today encompasses three main areas: (i) physics of neutron scattering, in particular aimed at code development (e.g. McStas[10], Monte Carlo particle list, NCrystal[11]), neutron scattering kernels and neutronics support for neutron scattering experiments; (ii) neutronics design, aimed at designing and optimizing nuclear scattering instruments, neutron optics, moderators and nuclear diagnostics; and (iii) reactor physics. Activities within all these fields are carried out in close collaboration with external partners, in particular with the European Spallation Source for a target monolith design including the moderator reflector system and target wheel diagnostics, Fusion for Energy for a neutronics design of the collective Thomson scattering diagnostic for ITER and Seaborg Technologies for MSR development projects.

Research on reactor physics focuses on the development of models for MSR designs. In its main design version, the fuel is dissolved in the molten salt, resulting in strongly coupled neutronic and thermohydraulic effects, further complicated by the transport of delayed neutron precursors by the circulating fuel. In an ongoing PhD project, different coupling schemes of neutronic–thermohydraulic single physics codes are being investigated, using open source software (e.g. OpenFOAM CFD) and restricted neutronics codes (e.g. SERPENT, Monte Carlo N-particle). The different coupling approaches are benchmarked against historical data from the MSRE as well as data from molten salt experimental loops, such as the DYNASTY (dynamics of natural circulation for molten salt internally heated) loop at Politecnico di Milano, Italy.

In the SEALION collaborative project (with Seaborg Technologies, Chalmers University of Technology in Sweden and Development of Advance

[10] General tool for simulating neutron scattering instruments and experiments.

[11] A library for thermal neutron transport in crystals and other materials.

Engineering Solutions (DAES), a Swiss startup), regulatory compliant multiphysics software has been developed to facilitate the process of licensing MSR designs. The approach applies an external coupling scheme between the neutronic and thermohydraulic codes, leaving these codes of single physics unaltered. The coupled codes are benchmarked against MSRE data.

A new reactor technology project started in 2020 at DTU, also in collaboration with Seaborg Technologies, to develop designs for freeze valves. The aim of the project is to develop and apply CFD models for the optimization and safety analysis of these designs.

Copenhagen Atomics is a Danish molten salt technology and reactor company that aims to develop thorium breeder MSRs that are mass manufacturable. The company was formed in 2014, incorporated in 2015 and was the first nuclear reactor company in Denmark. Copenhagen Atomics chose a unique approach to developing a reactor, in that all components of the design are developed, assembled and tested in house, and the systems as a whole are tested with non-nuclear salts. This approach allows for rapid iterations of testing to ensure the reliability of the systems before seeking nuclear regulatory approval of a design that has never been built and tested. Copenhagen Atomics promotes open collaborations with other companies, universities and national laboratories and offers many of its test systems for purchase and for academic research. The objective is to accelerate the development of molten salt technologies and help talented people gain hands-on experience with molten salt systems.

Copenhagen Atomics is actively developing components considered to be state of the art for MSRs, including a canned rotor molten salt pump with electromagnetic bearings, molten salt valves, molten salt filters, molten salt flowmeters, qualification of commercial heat exchangers for use in MSRs, salt spraying off-gas systems, qualification of composite materials as salt wetted structural material in MSRs, fluoride salt purification, on-line molten salt redox meter, on-line laser breakdown spectroscopy meter, autonomous reactor control systems and software, assembly line production of molten salt test loops and MSRs that are the size of a shipping container. Some of these research activities are carried out with university and industry partners. The company has also received approval by a national nuclear regulatory body to work with a salt containing fertile uranium and thorium at a facility outside of Denmark, and has begun to run static salt corrosion experiments and pumped salt loop tests of long duration with fertile bearing salts.

Copenhagen Atomics is developing a 100 MW(th) breeder MSR that is moderated with heavy water (thermal spectrum) and that fits inside a leaktight containment, which is the same size as a shipping container and has a length of 12.2 m. The heavy water moderator is unpressurized, thermally insulated from the fuel salt that can reach up to 700°C, continually circulated and cooled, and passively drained from the core in case a loss of power or abnormal condition is detected. Materials that are highly reliable with low absorption of neutrons for separating and insulating the fuel salt and moderator are the focus of ongoing research at Copenhagen Atomics, which is being carried out in collaboration with university and industry partners.

In 2014, Seaborg Technologies proposed the Molten Salt Thermal Wasteburner, or MSTW, a thermal spectrum, graphite moderated MSR with a single salt, fuelled with a combination of spent nuclear fuel and thorium [195]. It was envisioned to produce 115 MW(e) with a two stage turbine, from 270 MW(th). The design philosophy was to create a reactor that was small, modular and possible to manufacture in large quantities, with a lifetime of about seven years. The design developed into the current Compact Molten Salt Reactor (CMSR), a modular reactor generating 250 MW(th) and placed on a floating barge (see Appendix V, Section V.2.8 for specifications). The conceptual design phase of the CMSR was completed with the first regulatory approval phase in 2020. The current focus of the company is on molten salt experiments, building and testing equipment prototypes, modelling and simulation and developing the basic engineering design.

5.6. RESEARCH AND DEVELOPMENT ACTIVITIES BY THE EUROPEAN COMMISSION

5.6.1. Introduction and main achievements

5.6.1.1. History

The research efforts of the European Commission's JRC on the technology of MSRs started around 2004. At that time, a group of research organizations in Europe began collaboration on the topic of MSRs, which was selected by the Generation IV International Forum. This collaboration resulted in the project on molten salt reactor technology (the MOST project) in the Fifth Framework Programme of the European Atomic Energy Community (Euratom). The JRC activity started at the Karlsruhe site in Germany, a research installation with a strong focus on fuel and fuel cycle issues. Naturally, the JRC started to address the chemistry of the molten salt fuel.

The liquid MSR fuel is a liquid solution in which — unlike solid oxide fuel — chemistry plays a much more important role and thermomechanics a minor role. The temperature distribution in the liquid is relatively homogeneous, and the mutual solubility in the liquid of the salt and the fission products is high. As a result, the chemistry and some of the properties of the liquid MSR fuel can be described very well by equilibrium thermodynamics. This was the main focus of JRC research in the first period (2002–2006).

From the beginning, a combination of thermodynamic modelling and experimental studies has been the goal. The extensive literature on phase diagrams and thermodynamic properties of molten salt systems from the MSRE and MSBR period at ORNL provided a solid base from which to develop an extensive thermodynamic model of MSR fuel systems with the calculation of phase diagrams (CALPHAD) approach. In this approach, the experimental data are fitted to semi-empirical models for the various phases (i.e. solids, liquids, gases) to obtain a consistent description of multicomponent systems. Such a model, if based on sufficient and reliable experimental data, has a strong predictive capacity and can be used for design and safety analysis. The focus was on the $LiF–BeF_2–ThF_4–UF_4$ system.

Initially, the experimental work aimed to validate data from ORNL and other sources, and was extended when needed to improve the reliability of the models. However, the experimental work quickly proved to be complicated. Fluoride salts are corrosive, reactive and although only slightly volatile, the experimental equipment needs appropriate protection. Moreover, methods needed to be developed to work with small quantities (not the gram quantities used by ORNL), and it quickly became clear that the samples of ThF_4 and UF_4 that were made available to the JRC contained impurities that affected the measurements. The experimental work therefore originally focused on two areas: the synthesis of pure materials (see Section 5.6.2.1) and the development of the appropriate experimental tools for accurately measuring physical properties at high temperatures (see Section 5.6.2.4).

As a result, the JRC in Karlsruhe now has a collection of instruments and facilities suitable for investigating molten salt fuels. These facilities have been used extensively in a series of Euratom indirect action funded projects, as described in Section 5.6.1.2. In addition, a comprehensive thermodynamic database has been compiled, the Joint Research Centre Molten Salt Database (JRCMSD), which is nowadays available through the Thermodynamics of Advanced Fuels – International Database, or TAF-ID, project organized by the Nuclear Energy Agency of the Organisation for Economic Co-operation and Development. The experimental and modelling studies are well embedded in collaborations with European academic, research and industrial organizations, but also outside of the European Union, and address a wide range of different fuels and coolants, including fluoride and chloride salts. The ongoing work has evolved/extended to irradiation experiments in the High Flux Reactor (see Section 5.6.3), in which molten salts are irradiated in a reactor for the first time in several decades. These experiments, which are described in Sections 5.6.2.4 and 5.10.1.6, will provide further samples to test the thermochemical models and will

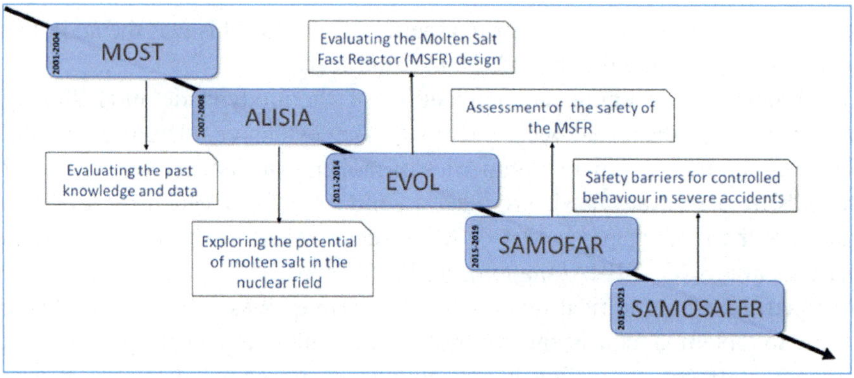

FIG. 18. Timeline of European Union projects on the R&D of MSRs (courtesy of O. Benes, Joint Research Centre of the European Commission).

provide insight into the fuel behaviour and how the fuel chemistry model will need to be developed further.

5.6.1.2. European framework projects

In the framework of the Euratom Research and Training Programme, a series of projects on MSR technology have been supported since 2004, as shown in Fig. 18. These projects showed an evolution from research on the ORNL concepts (e.g. the MOST project), to a broader review of molten salt applications in nuclear technology, to a focus on the MSFR design by the CNRS in France. Research within these projects is being carried out in various countries of the European Union in the framework of national research programmes combined with European Union Research Framework Programmes.

(a) Past projects

After completion of the MOST project, the project for the assessment of liquid salts for innovative applications (the ALISIA project) was funded for one year and was dedicated to exploring the potential of MSR technology in the nuclear field. Furthermore, it focused on the selection of the fuel salt and the design for a European MSR.

The project on the evaluation and viability of the liquid fuel fast reactor system (the EVOL project) focused on the integral evaluation of the MSFR design (which at that time utilized both core fuel and blanket salt that were rich in thorium to breed fissile uranium), including reactor and cleanup facility, and led to a buildup of necessary computational tools in Europe. It was performed in close collaboration with the sister project on minor actinides recycling in salts (MARS, 2010–2013), which was sponsored by the State Atomic Energy Corporation "Rosatom", in the framework of the collaboration between the European Union and the Russian Federation in the field of nuclear energy. The conclusions of the EVOL project can be summarized as follows [196]:

— Thermohydraulic and neutronic calculations followed by safety analyses were conducted to propose an innovative design. Previous studies emphasized that it is necessary to actively circulate and cool the blanket salt. In the current MSFR design option, the blanket salt flows in an external circuit adjacent to the salt container. This arrangement ensures a desired neutron economy because most of the neutrons escaping from the reactor core will be captured in the surrounding blanket. Moreover, the blanket salt follows a pathway that improves the cooling of the main metallic structures of the fuel circuit.

— Since ^{233}U is not available in the reactor, for the startup, the MSFR needs another fissile material (i.e. ^{235}U and/or ^{239}Pu) or TRU elements. Two options for the fuel composition have been proposed that meet the requirements from the points of view of neutronics, physicochemical properties (such as melting temperature, viscosity, density and solubility of elements), redox potential (required for material compatibility), fuel reprocessing and economy. The two options retained are $LiF–ThF_4–UF_4–PuF_3$ (78.6–12.9–3.5–5 mol%) and $LiF–ThF_4–UF_4–(TRU)F_3$ (77.5–6.6–12.3–3.6 mol%). The two options require an enrichment of ^{238}U into ^{235}U of 20% and 13%, respectively.

— The fuel reprocessing scheme has been demonstrated and the chemical database required for calculations of efficiency of the involved processes was established in the framework of the project. New steps have also been proposed for the fuel reprocessing, leading to an efficient recovery of the fuel components. The behaviour of all relevant fission products in the reprocessing has been evaluated.

— Three types of NiWCr alloys were manufactured in the project. The optimized composition for the MSFR was found to be (in wt%) 21.7 tungsten, 7.5 chromium, <0.02 manganese, 0.22 aluminium, 350 ppm zirconium and 0.22 titanium. The corrosion studies have demonstrated the role of oxygen and of the ratio of uranium (IV) to uranium (III) to decrease corrosion reactions. Dynamic studies have been performed through the collaboration between Rosatom and Euratom.

In the SAMOFAR project, the key safety features of the MSFR were analysed in depth, based on substantial improvement of the computational tools, particularly in the field of multiphysics analysis and experimental studies. The conclusions of the SAMOFAR project can be summarized as follows:

- — Initiating events have been identified and classified;
- — Properties of the MSFR fuel have been well defined;
- — Valuable experimental tools for validating simulation tools have been implemented;
- — Enormous progress has been made in the field of multiphysics modelling of the core;
- — The potential for safety barriers to be implemented has been further substantiated.

(b) Current project

The SAMOSAFER project on severe accident modelling and safety assessment for fluid-fuel energy reactors is the European Commission's latest project and a logical follow-up to the previous R&D projects funded by the European Union as part of Horizon 2020. The four year project started in 2019 and will build upon the experience and knowledge from previous projects funded by the European Union, with the aim of ensuring that MSR technology (not only limited to the MSFR concept studied in the past in the European Union) can fully comply with the more stringent safety requirements expected in the next 30 years.

In particular, the SAMOSAFER project will focus on the following phenomena important for safety assessments [197]:

- — Freezing of the fuel salt against cold walls and subsequent remelting of the salt;
- — Internal heating of the fuel salt causing lower natural circulation and local overheating;
- — Overheating of the fuel salt in the MSFR core during transients and in the drain tanks during storage;
- — Effects of transients on the thermomechanical integrity of the primary circuit;
- — Redistribution of the inventories of radionuclides in the fuel treatment unit via gas bubbling, fluorination and chemical extraction leading to changes in chemistry and mobility of radionuclides.

The chemistry and physics needed to increase the safety of the MSFR, and of an MSR in general, will also be modelled in the SAMOSAFER project. These include the following:

— Thermochemical modelling to evaluate the properties of the fuel salt to retain fission products, and the effects of fission products and corrosion products on the fuel salt's thermophysical properties (e.g. melting point, heat capacity, vapour pressure, viscosity, thermal conductivity);
— Radiation heat transfer to accurately calculate the decay heat removal from the emergency draining system;
— Predictive reactor control strategies to reduce the number of draining events, to avoid accidents and to mitigate their consequences;
— Redox control of the fuel salt to reduce corrosion in the primary circuit;
— Reactor scaling effects on the safety of nuclear reactors in general;
— Uncertainty quantification methods based on non-intrusive polynomial chaos expansion and reduced order methods (important for licensing).

New safety features to prevent and mitigate severe accidents will also be designed, such as an iodine trapping system, an off-gas system and a gas holdup system of the gaseous fission products in the fuel treatment unit, and freeze plugs and drain tanks (for both emergency and normal conditions).

For the experimental validation of the new simulation methods and tools, the following facilities will be used:

— *DYNASTY*. This is a facility at Politecnico di Milano with a height of 10 metres to study the flow dynamics of internally heated salts. Two versions exist: a single loop system and two connected loops. The latter set-up simulates the primary salt circuit (core region) connected to the salt flow in the intermediate circuit. DYNASTY is fully instrumented and can be used to study decay heat removal from the core region by natural circulation. It has been modified in the SAMOFAR project to fully meet the specific needs of MSRs.
— *ESPRESSO*. This is a new facility at TU Delft to investigate melting and solidification in conjunction with flow. It consists of a transparent container with cooling options at the bottom. Both the container and the fluid are transparent, which allows a visual determination of the local velocities in the salt with laser methods such as laser Doppler anemometry and particle image velocimetry, and the development of a solid layer near the cooled surface.
— *High Flux Reactor*. This is a materials testing reactor in Petten, the Netherlands, capable of generating 40 MW(th) and being used to irradiate

samples of fuel salt consisting of ThF_4 mixed in LiF. After irradiation for one or two years, the samples will be analysed both at NRG and at the JRC in Karlsruhe, with a focus on the composition of fission products in the samples, redox potential, and the interactions between the fuel salt and the graphite crucibles and between the fuel salt and the metal encapsulation.

— *SWATH-S.* This is a facility at the CNRS consisting of two vessels, one of which is filled with liquid salt (FLiNaK). The salt can flow by pressure from one vessel to the other, thereby passing through an experimental station in a glovebox. The experimental station can be adapted to the needs of the experiments, such as the measurement of the flow and temperature profiles of the salt in turbulent and laminar conditions or the freezing phenomena of the salt against cold walls.

— Experimental facilities of the JRC (see Section 5.6.2.4) will be used to provide novel data on the physicochemical properties of MSR fuels.

(c) Simulation tools used in the SAMOFAR and SAMOSAFER projects

For the safety assessment of the MSFR in various transient scenarios, multiphysics tools are necessary to capture the essential physical phenomena in the reactor system. Classical reactor physics computational codes cannot be used as they do not allow for the key features of the MSFR. The key features of the MSFR include the movement of precursors with the moving fuel, the strong coupling between the neutronic and the thermohydraulic aspects due to the use of molten salt fuel, the internal heat generation and the shape of the core having no fuel pins as a repeated structure. These features cause a variety of phenomena occurring during transients that are specific to the MSFR for which dedicated tools have been developed.

This subsection presents a basic description of the tools used in the SAMOFAR and SAMOSAFER projects. The capabilities of these codes include detailed CFD analysis (two dimensional, three dimensional, geometrically flexible), detailed neutronics (diffusion, transport) and the incorporation of the melting and solidification of the fuel salt as well as complex physics such as gas bubbling. Specifically, the code systems of the CNRS, Karlsruhe Institute of Technology/EDF, TU Delft, Politecnico di Milano and Paul Scherrer Institute are described below.

— *Karlsruhe Institute of Technology (SIMMER-III).* SIMMER-III is a two dimensional, multivelocity field, multiphase, multicomponent, Eulerian fluid dynamics code coupled with a fuel pin model and a space and energy dependent neutron kinetics model [198]. In this code, phase change (melting/freezing; evaporation/condensation) models are available.

This code consists of three basic elements: (i) a fluid dynamics model, (ii) a structure (fuel pin) model and (iii) a neutronics model. The fluid dynamics model (the main part of the whole code) is interfaced with the structure model through heat and mass transfer at structure surfaces. The neutronics model provides nuclear heat sources based on the time dependent neutron flux distribution consistent with the mass and energy distributions. The SIMMER-III code models incorporate five basic materials in the reactor core: fuel, steel, coolant, control and fission gas. A material can exist in different physical states. For example, fuel can be represented as fabricated pin fuel, liquid fuel, a crust refrozen on structure, solid particles, broken fuel pellets (called chunks) and fuel vapour, whereas fission gas exists only in a single gaseous state. The fuel includes fertile and fissile components to represent different enrichment zones in the core. Material density and energy component distributions are obtained through mass, momentum and energy conservation equations. Multivelocity field formulation and the fluid convection solution algorithm are based on a time factorization approach.

— *Politecnico di Milano (OpenFOAM).* A multiphysics solver for the analysis of the accidental scenarios of the MSFR has been developed and extended, starting from a previous tool based on the OpenFOAM library [199]. OpenFOAM is an open source library for numerical simulation in continuum mechanics using the finite volume method. The toolkit is very flexible because of the object oriented programming, allowing users to customize, extend and implement a complex physical model [200]. OpenFOAM facilitates the parallelization of the developed solvers owing to dedicated routines based on geometrical domain decomposition. The solver allows for the transient analysis of the MSFR, describing the neutronics and the thermohydraulic behaviour of the reactor, the movement of the delayed neutron precursors, and the decay heat. The solver was extended from the previous version by introducing new features for neutronics and thermohydraulics. For neutronics, a multigroup diffusive model has been adopted along with acceleration routines in order to speed up the convergence rate of the multigroup neutronics solution. This improvement was achieved by implementing the neutronics module of the GeN-Foam multiphysics platform [201, 202], which is a general purpose solver for transient analysis of nuclear reactors, in the Politecnico di Milano solver. This is possible because of the object-oriented programming of OpenFOAM. For thermohydraulics, the code was extended by introducing the compressibility, which could have an important effect on the simulation of fast transients [203]. In addition, the transport of gas bubbles in the salt mixture was added to simulate the effect of increasing/decreasing bubbling injection on the reactor dynamics. Besides the removal of fission products, the bubbling system can be adopted for reactivity control in the MSFR.

— *Paul Scherrer Institute (GeN-Foam).* GeN-Foam is a reactor modelling code based on OpenFOAM and is a library written in C++ used to create executables that solve problems of continuum mechanics or perform data manipulation. It is a multiphysics code capable of modelling neutronic, thermohydraulic and thermomechanical aspects in steady state and transient conditions. Features of GeN-Foam have already been described in detail in journal publications from the main code developer [201, 202].

— *TU Delft (PHANTOM-SN and DG-Flow).* The tools in the area of neutronics have evolved into full transport codes based on discontinuous Galerkin finite elements at TU Delft. The in-house code based on this principle is PHANTOM-SN and it can be considered a state of the art solver technology. A similar CFD code based on the same principle has been developed and is called DG-Flow. It combines the advantage of local conservation of finite volumes with the geometric flexibility and high order accuracy of finite elements [204]. The coupling of these two tools is capable of high accuracy simulation of transients in the MSFR, including the movement of the precursors. In the future, additional physics will be included, such as the tracking of bubbles and the incorporation of the bubbles' effect on the fluid phase.

  - PHANTOM-SN is a radiation transport tool capable of solving the linear Boltzmann radiation transport equation. The code has extensive capabilities, such as solving the steady state and time dependent linear Boltzmann equation, both principal and multimodal calculations of various eigenvalue types (criticality and time eigenvalues) and extensive perturbation analysis (both regular and generalized).
  - DG-Flow is a Navier–Stokes solver based on the discontinuous Galerkin approach. Both standard Navier–Stokes as well as turbulent flow can be handled by using Reynolds-averaged Navier–Stokes models.

— *CNRS (Transient Fission Matrix — OpenFOAM).* The transient fission matrix (TFM) approach [205, 206], described in Section 5.7.2.2, has been developed specifically as a neutronic model to account for the phenomena associated with precursor motion and to perform coupled transient calculations close to the Monte Carlo accuracy for the neutronics while sustaining a low computational cost. This approach is based on a pre-calculation of the neutronic reactor response through time prior to the transient calculation. The results of the calculations using the SERPENT Monte Carlo code [207] are condensed in fission matrices, keeping the time information. These matrices are interpolated to account for the effects of local Doppler and density thermal feedbacks due to temperature variation in the reactor system. This approach provides very quickly an estimation of the neutron flux variation for any temperature and precursor distribution in the reactor.

5.6.2. Fuel and coolant chemistry and supporting technology

5.6.2.1. Synthesis of actinide halides

Basic thermodynamic and electrochemical data of actinide fluorides and chlorides are required for the design and safety assessment of any MSR concept. Since these compounds are usually not available commercially, or not in the required purity, they have to be prepared from the available input materials, typically oxides. This subsection summarizes activities performed at the JRC in Karlsruhe pertaining to the synthesis and characterization of actinide fluorides and chlorides.

(a) Synthesis of actinide fluorides

Equipment for the synthesis of pure actinide fluorides has been designed and installed at the JRC in Karlsruhe, and it uses pure HF gas as the fluorination agent. It consists of a glovebox kept under purified inert argon gas atmosphere with oxygen and moisture content less than 1 ppm, a fluorination reactor at high temperature and a HF supply gas line connected to the glovebox. The fluorination apparatus is a horizontal tube reactor made of Inconel X-750 inserted in a resistance furnace enabling work at temperatures up to 1200°C. A schematic and a photograph of the fluorination reactor are provided in Fig. 19 and the equipment is described in detail in Ref. [208].

The syntheses are based on a solid–gas reaction between the stoichiometric oxides with pure HF gas at elevated temperatures. The oxide powder in batches of up to 15 g is inserted into a boron nitride boat and placed in the fluorination reactor, which is heated to the desired temperature. Hydrogen fluoride gas is then led directly above the boat to convert the oxide to fluoride. The HF gas molar excess over the reaction stoichiometry was usually 3–4. The non-reacted HF gas and the possibly formed gaseous products are removed to the off-gas treatment unit by a stream of pure argon carrier gas, introduced to the reactor through a port in a front flange. If needed, the synthesis can be completed by consequent reduction of the fluorination product by a mixture of 6% H_2 in dry argon gas at the desired temperature. A summary of the conditions for the established syntheses is given in Table 11, photographs of the final products are shown in Fig. 20 and detailed descriptions can be found in Refs [208–210].

A combination of the following techniques is used for the quality control of the synthesized fluorides: gravimetric mass balance of the reaction, X ray diffraction for the qualitative phase analysis, differential scanning calorimetry for the melting temperature determination and inductively coupled plasma mass spectrometry for the metal bases impurities. The compounds ThF_4, UF_4 and PuF_3 were synthesized at the JRC in Karlsruhe with a phase purity 100 ± 0.5% (uncertainty of the methods) and with a metal base purity greater than 99.8% [208, 209]. It was found that UF_3 had a phase purity of 99.6% ± 0.5% based on X ray diffraction, while differential scanning calorimetry and inductively coupled plasma mass spectrometry analyses were not available.

The irradiation experiments SALIENT-01 and SALIENT-03, which are described in Sections 5.10.1.6 and 5.6.3, respectively, are being carried out in a collaboration between NRG and the JRC in Karlsruhe. The actinide fluoride mixtures required for these experiments were synthesized at the JRC in Karlsruhe by the methods described above, as summarized in Table 12.

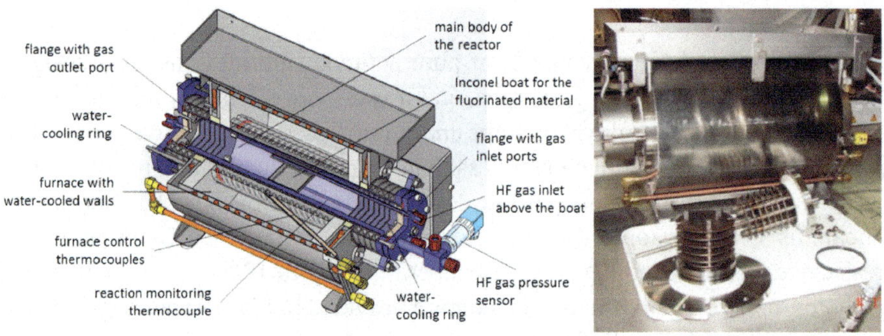

FIG. 19. A schematic and a photograph of the fluorination reactor installed at the JRC in Karlsruhe (courtesy of O. Benes, Joint Research Centre of the European Commission).

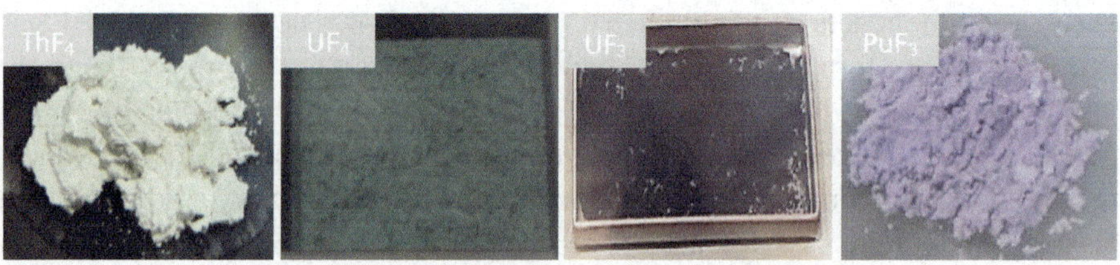

FIG. 20. Appearance of the pure actinide fluorides synthesized at the JRC in Karlsruhe (courtesy of O. Benes, Joint Research Centre of the European Commission).

TABLE 11. SYNTHESES OF ACTINIDE FLUORIDES ESTABLISHED AT THE JRC IN KARLSRUHE

Product	Chemical reaction	Temperature
ThF_4	$ThO_2(s) + 4HF(g) \rightarrow ThF_4(s) + 2H_2O(g)$	600°C
UF_4	$UO_2(s) + 4HF(g) \rightarrow UF_4(s) + 2H_2O(g)$	450°C
UF_3	$2UF_4(s) + H_2(g) \rightarrow 2UF_3(s) + 2HF(g)$	800°C
PuF_3	$PuO_2(s) + 4HF(g) \rightarrow PuF_4(s)$[a] $+ 2H_2O(g)$	550°C
	$2PuF_4(s) + H_2(g) \rightarrow 2PuF_3(s) + 2HF(g)$	600°C

[a] PuF_4 cannot be obtained as a pure product, but as a mixture of PuF_4 and PuF_3 owing to observed partial reduction during the fluorination process, likely due to traces of hydrogen in the HF gas that was used.

TABLE 12. MATERIALS SYNTHESIZED AT THE JRC IN KARLSRUHE FOR SALIENT IRRADIATION EXPERIMENTS

Salt mixture	Composition (mol%)						Mass (g)	Experiment
	^{7}LiF	LiF	ThF_4	UF_4	UF_3	PuF_3		
$^{7}LiF–ThF_4$	78.0	n.a.[a]	22.0	n.a.	n.a.	n.a.	28.6	SAL-1
$^{7}LiF–ThF_4–UF_4–PuF_3$	75.0	n.a.	18.7	6.0	n.a.	n.a.	28.0	SAL-2
$^{7}LiF–ThF_4–UF_4–UF_3–PuF_3$	75.0	n.a.	18.7	5.7	n.a.	0.3	15.0	SAL-2
$^{7}LiF–ThF_4–UF_4–CrF_3–PuF_3$	74.7	n.a.	18.6	6.0	n.a.	0.3	7.0	SAL-2
$LiF–ThF_4–UF_4–PuF_3$	n.a.	74.9	23.0	2.0	0.1	n.a.	50.0	SAL-2

[a] n.a.: not applicable.

(b) Synthesis of actinide chlorides

Equipment for the synthesis and purification of actinide chlorides was designed and installed at the JRC in Karlsruhe in 2019, and it consists of two connected gloveboxes, a chlorination reactor at high temperature, a Cl_2/HCl gas line connected to the main glovebox and an off-gas treatment unit in the auxiliary glovebox. The chlorination reactor is a gastight quartz tube closed by Inconel flanges inserted in a horizontal resistance furnace enabling work at temperatures up to 800°C. A schematic and a photograph of the reactor are shown in Fig. 21. The reactor is intended for a solid–gas reaction. The starting powdered material is inserted into a quartz boat and placed in the reactor, and the powder stays in the boat during the complete reaction. The set-up enables flow-through chlorination, while Cl_2 or HCl gas is led directly above the boat and the second reactive gas (e.g. CCl_4 vapours in argon, H_2 (6%)/Ar mixed gas or an inert carrier gas) are introduced into the reactor through the flange inlet. The excess reaction gases and the possibly formed gaseous products are removed to the off-gas unit. The equipment is described in detail in Ref. [211].

Procedures for syntheses of actinide chlorides are being investigated and optimized. A novel method for preparation of UCl_4 from stoichiometric UO_2 using a combination of Cl_2 and CCl_4 gases was proposed and tested, while syntheses of other actinide chlorides are in the preparatory phase. As reported and confirmed by thermodynamic calculations, UCl_4 cannot be formed by a simple reaction of UO_2 with Cl_2 or HCl gases at temperatures up to 1000°C, as $UOCl_2$ is always predominantly formed. In addition, the reaction of UO_2 with CCl_4 gas is efficient only at elevated temperatures, when higher chlorides (i.e. with higher valency) are formed together with UCl_4. The proposed method is therefore based

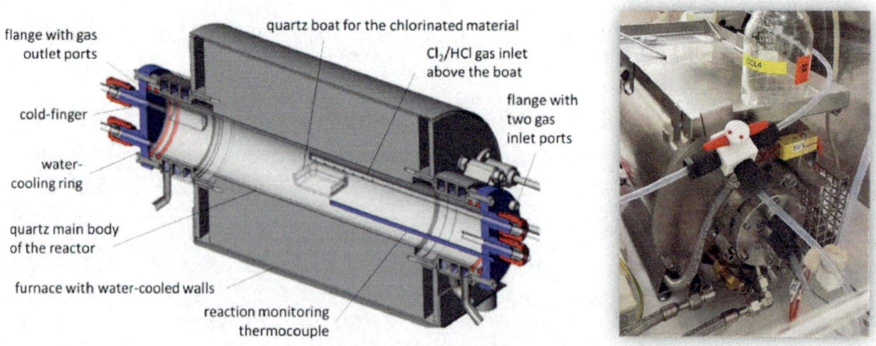

FIG. 21. A schematic and a photograph of the chlorination reactor installed at the JRC in Karlsruhe (courtesy of O. Benes, Joint Research Centre of the European Commission).

on combining Cl_2 and CCl_4 gases as a chlorination agent and a source of carbon to convert UO_2 to UCl_4 at temperatures lower than needed for the reactions with pure CCl_4.

A phase pure UCl_4 product was obtained using two step chlorination at a temperature of 175°C, followed by purification of the intermediate chlorination product composed of a UCl_4/UO_{2+x} mixture by sublimation at 600°C and recovery of the final UCl_4 product at a cold-finger at 400°C. The quality control of the obtained products is based on a combination of the same methods as described above for the synthesis of actinide fluorides. Photographs of the initial material, the chlorination and final products, and the X ray diffraction pattern of the recovered phase pure UCl_4 are shown in Fig. 22, and a detailed description of the method and the results have been published in Ref. [211].

5.6.2.2. *Purity control of the samples*

The purity of the samples needs to be checked after every synthesis. Two types of purity — phase purity and chemical purity — are generally considered. It has been demonstrated in the past that four independent methods are used to verify the sample's purity. The first purity check is the mass balance check with knowledge of the chemical state (and purity) of the initial material and the chemical state of the product. This step is accompanied by a visual check of the formed product (such as colour, homogeneity of colour and powder form).

The second purity check uses X ray diffraction analysis for phase purity including a phase quantification by Rietveld refinement. Figure 23 shows two X ray diffractograms of synthesized UF_4: one containing residual UO_2 (as initial

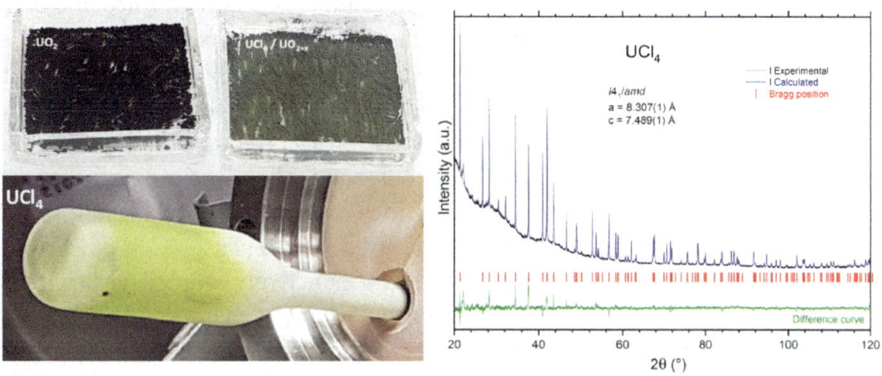

FIG. 22. Left: photographs of the initial material used for the synthesis of UCl_4, the chlorination and the final products; right: an X ray diffraction pattern of the phase pure UCl_4 final product (courtesy of O. Benes, Joint Research Centre of the European Commission).

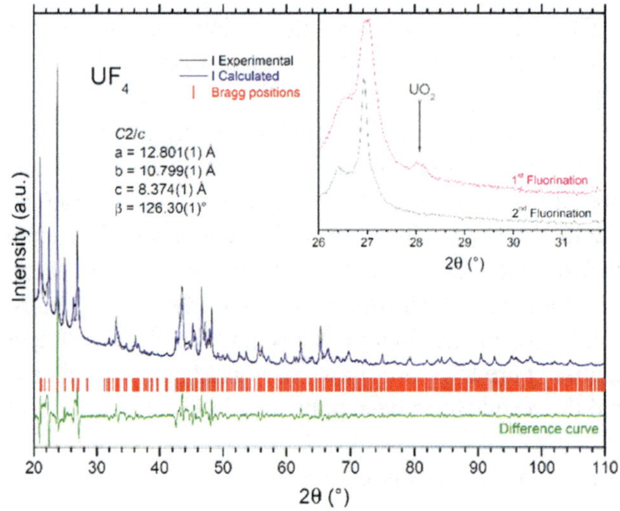

FIG. 23. X ray diffraction analysis of purified and unpurified UF_4 (reproduced from Ref. [208]).

material before fluorination) and one identified as UF_4 that is phase pure. The diffractogram that identified a phase pure UF_4 does not contain evidence of the UO_2 related peak.

It was, however, observed in the past that even samples that were considered as X ray diffraction pure (i.e. showing no evidence of other phase impurities) still contained impurities such as oxyfluorides or oxides that could significantly influence their behaviour. Therefore, a third purity check — a melting point determination using differential scanning calorimetry — is performed for each compound. This method has been demonstrated to be a very sensitive technique for the detection of impurities. Figure 24 shows two melting point determinations of purified and unpurified ThF_4. Only one peak is observed on the heat flow signal of differential scanning calorimetry with a correct melting point (not quantified in the figure) for the pure ThF_4 compound. The heat flow signal for the ThF_4 compound containing traces of impurities revealed two peaks and a lower melting point, clearly indicating the presence of a second phase.

The fourth purity check is performed using inductively coupled plasma mass spectrometry and is in place to quantify concentrations of chemical elements.

All four methods are part of the quality assurance protocol used at the JRC for safety studies and can be used for presenting to regulators.

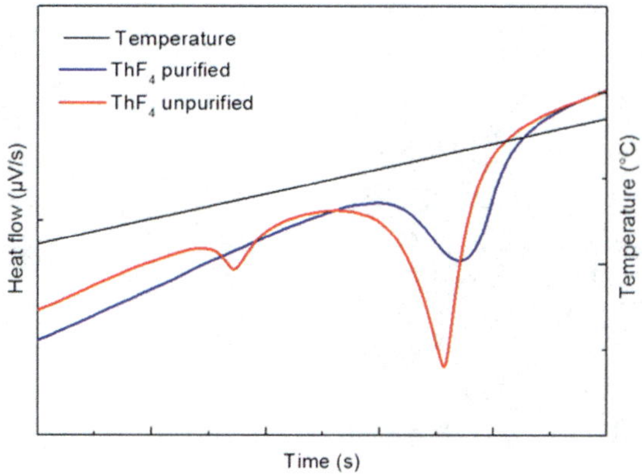

FIG. 24. Melting point determination of purified and unpurified ThF_4 determined by differential scanning calorimetry (courtesy of O. Benes, Joint Research Centre of the European Commission).

5.6.2.3. Electrochemistry of molten salts

(a) Electrochemical characterization of chlorides

A systematic experimental study was carried out at the JRC in Karlsruhe to obtain accurate knowledge of the thermodynamic and electrochemical properties of actinides dissolved in a molten LiCl–KCl eutectic salt [212–217]. Apart from data on the general electrochemical behaviour (such as electrode potentials and reaction mechanisms), thermochemical data (activity coefficients, enthalpies and entropies of formation) and kinetic data (diffusion coefficients) were evaluated for all actinides up to americium. Their electrochemical properties were measured on different inert and reactive working electrode materials.

Dedicated equipment for electrochemical measurements in molten chloride media was designed and installed at the JRC in Karlsruhe. It consists of a glovebox under purified argon atmosphere, equipped with a vertical well-type oven with an electrolyser made of stainless steel inside. The electrochemical cell is an alumina or quartz crucible containing the measured melt and three electrodes, positioned through a water cooled top flange of the electrolyser. A working electrode is typically a metallic wire with a diameter of 1 mm, a counter electrode is made from a molybdenum wire bent into the shape of a spiral and a reference electrode is a silver wire dipped into a LiCl–KCl–AgCl (1 wt%) melt in a Pyrex glass tube. The glovebox, schematic and photograph of the set-up are shown in Fig. 25. The equipment is described in detail in Refs [213, 215].

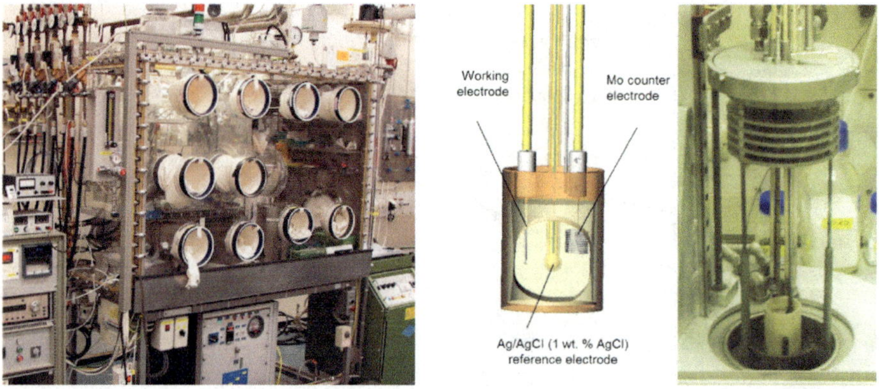

FIG. 25. Left: glovebox for electrochemistry in molten chlorides installed at the JRC in Karlsruhe; right: a schematic and a photograph of an electrochemical cell (courtesy of O. Benes, Joint Research Centre of the European Commission).

Transient electrochemical techniques, such as cyclic voltammetry, chronopotentiometry, square-wave voltammetry and similar were used to measure the properties listed above. The selected properties of actinides in molten eutectic LiCl–KCl measured on inert working electrodes are summarized in Table 13, which is compiled from Refs [212–219]. In addition, the electrochemical formation of actinide alloys with reactive materials of different working electrodes was investigated using mainly solid aluminium and liquid cadmium electrodes. A comparison of measured apparent standard potentials of selected actinides and lanthanides on inert, solid reactive and liquid reactive electrodes is illustrated in Fig. 26 [218].

Results of these electrochemical studies supported the design of an electrorefining process for group selective recovery of actinides from nuclear fuel using reactive solid aluminium cathodes [220, 221]. In this process, electroseparation of actinides from fission products is carried out in LiCl–KCl eutectic molten salt at a temperature of 450°C by applying a constant current between a metallic fuel contained in an anodic basket and an aluminium cathode. During the process, actinide cations arising from the anodic oxidation of the fuel are electrotransported and deposited onto the aluminium cathode where they form An–Al alloys, while fission products either stay in the melt or are not dissolved from the fuel. The process was intensively studied at the JRC in Karlsruhe and successfully demonstrated in a laboratory scale using METAPHIX, an irradiated experimental metallic fuel [222, 223]. An excellent performance of the solid aluminium cathode was shown, as very high selectivity of actinides over

TABLE 13. SELECTED PROPERTIES OF ACTINIDES IN MOLTEN LiCl–KCl EUTECTIC AT $T = 450°C$ EXPERIMENTALLY DETERMINED AT THE JRC IN KARLSRUHE [212–219]

Actinide	Th	U		Np		Pu	Am	
Reduction mechanism	1 step	2 step		2 step		1 step	2 step	
Redox systems	Th^{4+}/Th^0	U^{4+}/U^{3+}	U^{3+}/U^0	Np^{4+}/Np^{3+}	Np^{3+}/Np^0	Pu^{3+}/Pu^0	Am^{3+}/Am^{2+}	Am^{2+}/Am^0
Exchanged electrons, n	4	1	3	1	3	3	1	2
App. standard potential, E^{0*} (V vs Ag/AgCl 1 wt%)	−1.36	−0.23	−1.28	+0.46	−1.50	−1.57	−1.47	−1.67
Gibbs free energy form ΔG^{∞}_{MClx} (kJ/mol)	−987.6	−137.8	−726.2	−69.87	−776.4	−801.9	−816.6	−558.3
Species	Th^{4+}	U^{4+}	U^{3+}	Np^{4+}	Np^{3+}	Pu^{3+}	Am^{3+}	Am^{2+}
Activity coefficient, γ^0	4.6×10^{-4}	1.5×10^{-2}	5×10^{-4}	8.5×10^{-3}	3×10^{-5}	$\sim 1 \times 10^{-2}$	4.7×10^{-3}	2.7×10^{-2}
Diffusion coefficient, D (cm^2/s)	3.2×10^{-5}	2.1×10^{-5}	2.7×10^{-5}	2.1×10^{-5}	1.9×10^{-5}	1.6×10^{-5}	2.4×10^{-5}	1.2×10^{-5}

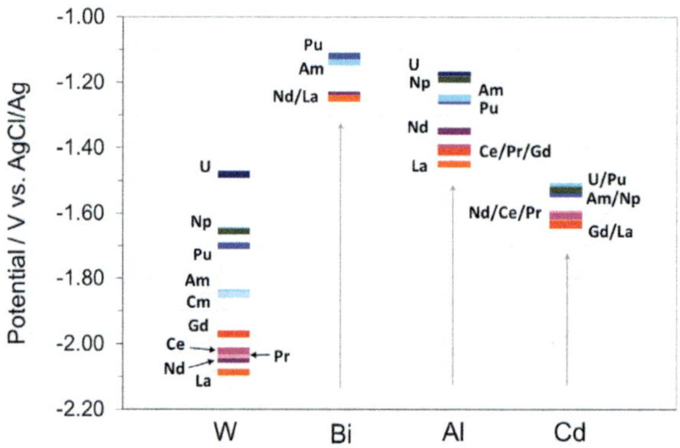

FIG. 26. Apparent electrochemical potentials measured at the JRC in Karlsruhe for various actinides and lanthanides using different working electrodes in LiCl–KCl eutectic at 450°C. The represented values are based on the measurements taken from Ref. [218] (courtesy of O. Benes, Joint Research Centre of the European Commission).

lanthanides was reached, as well as sufficient efficiency of the process and high capacity of aluminium to take up actinides.

(b) Electrochemical characterization of fluorides

The electrochemistry of actinides in molten fluoride salts is studied at the JRC in Karlsruhe within the framework of research on properties of fuel candidates for MSRs. The experimental set-up that was initially developed consists of a corrosion resistant Inconel electrolyser inserted in a vertical furnace enabling a maximum working temperature of 1200°C, built in at the bottom of the same glovebox as used for the synthesis of actinide fluorides (see Section 5.6.2.1(a)). The lid of the electrolyser is equipped with ports for precise guiding and positioning of the electrodes attached to Inconel holders. The electrochemical cell consists of a glassy carbon crucible containing the measured molten salt and three electrodes. A working electrode is typically 0.5–1 mm metallic wire (such as tungsten, molybdenum, gold, nickel, tantalum and silver), a platinum wire (1 mm diameter) serves as a PtO_2/O^{2-} quasi-reference electrode, and a glassy carbon rod or a tungsten wire is connected as an auxiliary electrode. Figure 27 shows the electrochemical set-up and details of the complete installation can be found in Ref. [224]. Since the electrode holders can be fully removed from the ports and the electrolyser can be closed gastight, the HF gas line connected to the glovebox described above allows for purification of liquid fluoride melts by bubbling pure HF gas at high temperatures.

A method for purification of a LiF–CaF_2 (80.5–19.5 mol%) eutectic carrier melt was established based on fluorination using HF gas bubbled directly into the melt through a nickel tube for 120 min at 850°C with a flow rate of 6 cm^3/min. The procedure was proven sufficient to remove dissolved oxygen from the melt, as shown in Fig. 28, which displays the electrochemical response of oxygen in the melt before and after the purification. The details of the technique are published in Refs [224, 225].

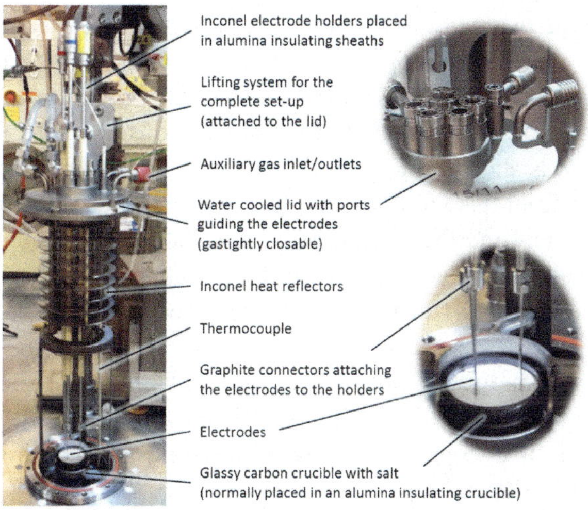

FIG. 27. A photograph of the electrolyser installed at the JRC in Karlsruhe with a description of the most important parts of the set-up and electrochemical cell (reproduced from Ref. [224]).

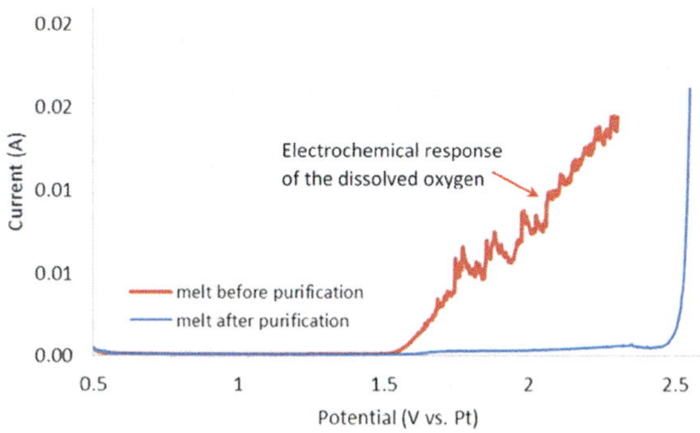

FIG. 28. Linear sweep voltammetry of LiF–CaF_2 eutectic melt before and after purification by hydrogen fluoride gas. Working electrode: gold wire (cross-section = 0.08 cm^2), counter electrode: glassy carbon rod, reference electrode: platinum wire, scan rate: 10 mV/S, temperature: 850°C (courtesy of O. Benes, Joint Research Centre of the European Commission).

The electrochemical properties of thorium were studied in the melts LiF–CaF_2–ThF_4 (0.1–0.3 mol% ThF_4) and LiF–ThF_4 (22 mol% ThF_4) [224, 225]. Verification of the published data on the electrochemical properties of thorium was carried out and a novel method was proposed for the determination of the activity coefficients of compounds dissolved in molten fluoride media using electrochemical techniques. The approach overcomes both the absence of a thermodynamic reference electrode in molten fluoride salts at high temperature and the usually unknown value of the activity of free fluorine ions in the studied media. It is based on measurements of the difference between the electrode potentials of the studied compound and the least stable carrier melt constituent (melt decomposition potential). An experimental determination of this potential difference using the cyclic voltammetry technique of the LiF–CaF_2–ThF_4 melt (0.18 mol% ThF_4) is shown in Fig. 29. The activity coefficient of the compound can be calculated from this value, assuming that the standard electrode potentials of the compound and the least stable solvent constituent are known or can be calculated, as well as the activity of the least stable solvent constituent. The activity coefficient of ThF_4 in the above mentioned LiF–CaF_2–ThF_4 melt was determined to be 3.28×10^{-3} [224].

Experimental studies on electrochemical and thermodynamic properties of other actinides in selected fluoride melts of interest are planned at the JRC in Karlsruhe. The aim is to create a database of properties that are important for the fuel chemistry of MSRs.

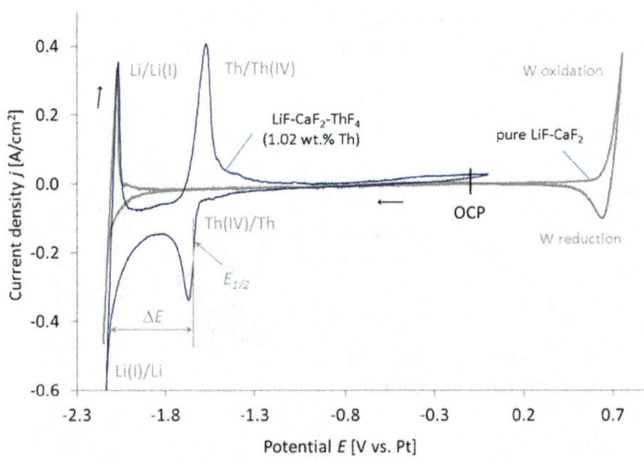

FIG. 29. Cyclic voltammetry of pure LiF–CaF_2 eutectic melt (thick line) and LiF–CaF_2–ThF_4 (0.18 mol% ThF_4) melt showing graphical evaluation of the potential difference between ThF_4 and the least stable carrier melt constituent, LiF (working electrode: tungsten, counter electrode: tungsten, reference electrode: platinum, temperature: 850°C, scan rate: 100 mV/s) (reproduced from Ref. [224]).

5.6.2.4. Properties at high temperature

The JRC in Karlsruhe has been significantly involved in experimental studies on properties at high temperature of molten fluoride salts for more than 15 years. A huge step towards achieving reliable results was accomplished by establishing a suitable and cost efficient encapsulation technique that prevents vaporization of samples during measurements at high temperature and avoids chemical interaction between the material and the salt. Furthermore, the encapsulation technique needs to be selected for the purpose of the measurement and needs to be compatible with the measuring device. Two types of equipment have been suggested: one rather compact in size and based on laser welding of nickel capsules [226], and the other using a firm stainless steel crucible with inert liner insets, which can be made from nickel metal, boron nitride (oxygen free), graphite or any other inert material with respect to the sample [227]. Inner coatings are possible as well. Both types have been described in detail in earlier studies [225, 227] and are shown in Figs 30 and 31.

(a) Melting point determination and phase diagram investigations

One of the key properties of the MSR fuel is the melting temperature, which determines the lower limit of the operating temperature of the reactor. Expertise in measuring the melting points of halide salts has been developed at the JRC and is available under the plan for quality assurance required by licensing authorities. Differential scanning calorimetry is used for melting point determination by

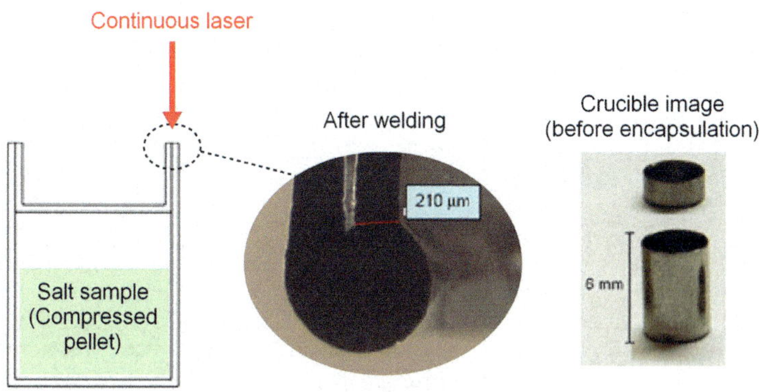

FIG. 30. A schematic representation and a photograph of the nickel crucible designed for the encapsulation of fluoride and chloride salt samples (wall thickness of ~0.2 mm); centre: metallography of the welded edge of the crucible (courtesy of O. Benes, Joint Research Centre of the European Commission).

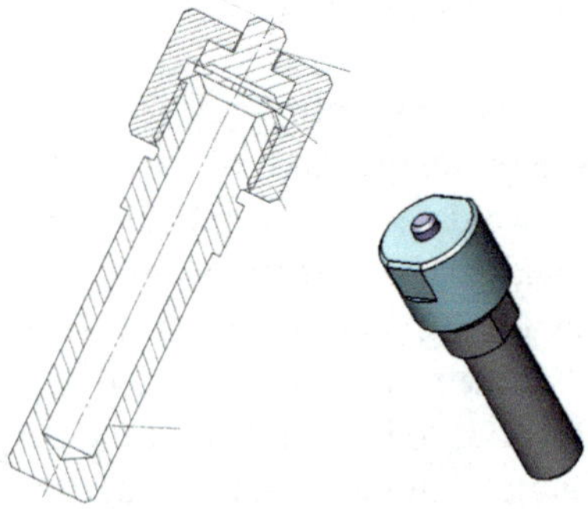

FIG. 31. A schematic representation of the differential scanning calorimetry crucible developed at the JRC for measurements of halide samples (courtesy of O. Benes, Joint Research Centre of the European Commission).

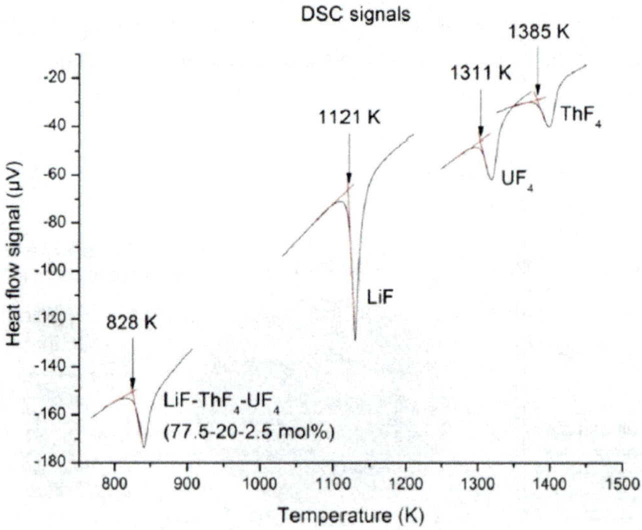

FIG. 32. A differential scanning calorimetry output of the melting point determination of LiF, UF_4 and ThF_4 end-members and the LiF–ThF_4–UF_4 (77.5–20–2.5 mol%) fuel for the MSFR (reproduced from Ref. [230] with permission).

applying the encapsulation technique shown in Fig. 31. The focus is given not only on melting point determination of pure fuel components or specific mixtures, but also on systematic investigations of key phase diagrams [203, 210, 228, 229]. These provide solid bases for the validation and extension of the JRCMSD thermodynamic database, an extensive database owned by the JRC that describes the key MSR systems studied around the world, including chloride and fluoride salt media. Both examples are illustrated in the following figures. Figure 32 shows the melting point measurement — determined as 828 K — of the MSFR fuel of the LiF–ThF_4–UF_4 (77.5–20–2.5 mol%) composition performed within the framework of the SAMOFAR project, as published by Tosolin et al. [230]. Figure 32 further summarizes the melting point determination of all three LiF, UF_4 and ThF_4 constituents as part of the purity analysis of the starting materials. Figure 33 shows a recent study of the phase equilibrium points of the ThF_4–PuF_3 system, indicating three phase equilibria: (i) the liquidus temperature at which the first crystal precipitates upon cooling; (ii) the solidus temperature, represented by eutectic and peritectic melting; and (iii) the decomposition of the $PuTh_2F_9$ intermediate phase at high temperature.

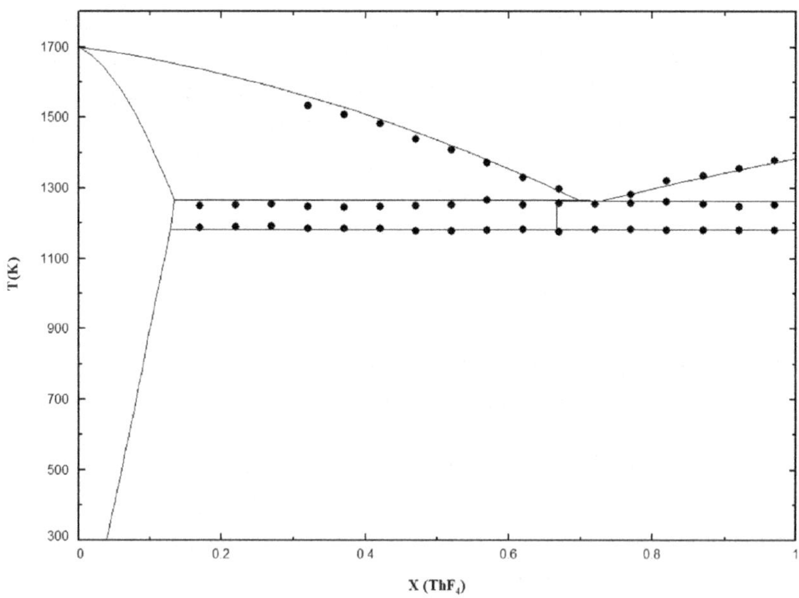

FIG. 33. The assessed ThF_4–PuF_3 phase diagram. The points are phase equilibrium points obtained by differential scanning calorimetry (courtesy of O. Benes, Joint Research Centre of the European Commission).

(b) Heat capacity

Two techniques are used at the JRC Hot Cell Laboratory[12], differential scanning calorimetry and drop calorimetry, for determining the heat capacity of molten salts. Both techniques provide reliable results and are also suitable for fusion enthalpy or transition enthalpy evaluations, as discussed in the next subsection. The use of two independent methods allows for the complementary determination of the fuel salt heat capacity required for quality assurance purposes. The focus is on three main domains at the JRC:

- Systematic investigation of multicomponent halide mixtures. Earlier studies from the JRC [231, 232] show that heat capacity of a complex halide liquid solution does not follow ideal behaviour, that is, it cannot be simply estimated from the end members (individual components forming the mixture) using the so-called Neumann–Kopp rule, but rather shows significant departures from this ideal trend. It has been further shown that the excess heat capacity is proportional to the cationic radius difference of mixing cations. The summary of the thus determined excess heat capacity for the liquid phase of the LiF–AlkF (where Alk is sodium, potassium, rubidium or caesium) binary pairs is shown in Fig. 34.
- Provision of novel data to fill relatively large gaps existing for fundamental data on actinide halides, and re-evaluation of existing data with current techniques, which meet the required standards. An example of novel measurement of the ThF_4 heat capacity for both solid and liquid phases utilizing complementary results obtained by differential scanning calorimetry and drop calorimetry is given in Fig. 35 [233].
- Measurements of the heat capacity of specific fuel composition.

(c) Fusion and mixing enthalpy measurements

As briefly mentioned in the above subsection, differential scanning calorimetry and drop calorimetry techniques are excellent tools to determine the enthalpies of phase transitions, including the fusion enthalpy. Figure 36 shows fusion enthalpy determination of the pure ThF_4 compound, directly obtained from enthalpy increment data of the solid and liquid phases [233]. The thus determined value is in perfect agreement with a value measured by differential scanning calorimetry, that is, 41.9 kJ/mol (not shown in the figure).

---

[12] The hot cell laboratory in Karlsruhe consists of hot cells (shielded nuclear radiation containment chambers) where highly radioactive materials can be received, handled, examined and returned to their owners.

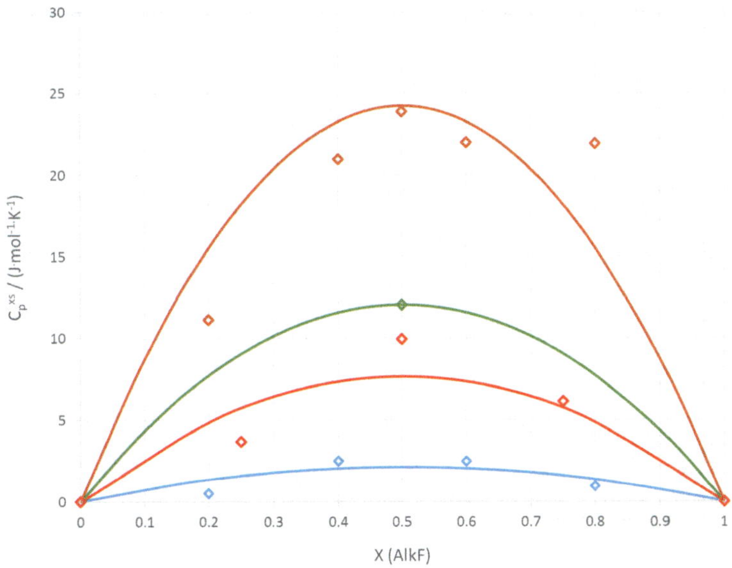

FIG. 34. Experimentally determined excess heat capacity as a function of molar ratio of mixed higher cation X(AlkF) for the LiF AlkF binary liquid solutions: LiF–NaF (light blue); LiF–KF (red); LiF–RbF (green); LiF–CsF (orange); the experimental data are taken from Ref. [231] (courtesy of O. Benes, Joint Research Centre of the European Commission).

Furthermore, the enthalpy of mixing of binary (and higher order) halide systems is determined using a differential scanning calorimetry technique. This value is a measure of the liquid phase stability and gives a valuable basis for phase equilibria modelling (described in Section 5.6.2.4(a)). Figure 37 gives an example of JRC's earlier measurement of enthalpy of mixing of a binary LiF–ThF_4 system [228], indicating a relatively strong exothermic effect during mixing of the end members.

(d) Vapour pressures

A Knudsen effusion mass spectrometry, or KEMS, device as shown in Fig. 38 is used for determining the vaporization behaviour of MSR fuels and coolants. The whole assembly is placed in a glovebox that is alpha tight and gamma shielded, thereby allowing handling of radioactive materials, including irradiated materials.

Knudsen effusion mass spectrometry is a powerful technique for the determination of the thermodynamic properties at high temperatures based on the equilibrium between condensed and gaseous phases. It can be used to identify the gas composition and its vaporizing behaviour as a function of

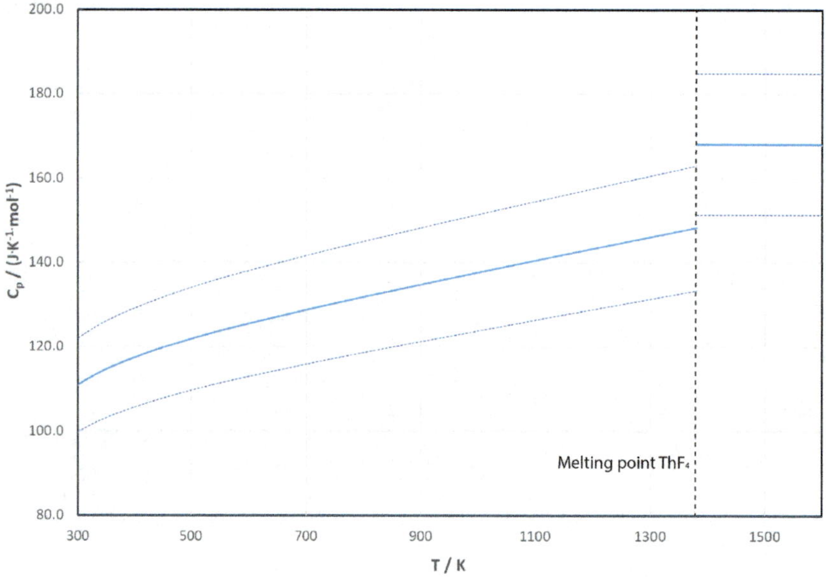

FIG. 35. Heat capacity of solid and liquid phases of ThF_4 obtained by differential scanning calorimetry and drop calorimetry (reproduced from Ref. [233] with permission).

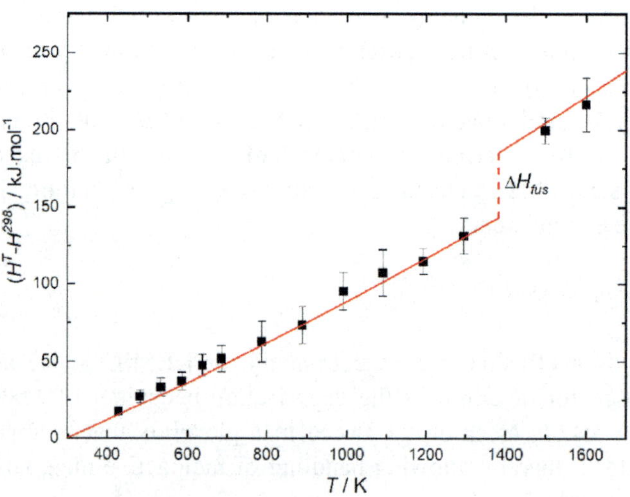

FIG. 36. Enthalpy increment data measured by drop calorimetry for ThF_4 solid and liquid phases as a function of temperature (T) (in K) (reproduced from Ref. [233] with permission).

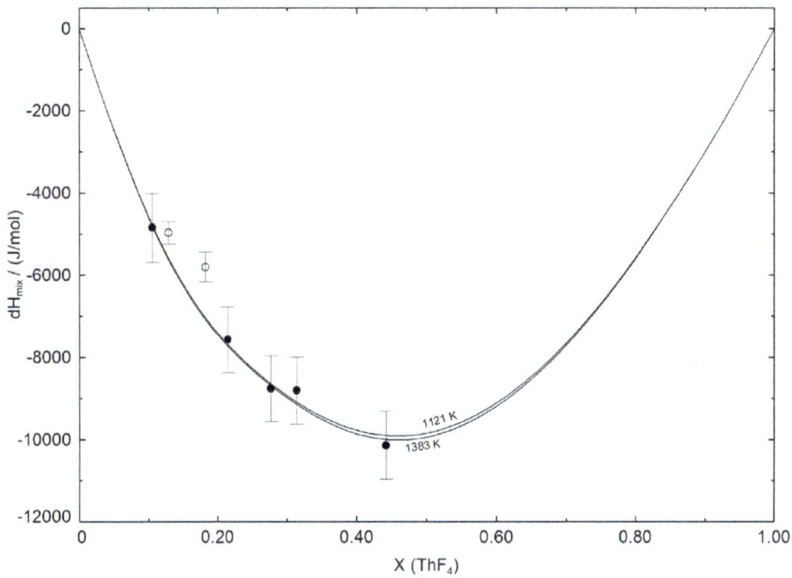

FIG. 37. Mixing enthalpy of the LiF–ThF_4 binary system as determined by differential scanning calorimetry; data are measurements taken from Ref. [228] (courtesy of O. Benes, Joint Research Centre of the European Commission).

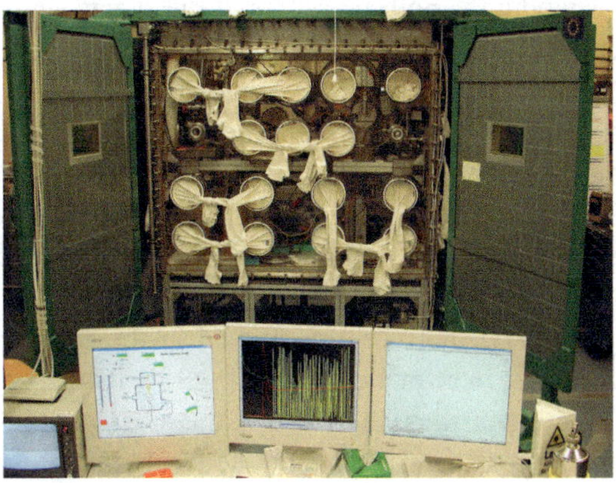

FIG. 38. Device for Knudsen effusion mass spectrometry installed in an alpha-tight glovebox with external gamma shielding at the JRC in Karlsruhe (courtesy of O. Benes, Joint Research Centre of the European Commission).

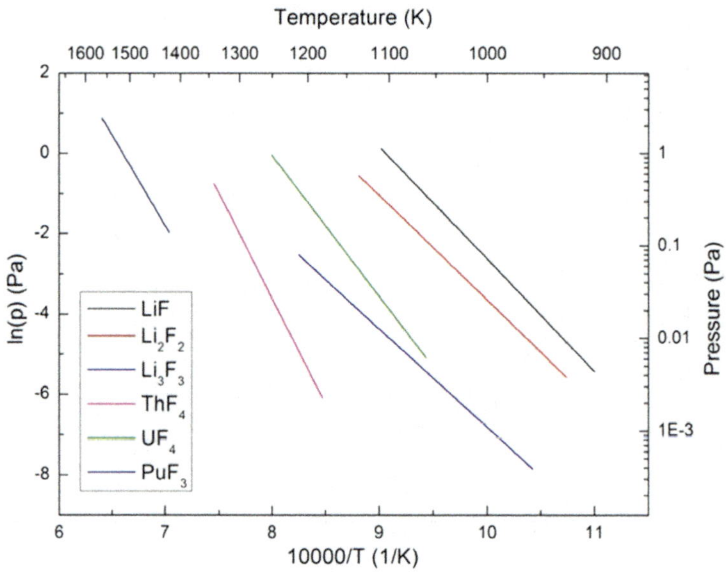

FIG. 39. Partial and total vapour pressures for the mixture LiF–ThF$_4$–UF$_4$–PuF$_3$ (77.5–6.6–12.3–3.6 mol%). The relative uncertainty on vapour pressure measurements is 50% (reproduced from Ref. [234] with permission).

temperature. This is done by conversion of the detected spectral ion intensity data into the corresponding vapour pressures of the species in the gaseous phase. This method can further be used to determine other thermodynamic properties, such as sublimation enthalpies, ionization energies of gaseous species, activity coefficients or to determine the retention capacity of molten salt with respect to selected (caesium, iodine) fission products. This is a reliable method for determining the boiling points of MSR fuels.

An example of measurement of vapour pressures of MSR fuel is given in Fig. 39, which shows partial vapour pressures of the LiF–ThF$_4$–UF$_4$–PuF$_3$ (77.5–6.6–12.3–3.6 mol%) fuel mixture studied in the SAMOFAR project and published in Ref. [234]. It is evident from Fig. 39 that each fuel component evaporates at different intensities. In this case, the most volatile is the LiF matrix component, followed by UF$_4$ and ThF$_4$, and the least volatile is PuF$_3$. The partial vapour pressures of the homogeneously mixed fuel are thus proportional to their concentrations. The sum of the partial vapour pressures gives the total vapour pressure of the fuel, which is used to extrapolate the boiling point of the fuel, which in this case is determined as 1896 ± 10 K.

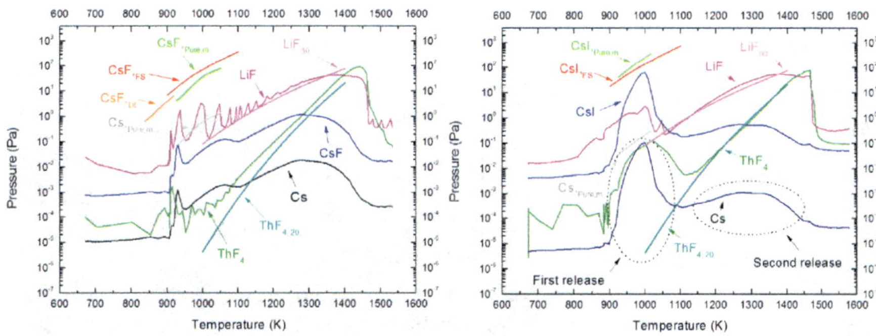

FIG. 40. Left: volatility of CsF from the LiF–ThF$_4$ eutectic; right: volatility of CsI from the LiF–ThF$_4$ eutectic (reproduced from Ref. [235] with permission).

(e) Fission product release — retention capacity of the fuel

As mentioned above, KEMS is a technique that can measure the retention capacity of the MSR fuel for fission products. Understanding whether a fission product will dissolve in the fuel matrix or remain separated is very important, particularly in the case of a volatile fission product like caesium or iodine. Recently, a series of experiments has been performed at the JRC to understand the behaviour of these two major fission products and the effect of their chemical form on the volatility behaviour. Figure 40 shows the vaporization behaviour of caesium fluoride (CsF) and caesium iodide (CsI) compounds mixed with LiF–ThF$_4$ eutectic solvent [235]. It was found that CsF is completely dissolved in the fuel matrix, and thus its volatility is very low. However, as iodides are not very soluble in fluoride melts, most of the CsI remained undissolved and consequently the volatility of CsI remained almost as high as for the pure compound. This is evident from the first release of CsI from the fuel, as indicated in Fig. 40.

(f) Thermal conductivity

The thermal conductivity measurements are carried out using the laser flash device. The device was fully designed and constructed at the JRC in Karlsruhe, as described in detail in Ref. [236], and is shown in Fig. 41 [237]. To measure radioactive elements, the laser flash device is installed in a glovebox, which is surrounded by lead shielding (that allows thermal conductivity measurements of irradiated fuel materials) and equipped with two manipulators. Two types of measurement are performed on a halide salt system: (i) the solid phase to understand thermal transport through, for example, a solid crust layer precipitating on the inner walls of structural vessels, and (ii) the liquid phase.

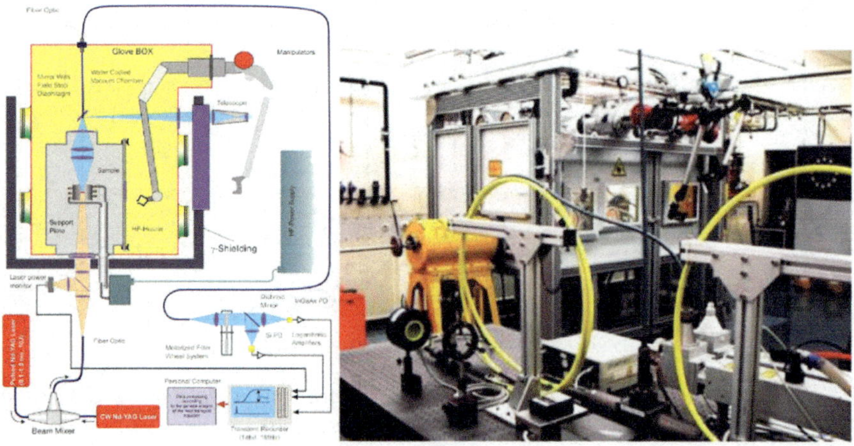

(a) Schematic setup (b) Photo of LAF

FIG. 41. Left: schematic representation of the laser flash, or LAF, device used at the JRC for measurements of thermal conductivity (reproduced from Ref. [237]). Right: LAF installed in the Hot Cell Laboratory (courtesy of O. Benes, Joint Research Centre of the European Commission).

The measurements of solid bulk are performed on small sized discs with a diameter of approximately 5 mm and a thickness of approximately 1 mm. It is important that the sample has as low porosity as possible. For this reason, most of the measurements are performed on pre-molten samples that upon cooling solidify into a disc shape.

For the thermal conductivity measurements of liquid salts, a specially designed, laser welded, hermetically closed crucible is used to keep the liquid phase in place during the measurements at high temperature.

(g) Post-irradiation examination

Because of its capability to work with molten salts containing actinides, the JRC became a key partner for the Dutch national programme on molten salt irradiation, named the SALIENT project, led by NRG. The fuel for the SALIENT-01 irradiation experiment has been synthesized at the JRC premises in Karlsruhe and, in the near future, part of the irradiated fuel will be transported back from NRG for post-irradiation examination. This will include fuel examination by Knudsen effusion mass spectrometry to study the volatility of the formed fission products and experiments performed by transmission electron microscopy to investigate the formation and size distribution of metallic precipitates. A benchmark for observations made to solid oxide fuels will be made. Currently, the JRC is involved in the synthesis of fuel containing plutonium (75LiF–18.7ThF_4–6UF_x–0.30PuF_3 mol% composition) for the SALIENT-03 irradiation experiment, which will be hermetically sealed by qualified welding using alloys based on nickel, such as Hastelloy N (USA) and GH3535 (China) alloys. More information on the Dutch programme on molten salt irradiation is given in Sections 5.6.3, 5.10.1.3, 5.10.1.4 and 5.10.1.6.

5.6.2.5. JRCMSD thermodynamic database

The JRC is the owner of an extensive thermodynamic database, known as the JRCMSD, which describes the key MSR systems studied around the world, including chloride and fluoride salt media. The database is continuously developed in collaboration with other partners (e.g. TU Delft) by novel assessments and data reviews addressing not only fuel behaviour in the so-called fresh state, but also fission product behaviour and corrosion interactions. The database has been recently shared with members of the project to develop the Thermodynamics of Advanced Fuels – International Database, known as TAF-ID, organized by the Nuclear Energy Agency of the Organisation for Economic Co-operation and Development. A summary of the described binary systems from the JRCMSD is given in Fig. 42. The database provides a complete thermodynamic description of the following key systems (with the main references included):

- LiF–NaF–BeF_2–PuF_3 system: fuel for an actinide burner with BeF_2 as a component [238].
- LiF–NaF–KF–RbF–CsF–LaF_3–PuF_3 system: BeF_2–free fuel for an actinide burner with consideration of KF and RbF as matrix alternatives and CsF and LaF_3 as fission product representatives [239].
- LiF–NaF–BeF_2–UF_4–ThF_4 system: fuel for a breeder reactor using molten salt with NaF as an alternative candidate for the matrix component [240].
- LiF–NaF–UF_3–UF_4 system: fuel system showing the influence of UF_3 content on, for example, melting behaviour of the fuel. UF_3 is considered a redox control additive [241].
- LiF–ThF_4–CeF_3–PuF_3 system: fuel for a non-moderated breeder reactor using molten salt with CeF_3 to represent the fission product (and as a proxy to PuF_3 at the early stage of the database development) [229].
- LiF–CsF–ThF_4–LiI–CsI–ThI_4 system: fuel for a key MSFR system with addition of fission products CsF and CsI [242].
- NaCl–$MgCl_2$–UCl_3–$PuCl_3$ system: breeder and burner concept utilizing chloride salts [243].

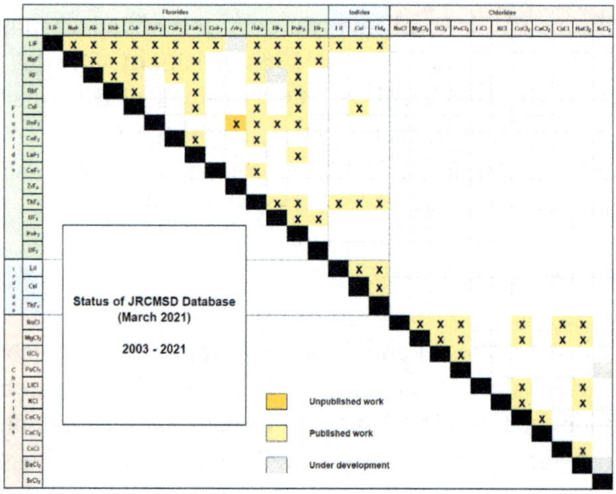

FIG. 42. Status of binary systems in the JRCMSD thermodynamic database (courtesy of O. Benes, Joint Research Centre of the European Commission).

The database is a very good tool for predicting some of the key properties of the MSR fuel, such as melting point, chemical nature of precipitates, heat capacity, vapour pressure, heats of fusion or boiling points. It was also found to be a very useful tool for optimizing the composition of an MSR fuel, mainly because of the multicomponent nature of the fuel, and the associated database ability to simulate the properties of any composition needed. Figure 43 shows examples of calculated melting behaviour (the liquidus projection) of two selected systems: the $LiF–ThF_4–PuF_3$ and the $NaCl–UCl_3–PuCl_3$ systems.

5.6.3. Component and technology development

The High Flux Reactor, or HFR, is a materials testing reactor that is located in Petten, owned by the European Union and operated by NRG. Since 1961, the reactor has been employed to conduct irradiation experiments to investigate the behaviour of both structural materials and fuel exposed to high thermal neutronic flux. From 2017 to 2019, it was used to irradiate molten salt within the framework of a bilateral collaboration between the JRC and NRG. The first experiment conducted and the second experiment being carried out are called SALIENT-01 and SALIENT-03, respectively. Since the deployment of SALIENT-01 (see Section 5.10.1.6), which used capsules made of graphite containing a mix of salts based on fluoride ($78LiF–22ThF_4$ mol%) [244], further experience has been gained and new challenges resulting from the handling of MSR fuel under irradiation have been successfully tackled.

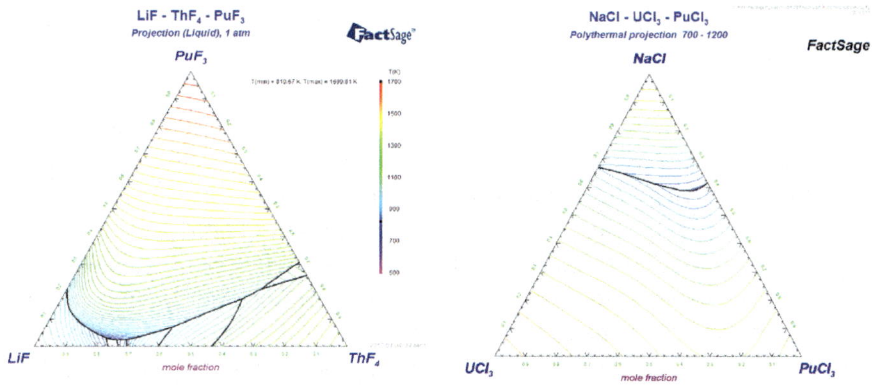

FIG. 43. The calculated liquidus projection of the LiF–ThF$_4$–PuF$_3$ and NaCl–UCl$_3$–PuCl$_3$ systems (left and right, respectively) (courtesy of O. Benes, Joint Research Centre of the European Commission).

Although the compatibility between molten salt based on fluoride and nuclear grade graphite is already well documented, the second experiment (SALIENT-03) was conducted encapsulating molten salt (75LiF–18.7ThF$_4$–6UF$_x$–0.30PuF$_3$ mol%) in alloys based on nickel (e.g. Hastelloy N and GH3535). The experiment allowed a more precise understanding of the fission products released and trapped in molten salt, as well as a study of the interaction between the salt and the metallic structural material [245]. The SALIENT-03 experiment has the following objectives:

(a) To study the in-pile corrosion of the MSR containment material Hastelloy N and also GH3535 when they are in contact with molten fluoride fuel salt, as well as under the following scenarios:
   (i) In a strong, stable temperature gradient (>100°C);
   (ii) As a function of the redox potential of the salt;
   (iii) As a function of the concentration of the corrosion products in the salt;
   (iv) In the presence of corrosive fission products generated during irradiation (notably tellurium).
(b) To continue and improve the study begun in SALIENT-01 of fission product behaviour in molten fluoride salts. This behaviour is representative of the behaviours found in MSRs used for breeding thorium.

The design of the SALIENT-03 experiment has been tailored to achieve the following goals:

- A fission burnup of at least 2% fission of initial heavy metal atoms to be able to produce a sufficient quantity of fission products dissolved in the salt to be analysed during the post-irradiation examination phase.
- A constant irradiation temperature of the capsules containing the molten salt of at least 700°C maintained for 10 000 h to allow significant corrosion to occur. A simple correlation between the corrosion rate and the operating temperature can be determined.
- The irradiation of five capsules containing a slightly modified composition of salt to investigate the influence of redox potential on the corrosion products formed.
- The development of a sophisticated quality system based on redundant checks of the composition of the salt to assure its purity. The purity of the salt is key to understanding the behaviour of the salts under irradiation.
- The addition of electric heaters when the reactor is not operating, to avoid radiolytic production of F_2 gas and reduction of the salt samples [246, 247].

The SALIENT-03 experiment has been equipped with the following instrumentation to collect as much information as possible during in-pile irradiation:

- Forty-three thermocouples to map the temperature distribution of the irradiated capsules accurately;
- Fifteen flux monitors (three per capsule) to check the neutron fluence to which the salt has been exposed;
- A pressure transducer connected to the first capsule to check on-line the built-in pressure due to the generation and release of gaseous fission products during irradiation;
- Three electrodes connected to the first capsule to check on-line the variation of redox potential during irradiation.

The information acquired during irradiation will be complemented with a comprehensive post-irradiation examination of the capsules and the salt contained in them.

The SALIENT-03 experiment contains five capsules. Four of them are made of Hastelloy N and the fifth is made of GH3535. The composition of the salt inside each capsule is slightly different to study the corrosion of the most promising nickel based alloys under irradiation when exposed to different redox potentials and different CrF_x concentrations. Four of the five capsules irradiated

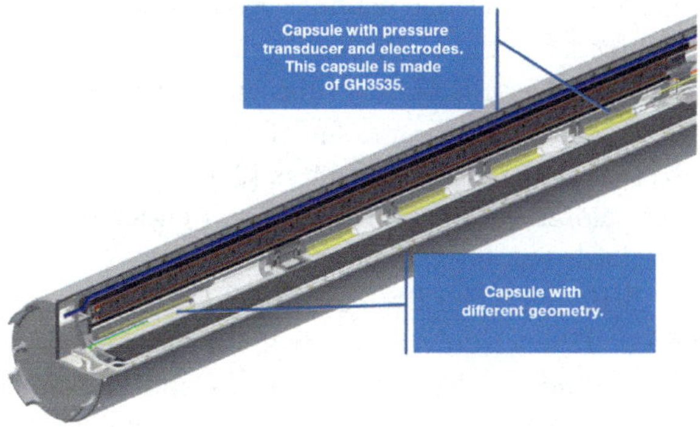

FIG. 44. Section of the SALIENT-03 experiment (courtesy of O. Benes, Joint Research Centre of the European Commission).

have the same geometry. The fifth is longer than the others, thus allowing buoyancy to increase in the salt and measurements to be taken of the temperature in the centre of the capsule and not only on its external wall. Figure 44 shows the capsules with their safety containments.

The different components of the salts have been synthesized at the JRC in Karlsruhe. Tests for the procedure of filling the capsules with salt and qualified welding were performed as well. The design of the experiment has been completed, and the fabrication of the capsules and sample holder is ongoing. The SALIENT-03 irradiation experiment is being carried out in the framework of a collaboration between NRG and the JRC.

5.7. RESEARCH AND DEVELOPMENT ACTIVITIES IN FRANCE

5.7.1. Introduction and main achievements

For more than 20 years, the CNRS has carried out R&D activities on the MSFR concept with national and European partners. The Generation IV International Forum selected the MSFR (representative of a reactor with molten salt as fuel and coolant) and the advanced high temperature reactor, or AHTR (representative of a reactor with molten salt as coolant only) because of their promising safety and design characteristics [248]. The so-called 'reference MSFR' design is a 3000 MW(th) reactor based on a fluoride fuel salt volume of 18 m^3 and operated at a mean fuel salt temperature of 700°C [249]. New research activities started in 2018 also on small modular versions of the MSFR (s-MSFR), to be operated as a breeder in the U–Pu fuel cycle or as an actinide burner and thus using a chloride salt. These R&D activities have been expanded in the light of a growing interest in France to assess the ability of these MSFR systems to satisfy the objectives of Generation IV reactors [250] in terms of sustainability, resource saving (e.g. closed fuel cycle, no uranium enrichment), safety, waste management (e.g. actinide burner) and non-proliferation.

From 2009 to 2019, public research on the MSFR in France was conducted mainly by the CNRS and French universities, with the participation of the Institute for Radiological Protection and Nuclear Safety (Institut de Radioprotection et de Sûreté Nucléaire, IRSN) and industrial partners, including Compagnie de Réalisation Industrielle de Simulateurs (CORYS), EDF, Framatome and Orano, as well as CEA. It was supported by French research programmes, including PACEN, NEEDS, the MSFR master project of the National Institute of Nuclear and Particle Physics (Institut national de physique nucléaire et de physique des particules, IN2P3; part of the CNRS), and the European EVOL, SAMOFAR and SAMOSAFER projects. During this period, the number of researchers varied, at 5–15 per year, with most of them working on basic data acquisition and physical phenomena modelling.

The R&D activities in France focused on the MSFR concept and are related to the following:

- *Calculations of material composition, nuclei evolution and burnup associated with MSRs*. Development and validation of the in-house REM neutronic code, and studies and optimization of different kinds of MSRs (fluoride and chlorides salts, fast and thermal spectrum, a cycle based on Th–^{233}U or on ^{238}U–Pu or versions for burning actinides).
- *Simulation studies during normal and accident conditions*. Development of two simulation tools for MSRs, namely the TFM-OpenFOAM code, a coupled CFD neutronics code that accounts for the delayed neutron precursor motion in the circulating liquid fuel and developed by the CNRS for transient analysis; and the LiCore power plant simulator under development by the CNRS and CORYS.
- *Safety-by-design for MSRs*. Safety evaluation of the 'reference MSFR' (a 3000 MW(th) MSFR based on the Th–U cycle) to achieve safety-by-design for a large MSFR, with the plan to apply the same approach for the s-MSFR in the coming years.
- *Measurement and study of the basic chemical properties of molten salts*. The main experimental activities on MSRs in France are conducted by LPSC in Grenoble and by the Laboratory of the Physics of the two Infinities Irène Joliot-Curie (IJCLab) in Orsay. LPSC has conducted the FFFER project

(devoted to testing gas dispersion and separation in circulating salts) and the SWATH experimental loops project (to validate heat transfer models in the presence of a solidified layer of salt in contact with a circulating liquid salt) using liquid LiF–NaF–KF eutectic (FLiNaK) at 550–750°C, and IJCLab has conducted chemical and material experiments (see Sections 5.7.4 and 5.7.5). Two other laboratories, the Chemical Engineering Laboratory (LGC) in Toulouse and Conditions Extrêmes et Matériaux: Haute Température et Irradiation (CEMHTI) in Orléans, are no longer supported in the MSR field but have expertise in these experimental activities.

— *Corrosion studies.* Focusing in particular on the influence of the salt redox potential on corrosion, these studies have been carried out with molten salts in contact with Hastelloy C276[13] and AISI 304[14].

These R&D activities will continue to address the same topics during the European SAMOSAFER project (2019–2023) and will focus on the 'reference MSFR', the fluoride breeder version using the thorium fuel cycle. Other studies have started since 2019 on alternative versions of the MSFR based on a chloride salt and designed as breeders in the U–Pu fuel cycle or as actinide burners. Some of these alternative versions are SMRs. All these versions (reference and alternatives) are characterized by a fast neutron spectrum and by an integrated design for the fuel circuit. The activities are carried out in the framework of national collaborations between academic partners, including the CNRS, Grenoble Institute of Technology and IMT Atlantique, and industrial partners, including Framatome, Orano, CORYS and EDF, together with CEA and IRSN.

5.7.2. Reactor physics

5.7.2.1. Simulation of material evolution during operation

The MSFR team at the LPSC laboratory in Grenoble has been developing, updating and validating the in-house REM neutronic code for more than 20 years [251, 252]. This code has been used to perform various MSR studies and optimizations of the MSFR concept based on the evolution of material composition during irradiation or in storage. The studies include burnup studies of different kinds of MSFR (fluoride and chloride salts, fast and thermal spectrum, and versions of MSRs as breeders and actinide burners using a cycle

---

[13] Hastelloy C276 is a nickel–molybdenum–chromium alloy designed to have excellent corrosion resistance in a wide range of severe environments.

[14] American Iron and Steel Institute (AISI) stainless steel 304 is the standard material of austenitic chromium–nickel steels.

based on Th–^{233}U [253] and based on ^{238}U–Pu). A neutronic benchmark [254] has been carried out in the framework of the European EVOL project. Finally, these burnup calculations have been used for applications such as proliferation evaluations [255, 256] and radioprotection issues in the framework of work package 5 of the SAMOFAR project between the Institute of Nuclear Physics (Institut de Physique Nucléaire, IPN) in Orsay, LPSC and the CEA's Institute of Research into the Fundamental Laws of the Universe (Institut de recherche sur les lois fondamentales de l'Univers, IRFU).

Dedicated studies were performed at the SUBATECH laboratory to evaluate the decay heat produced by the MSFR and to assess the potential need of new nuclear data for such a Generation IV concept. The decay heat is calculated by combining reactor simulations to estimate the fuel inventory with nuclear data (decay properties of fission products and actinides, cross-sections and fission yields) as inputs. The codes currently used are mainly validated for boiling water reactors or pressurized water reactors in the U–Pu cycle, but they are not yet validated for Generation IV reactor concepts. Moreover, some fission products in the decay data libraries have decay schemes that are biased because of the pandemonium effect. This pandemonium effect comes from the low efficiency of germanium detectors at high energy, resulting in an overestimation of the β^- contribution and an underestimation of the γ contribution in the decay heat. The SERPENT-2 code was used to carry out a simulation of the MSFR reactor core operated with a fluoride salt, based on the ^{232}Th–^{233}U cycle. The aim was to identify the main nuclei contributors to decay heat for different cooling times and to determine if some important fission products are also biased by the pandemonium effect and need to be remeasured with an alternative experimental technique based on the total absorption spectroscopy method. A preliminary list of ten potential pandemonium nuclei was established [257] and will be extended and compared with cases with chloride salts and the ^{238}U–^{239}Pu cycle, before being discussed with the collaboration on total absorption spectroscopy between SUBATECH, the University of Surrey and IFIC Valencia for a potential experimental proposal on an isotope separation on-line, or ISOL, facility.

In addition, work is also foreseen for the period 2020–2024 on uncertainty calculations of the decay heat of the MSFR, especially to determine the impact of nuclear data by simultaneously using approaches based on the total Monte Carlo method and the perturbation theory.

5.7.2.2. Simulation tool for transient analysis: TFM-OpenFOAM code

The safety and operating procedures for the MSFR, as a reactor with circulating liquid fuel, need to be newly defined. The intrinsic core stability is guaranteed by the reactor's excellent negative feedback coefficients. Owing to this stability, the power production may be driven only by the heat extraction, resulting in an interesting flexibility to follow the electric grid's load. Control rods in the core might then not be required, which eliminates a classical accident initiator. Since the heat is produced directly in the salt that circulates in the fuel circuit, the negative feedback coefficients act rapidly to stabilize the core, unlike reactors with solid fuel. The fuel salt itself is cooled in the heat exchangers. Because of these specificities, dedicated tools are being developed to simulate the reactor's behaviour for normal (e.g. load following) and accident (e.g. reactivity insertion) conditions, to develop, optimize and assess the MSFR operating procedures. The reactor modelling requires specific treatment to consider the phenomena associated with the liquid fuel circulation.

The study of MSR cores in such conditions requires a code coupling the neutronic and thermohydraulic evolutions. Accordingly, two tools for transient simulation have been developed at the CNRS: the TFM-OpenFOAM three dimensional coupled code and a simulator of an MSR power plant based on the LiCore code model (see Section 5.7.3.1).

Important aspects of such systems regarding the coupling are the delayed neutron precursor convection and a complex flow pattern in the core cavity. Thus, a multiphysics tool called TFM [205, 258], which is more generic than those existing for reactors with solid fuel, has been developed by coupling CFD and a green function approach based on Monte Carlo calculations. With CFD, the Reynolds averaged Navier–Stokes equation can be solved and a three dimensional flow description in the core is provided. The TFM approach is a time dependent version of the fission matrices characterizing the transport of a neutron from its birth position to its death or fission position. This includes the prompt as well as the delayed neutrons, irrespective of their location in the fuel circuit (in the critical zone or in the recirculation/cooling sectors). Then, using a technique of power iteration, all the generations of prompt neutrons are reconstructed and finally the reactor fission distribution, which accounts for the precursor transport, is obtained. The neutronics approach and this approach's coupling to OpenFOAM code is described in Refs [72, 206].

The TFM-OpenFOAM code can be used to calculate various normal (e.g. startup and load following procedures) and abnormal transients (e.g. reactivity insertion, loss of fuel flow, loss of heat sink, low power overcooling accident). Depending on the calculation options and mainly the refinement of the meshing, the calculation time for a given transient varies from several seconds (e.g. real time calculation) to one week. Finally, the TFM-OpenFOAM code developed initially for MSFR studies was also used for pressurized water reactor calculations [206] and was recently extended to perform transient calculations of sodium fast reactors [259–262] and of research reactors [263].

5.7.3. System behaviour and safety evaluation

Molten salt reactors with liquid fuel are flexible in terms of operation (such as the capability to follow the grid's load) and design choices (such as fuel composition, power level), and they differ considerably from reactors with solid fuel in terms of design, operation and safety approach.

5.7.3.1. Simulation tool for the MSFR power plant: LiCore code

Dedicated developments and studies have been performed in the framework of the European SAMOFAR project and in parallel in France involving the CNRS, CORYS and Framatome on the code called LiCore. This code is a power plant simulator based on basic principles and adapted for MSRs.

The MSFR is composed of three circuits: the fuel circuit, the intermediate circuit and the power conversion circuit. The fuel circuit, defined as the circuit that contains the fuel salt during power generation, includes the core cavity and the recirculation-cooling loops or sectors, which contain the inlet and outlet pipes, pumps and fuel heat exchangers. The neutronic model LiCore, at the centre of the simulator, corresponds to an improved model of point kinetics to account for the specificities of an MSR, notably the circulation of the delayed neutron precursors out of the core. Coupled to a piston model for the fuel motion in the core, this code can perform calculations faster than real time to simulate the behaviour of the fuel circuit. A simplified model of the intermediate circuit allows parametric studies of the MSFR fuel circuit during normal and accident conditions to be performed. Consistency of the results provided by the LiCore code with the three dimensional coupled neutronic–thermohydraulic TFM-OpenFOAM code has been checked [206, 264].

Since 2017, the CNRS has been collaborating with the CORYS company, a subsidiary of Framatome that develops simulators for trains and nuclear power plants. The LiCore code has been integrated successfully in ALICES (see Fig. 45), the integrated simulation toolset designed by CORYS for the development, maintenance and operation of major simulators, such as power plant simulators. Additional modules are being added to fully simulate the intermediate and energy conversion circuits. The idea is to add a simulation of the intermediate and energy conversion circuits. This integrated version allows the whole MSFR plant to be studied (see Fig. 46), and thus helps to define the operating procedures of the reactor. The next steps for developing this power plant simulator will be the addition of control–command devices and improvement of the modelling of components such as the turbine.

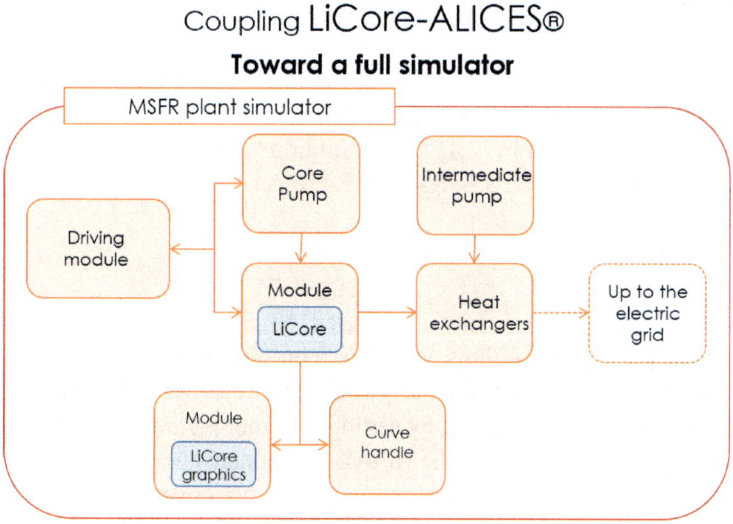

FIG. 45. Schematic of the LiCore-ALICES power plant simulator (courtesy of E. Merle, Laboratoire de Physique Subatomique et de Cosmologie).

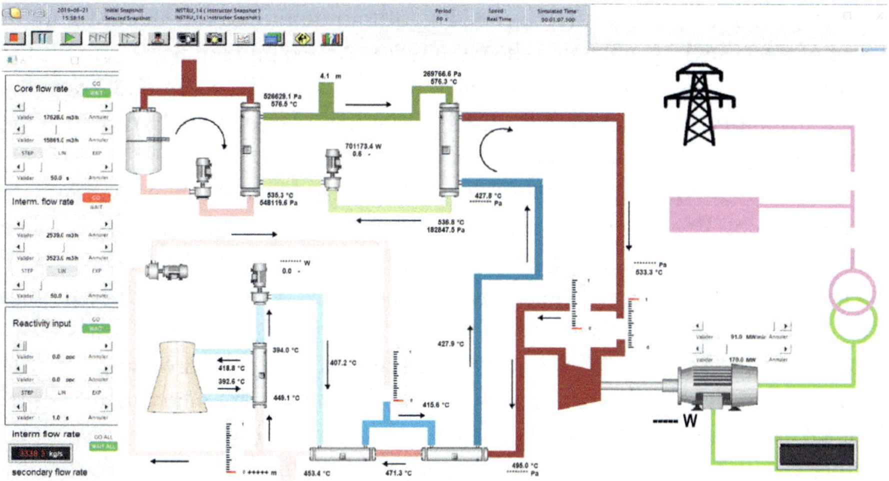

FIG. 46. Main screen of the LiCore-ALICES power plant simulator for the MSFR (courtesy of E. Merle, Laboratoire de Physique Subatomique et de Cosmologie).

5.7.3.2. *Procedures for normal operation*

Studies have been performed with the LiCore and the TFM-OpenFOAM codes to define a number of operating procedures (e.g. for startup [265], for following the electric load [72]). This definition of procedures continues in the framework of the European SAMOSAFER project.

5.7.3.3. *Geometry optimization of the fuel circuit*

A static system code that is focused on the MSFR's fuel circuit and connected to genetic algorithms has also been developed at LPSC. This code allows optimization studies of the fuel circuit. The objective is to obtain a globally optimized geometry of the circuit by setting different parameters and constraints, instead of separately optimizing the various components (such as heat exchangers and pipes). This is essential since some constraints cover several components (e.g. the fuel salt and coolant volume). A configuration that best satisfies these constraints can then be sought by adjusting a list of variable design parameters.

This code describes the system as an assembly of components, with each component having specific physical properties or parameters (e.g. length, hydraulic diameter, power), as shown in Fig. 47. It defines a set of constraints on each parameter (i.e. minimum and maximum values) that contributes to the figure of merit of the system. A genetic algorithm optimizes the global figure of merit of the system by changing the parameters. An example of application is the optimization of the fuel circuit to reduce the total fuel salt volume to lower the fissile inventory while keeping the fuel salt volume large enough to reduce the pressure drops in the pipes and heat exchangers.

Three types of heat exchangers have been considered following preliminary assessments by Framatome: corrugated heat exchangers using Hastelloy N, plate heat exchangers using SiC and channel heat exchangers. Calculations have been performed, for instance, with given inlet temperatures of the intermediate fluid in the heat exchangers and fuel temperature in the core, and with given mass flows of the salts in the intermediate and fuel circuits. The results are the geometric parameters of the heat exchangers (plate heat exchangers are better than channel heat exchangers) and the results show that the main risk is due to the lowest fuel temperature (freezing). Applications include the main temperatures (criticality temperature and wall temperatures) and hydraulic characteristics as functions of the core specific power, in the range of 50–600 MW/m^3.

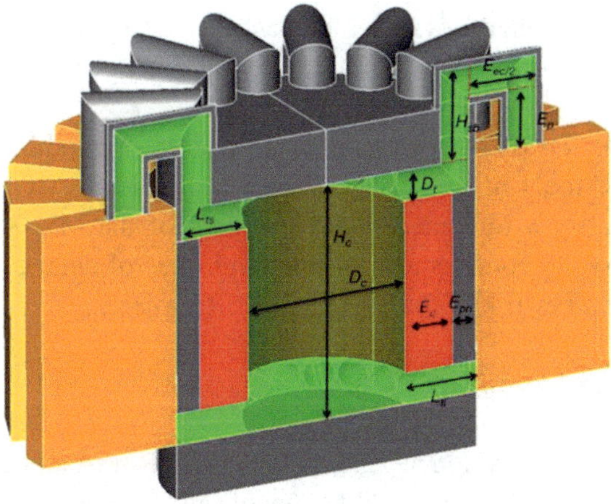

FIG. 47. Sizing parameters for the fuel circuit of the MSFR (courtesy of E. Merle, Laboratoire de Physique Subatomique et de Cosmologie).

5.7.4. Safety approach

A safety approach dedicated to the MSFR, or more comprehensively for reactors with fuel that is circulating in liquid state, is being developed. Some preliminary steps have been studied in the framework of the EVOL project as well as in M. Brovchenko's PhD thesis [266]. The approach for this development has been defined in the SAMOFAR project in a task led by IRSN and involving mainly the CNRS, Framatome and Politecnico di Torino (Italy). This approach aims at reaching a safety level that is 'built-in' and not 'added on' by applying it at the earliest stages of design. The approach is based on the Integrated Safety Assessment Methodology developed by the Risk and Safety Working Group of the Generation IV International Forum [267], coupled to usual risk analysis methods such as functional method analysis, master logic diagram and the line of defence method [268, 269].

Based on the new integrated design of the MSFR plant, abnormal situations of the fuel circuit and the emergency draining system have been identified and classified according to their initiating event (i.e. according to the phenomena involved), according to the following sources:

- — The analysis of the accident types identified for currently operating pressurized water reactors;
- — Deterministic calculations, such as criticality and thermal studies [265, 270], and calculations of transient and accident scenarios using multiphysics calculation tools [72, 206];

— Preliminary risk analyses taking into account design optimizations, together with the identification of the associated initiating events [268–270].

The application of the assessment methodology combining deterministic and probabilistic tools has been launched during the SAMOFAR project, for example, to identify a reference severe accident of the MSFR and to propose and evaluate possible confinement barriers and lines of defence. This work will continue in the framework of the SAMOSAFER project.

5.7.4.1. Abnormal transient calculations

As mentioned in Section 5.7.2.2, various MSFR transient calculations have already been performed with the TFM-OpenFOAM code, including parametric studies of overcooling and reactivity insertion transients. A parametric study of an overcooling transient is illustrated in Fig. 48 [72].

These transient calculations have highlighted the excellent stability of the MSFR core even in the case of a violent and quick perturbation. Parametric transient studies (such as the overcooling event at low power level; see Fig. 48) performed up to prompt critical regime [72] have also demonstrated that no cliff-edge effect occurs when this regime is reached; that is, no sudden violent behaviour is observed for the MSFR. This behaviour is important regarding safety considerations and analyses.

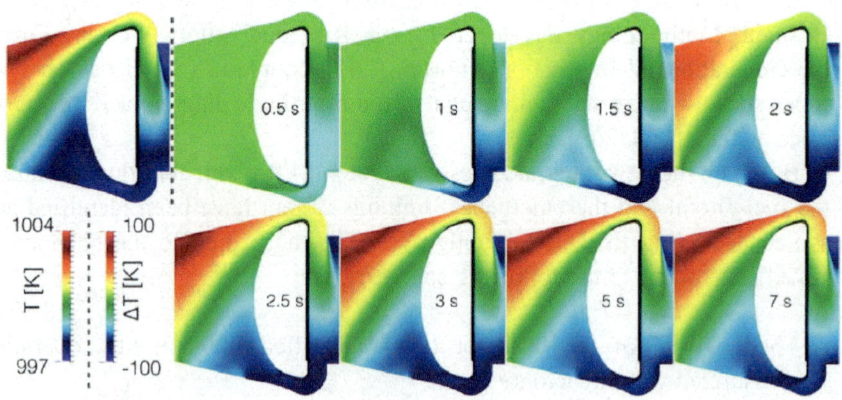

FIG. 48. Distribution of the fuel salt temperature at t = 0 (left) and its time variation T(t) − T(0) for a 100 MW–3 GW overcooling transient (reproduced from Ref. [72] with permission).

5.7.4.2. *Fuel and coolant chemistry*

These studies rely on the chemistry of fission products and actinides in the fuel salt. The fuel salt selected at the end of the EVOL project for the MSFR concept was LiF–ThF_4–UF_4 (77–19–4 mol%). The main experiments carried out at the Institute of Nuclear Physics in Orsay (now part of IJCLab) were focused on the solvent salt LiF–ThF_4 (77–23 mol%) to study the behaviour of uranium and fission products in this salt [271, 272]. The fission products considered were iodide, zirconium and lanthanides. The possibility of extracting iodide by forming iodine gas by fluorination with an efficiency higher than 98% was shown. Zirconium chemical analysis showed that Zr(IV) is the only soluble oxidation state stable in the molten salt; this is in disagreement with thermodynamic calculations that indicate a stability domain for ZrF_2 [273]. The chemistry of uranium was particularly well studied because (i) uranium is the fissile material of the MSFR concept and (ii) it can exist as two soluble states U(IV) and U(III). This characteristic is used to control the redox potential of the fuel salt to avoid the corrosion of the structural materials. Section 5.7.5 discusses this point further.

In the presence of less than 2 ppm of O_2 in the environment in equilibrium with the salt, an oxidation of iodide (I^-) to iodine (I_2) is observed with the formation of stable thorium oxyfluoride ($ThOF_2$), whose solubility has to be determined. The high stability of this compound leads to increasing the oxidizing power of O_2. When UF_4 is added in the molten salt without UF_3, a decrease of its concentration occurs with time due to the chemical oxidation of UF_4 by O_2 that in turn produces soluble UO_2F_2. The solubility of the oxyfluorides UO_2F_2 and $ThOF_2$ has not yet been evaluated to determine the concentration at which precipitation will occur.

The thermodynamic diagram of uranium in LiF–ThF_4 was drawn based on thermodynamic data and experimental results (see Fig. 49). In this figure, *E(V)* represents the potential of the salt and *pa*(Li_2O) the cologarithm of the activity of Li_2O, which represents the level of oxides in the salt. This kind of diagram is usually used to evaluate the thermodynamic stability of the salts [149]. The stability domain of each compound is presented by areas limited by straight lines. The red lines correspond here to the thorium compounds and the green lines to the uranium compounds.

These studies determined fundamental data such as activity coefficients and diffusion coefficients in LiF–ThF_4. These data are provided in Table 14.

The main steps of the processing scheme of the MSFR spent fuel are displayed in Fig. 50. The processing is composed of two parts: an on-line physical processing, which is performed by injecting a bubbling gas into the fuel salt, and an off-line chemical processing, which consists in removing 10–40 L of the fuel salt every day for batch processing of several chemical reactions based on the redox and acid base properties of elements in the fuel salt. Since the different elementary steps involved in off-line processing rely on the chemical properties of the elements composing the fuel salt, a considerable quantity of basic data is required to calculate each step efficiency.

Unlike the MSBR project conducted by ORNL in the 1960s [275, 276], which was based on thermal neutron spectrum, the removal of ^{233}Pa is not an issue

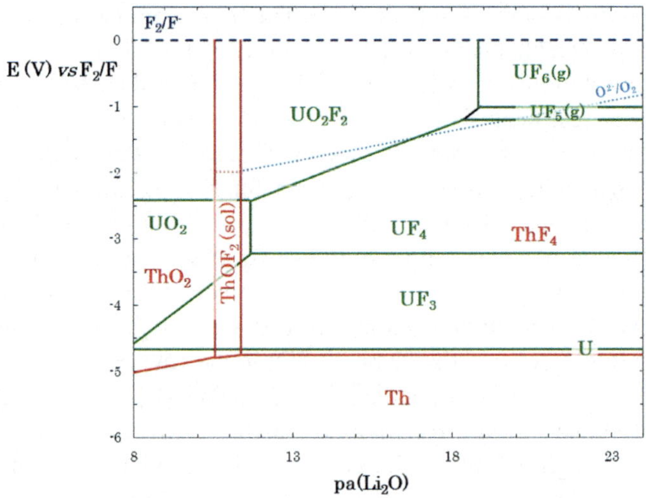

FIG. 49. Thermodynamic diagram of the stability of uranium in LiF–ThF$_4$ molten salt at a temperature of 650°C (P(F$_2$) = 1 atm, a(F$^-$) = 1, [UF$_4$] = 2.5 mol%) [272] (courtesy of E. Merle, Laboratoire de Physique Subatomique et de Cosmologie).

TABLE 14. DATA EXPERIMENTALLY DETERMINED IN LiF–ThF AT 600°C AND 650°C

	T (°C)	UF_4	UF_3	ThF_4	KI	ZrF_4
Activity coefficient, *lg* γ	600	−4.34	−0.54	−1.98	n.d.[a]	n.d.
	650	−4.05	−0.52	−1.96	−1.63	−4.33
Diffusion coefficient, *D* (cm²/s)	650	1.85 + 10^{-6}	2.5 + 10^{-6}	n.d.	n.d.	n.d.

[a] n.d.: not determined.

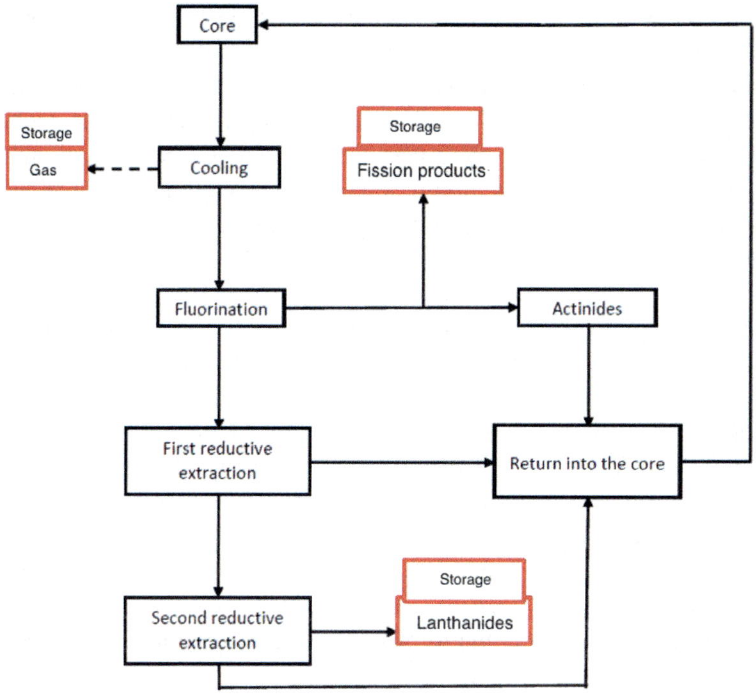

FIG. 50. Main steps of the chemical processing scheme of the MSFR fuel salt, including the physical (bubbling) processing on the top left (gas) [274] (courtesy of E. Merle, Laboratoire de Physique Subatomique et de Cosmologie).

in the MSFR concept because of its fast neutron spectrum. Owing to the high neutron captures by ^{233}Pa and fission products in a thermal neutron spectrum, the required rate of fuel salt processing was estimated at about 4000 L/day for the MSBR. The fast neutron spectrum of the MSFR concept enables a much slower processing rate of 10–40 L/day. No ^{233}Pa removal is required for the MSFR, as protactinium is managed with the other actinides in the chemical processing steps.

(i) *On-line processing.* This part of the processing that entails helium bubbling has two main objectives. First, the removal of gaseous fission products xenon and krypton, and second the removal of part of the noble metals produced in the core when they are in their metallic states. This process, based on a flotation process, has already been studied by ORNL. An amount of 15% of the gaseous fission products having a very short lifetime will decrease in the salt before extraction. The other gaseous fission products (85%) have lifetimes that are long enough to allow their extraction and treatment.

(ii) *Off-line processing.* The fluorination step has been studied at ORNL and the results obtained have been adapted for processing in the MSFR. The objective is to remove the elements with high gaseous oxidation states, such as niobium, tellurium, iodine, molybdenum, chromium and technetium for the fission products and uranium, plutonium and neptunium for the actinides. This technique involves oxidization of all the chemical elements contained in the salt to their higher oxidation states to produce gaseous elements, which are naturally separated from the salt.

A part of plutonium can also be oxidized in certain conditions [277]. The extraction by fluorination is possible by using small (around 100 μm) liquid salt droplets falling in a 50% F_2 atmosphere at a temperature in the range of 550–660°C. Up to 90% of plutonium was removed from the salt falling in a 1.3 m high fluorinator, and this efficiency is expected to be improved at higher temperatures and with smaller drop diameters. In the 1970s, the plutonium extraction that required crushing of solid salt was not feasible at a large scale. Today, granulation in a diameter range corresponding to that of the MSFR has reached an industrial stage for liquids at high temperature. This industrial technology can be adapted to the formation of molten fluoride droplets falling in a quasi-stagnant fluorine atmosphere.

The gaseous elements are then adsorbed on NaF traps. This adsorption reaction depends on the temperature of the traps. For instance, at 400°C, the adsorption of fission products (such as molybdenum, tantalum, niobium, tellurium, fluorides) occurs, but adsorption of UF_6 and NpF_6 does not occur. The adsorption of these two actinides occurs when the gas flows through a second trap heated at 25°C. The gas flows through several NaF traps heated at a given temperature to adsorb specific fission products or actinides and to perform their separation. In a second stage, the desorption of the elements occurs by heating the traps at a higher temperature. Fission products are reduced by hydrogen and are managed as waste. The objective of the last step is to reduce the actinides extracted by fluorination by using hydrogen gas in order to reintroduce them back into the fuel salt. Only a proportion of UF_4 is reintroduced. The remaining amount extracted is stored to feed new reactors since the MSFR is a breeder concept. Finally, the remaining gaseous elements removed by fluorination are also reduced to their solid state with hydrogen gas before storage. It appears that all the fission products are retained by absorption on the first trap (at 400°C) and all the actinides are absorbed on the second trap (at 25°C).

Concerning the reductive extractions step, a large number of results from ORNL can be used to define the MSFR processing concerning the steps of fluorination and helium bubbling previously described. As a consequence, in the framework of the first period of the EVOL project, the R&D activities on the processing scheme focused on the separation of actinides and lanthanides since no data were available for the calculation of the associated efficiencies. This separation was expected to be performed in two steps based on reductive extraction, with a 'first reductive extraction' dedicated to actinides and a 'second reductive extraction' for lanthanides. The reductive extraction is done through the contact between two liquid phases: the molten salt containing the soluble actinides and fission products that have to be extracted, and a liquid metal made of liquid bismuth, which contains lithium. It was demonstrated that the amount of lithium is the main parameter to control the selectivity and efficiency of the separation of actinide and lanthanide [278].

Reductive extraction requires the use of a Bi–Li liquid mixture of high purity. This mixture is prepared by cathodic electrolysis in the LiF–LiCl molten salt on a purified liquid electrode of bismuth [279]. Reductive extraction tests were performed at laboratory scale in LiF–ThF_4 containing uranium and neodymium to simulate actinides and lanthanides, respectively [280], in contact with Bi–Li (10 mol%). The efficiency measured was very low because it was shown that a solid alloy of Th_3Bi_4 and BiTh was formed at the interface of the liquid metal/molten salt, which blocks the transfer of uranium and neodymium from the molten salt to the liquid metal.

The final step concerns the back extraction of actinides and lanthanides. The back extraction of actinides is naturally achieved in the cleaned LiF–ThF_4 salt by contact with the bismuth pool containing actinides. This salt mixture is reintroduced in the reactor core after controlling the redox potential and adjusting the amount of fertile and fissile material (introduction of ThF_4 and UF_4, if necessary). The back extraction of lanthanides is accomplished in an inactive molten salt, possibly LiF–LiCl. The lanthanides are precipitated in oxide form by bubbling an inert gas saturated with a mixture of steam (pure water vapour) and other gases at a given temperature. In this way, the precipitation of the lanthanide oxide is attained, with the formation of HCl or HF gas [281]. The important aspect of this approach is to limit the volumes of effluents.

5.7.4.3. *Evaluation of the decay heat in the reprocessing plant and shielding requirement*

During the SAMOFAR project, an evaluation of the safety of the reprocessing plant was carried out by the CNRS and the CEA's Institute of Research into the Fundamental Laws of the Universe [282]. The first step consisted in providing the amount of each isotope in all places of the chemical plant. To perform this evaluation, it was necessary to know the initial inventory of isotopes and to introduce the efficiency of extraction of all the elements in

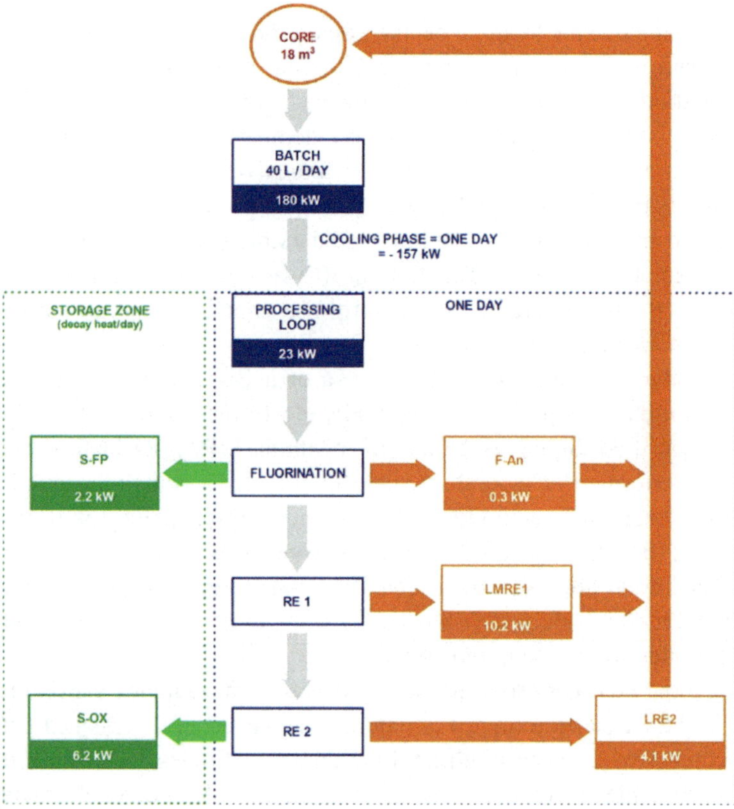

FIG. 51. Decay heat in each step of the processing scheme and in the storage zone (courtesy of E. Merle, Laboratoire de Physique Subatomique et de Cosmologie). An — actinides. FP — fission products. LMRE — liquid metal reductive extraction. OX — oxides. RE — reductive extraction. 'ONE DAY' in the dotted blue rectangle indicates the processing time (i.e. one day).

all the chemical steps. That was done considering both experimental results and thermodynamic data. Figure 51 shows the final scheme.

This work highlighted the necessity of defining one more step in the reprocessing scheme, which is indicated by the 'cooling phase'. This step consists in waiting for one day for the decay heat in the batch to decrease before starting the reprocessing. During this period, ^{233}Th decays to ^{233}Pa and the heat decreases by 157 kW.

By using the information in Fig. 51, the shielding requirement in the chemical plant was evaluated to be around 3 m of concrete in the thorium breeder 'reference MSFR'.

5.7.4.4. *Chemical experiment facilities for salts containing actinides in Orsay*

Several gloveboxes are installed at the Institute of Nuclear Physics in Orsay to perform experiments in active and inactive conditions. The gloveboxes are under inert gas, in depressurized conditions or in both conditions. Work can be carried out with about 100 g of ThF_4 and UF_4. Other techniques include thermogravimetric analysis and differential scanning calorimetry to measure melting points, inductively coupled plasma atomic emission spectrometry for elemental analysis in molten salts, X ray diffraction and scanning electron microscopy.

5.7.5. Performance of materials

5.7.5.1. *Influence of the redox potential on material corrosion*

The redox potential of the salt is the main parameter that controls the corrosion of the structural material. As shown in Fig. 49, the fuel redox potential can be controlled by regulating the ratio $[UF_4]/[UF_3]$ using the Nernst relation (see Eq. (1)).

$$E_{\mathrm{salt}} = E^{\circ}{}_{\mathrm{UF_4/UF_3}} + 2.3\frac{RT}{F}log\frac{[\mathrm{UF_4}]}{[\mathrm{UF_3}]} \tag{1}$$

where

R is the ideal gas constant,
T is the temperature (K),
and F is the Faraday constant.

This potential can be adjusted by adding uranium metal in the molten salt containing UF_4. In this case, the following chemical reaction occurs: $3UF_4 + U \rightarrow 4UF_3$, which is a fast and total reaction. Therefore, the mass of uranium metal added is directly related to the amount of UF_3 produced and the ratio $[UF_4]/[UF_3]$ can be adjusted in this way.

To test the influence of the redox potential on the corrosion, tests were carried out at 600°C over a period of 360 h with and without potential control on two alloys: Hastelloy C276 (a nickel based alloy) and AISI 304 (an iron based alloy). The tests highlighted the very important and efficient role of redox potential control to avoid corrosion in both cases [283].

Experiments in chloride media are also under way. Thermodynamic calculations evidence the risk of the formation of plutonium oxide (PuO_2)

and chlorine gas in the presence of oxygen traces. To prevent the formation of plutonium oxide, the addition of $MgCl_2$ is proposed. In this case, the oxides produced by oxygen reduction will be preferentially combined with Mg(II) to produce MgO. Magnesium oxide is a solid that can be filtered in the reactor core.

The redox potential in chloride media can be controlled, as in fluoride media, by the ratio $[UCl_4]/[UCl_3]$ (for instance, a ratio of 1/100 leads to the prevention of oxidation of nickel and iron), but chromium is oxidized to Cr(II) in chloride or oxide form. Chromium oxide is known to be highly protective of an alloy exposed to air, which is interesting for shaping phases. In molten salts, the stability of such a layer (of chromium oxide) has to be evaluated. The presence of $MgCl_2$ can decrease the stability of chromium oxide and increases the corrosion rate. These studies are under way at the Institute of Nuclear Physics in Orsay.

5.7.5.2. Component and technology development

This subsection deals with the pre-conceptual design activities carried out in France on the MSFR concept, and with the development and validation of technology and components related to experimental facilities.

(a) Pre-conceptual design activities on MSR systems

Several simplified designs of the 3 GW(th) MSFR with the Th–U cycle were proposed during the EVOL and SAMOFAR projects in order to integrate qualitative and quantitative studies related to safety analysis. Safety analysis of the functional design of the fuel circuit was conducted, and the results were used to propose changes to eliminate or reduce the consequences of accidents for each design option. This continuous process is ongoing to achieve safety-by-design, and its advancement depends on the sophistication of the available numerical modelling.

Some examples are given below on the conclusions of the SAMOFAR project for an updated MSFR design; these conclusions from 2019 are not publicly available. This design is limited to the functions to be performed by the components and no technical detail is given about them.

For the first time for the MSFR concept, an arrangement of the intermediate salt circuit is proposed to fit a selection of functions and recommendations and to allow a safety analysis. The general approach is the connection of several fuel loops and several loops for power conversion through common collector tanks.

Figure 52 shows the functional structure of the intermediate salt circuit between the intermediate heat exchangers and the conversion heat exchangers, the intermediate salt tanks (hot and cold tanks), and the position of the various elements with respect to the confinement barriers. The pumps are in blue. The startup and shutdown of the circuit is considered as well as the case of the reactor in standby mode (the heat is then removed by a 'dumping heat exchanger'). The size of the circuit and its fluid volume was calculated to provide damping of the power demand by the conversion that stabilizes the system. The intermediate circuit can be also used to adapt the operating parameters during load following.

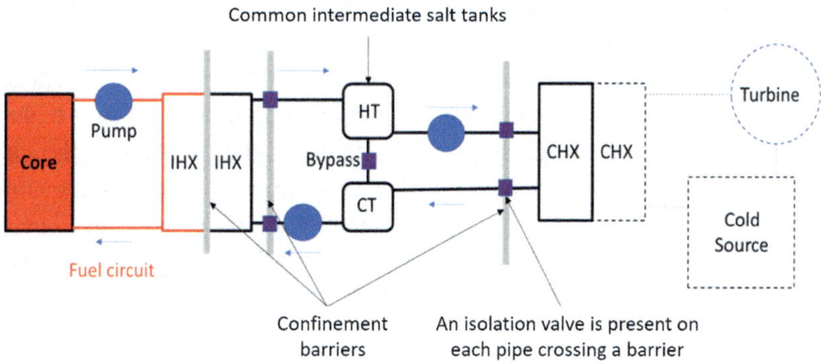

FIG. 52. Functional structure of the intermediate salt circuit of the MSFR (courtesy of E. Merle, Laboratoire de Physique Subatomique et de Cosmologie). IHX — intermediate heat exchanger. CHX — conversion heat exchanger. HT — hot tank. CT — cold tank.

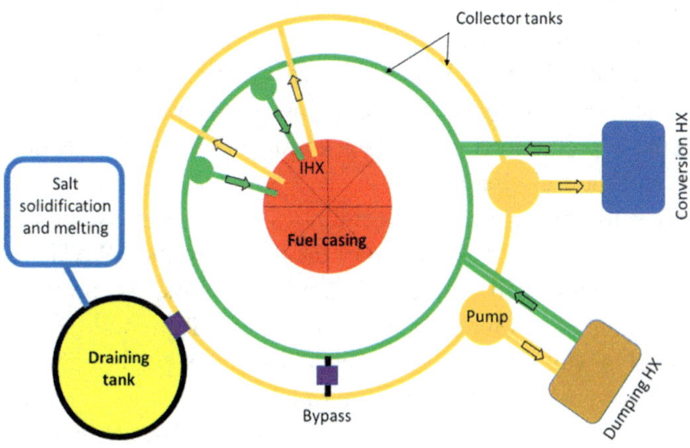

FIG. 53. Schematic presentation of the intermediate salt circuit of the MSFR (courtesy of E. Merle, Laboratoire de Physique Subatomique et de Cosmologie). HX — heat exchanger. IHX — intermediate heat exchanger.

A set of passive or partly active decay heat removal devices has been selected and studied. Two emergency draining containers were proposed: an emergency draining tank (relatively frequent draining events are expected because this design allows for reversible draining of the fuel salt) and a core catcher (in case of failure of the emergency draining tank). Both are passively cooled by gas through the second (core casing) and third (reactor casing) barriers. More detailed studies of this cooling are needed to assess its reliability because of the possibility of common cause failure. Figure 53 shows a schematic presentation of the intermediate salt circuit (in orange) in interaction with the fuel circuit (in green; circles are pumps), with several intermediate heat exchanger and conversion heat exchanger loops and with common collector tanks (for maintenance) and accessories. Figure 54 depicts a plant design in the case of absence of an emergency draining tank. Three confinement barriers are shown in this figure: the fuel casing, the reactor casing and the reactor building. The decay heat removal system uses natural gas convection inside the reactor casing and the reactor building for two cases: fuel remaining in the core and fuel drained into the core catcher. Approximate sizes were deduced from simplified calculations.

(b) SWATH facility and associated simulation tools

The SWATH experiments at LPSC developed during the SAMOFAR project aimed at improving and validating numerical thermohydraulic models of molten salt used for design and safety studies, and more specifically during the draining of the fuel salt. This involves convection (reinforced by turbulence mixing), conduction and radiation between different phases (solid, liquid, gas) present in some of the MSFR components for various states of the reactor. Owing to the relatively high Prandtl values of the molten salts, the thermal development lengths will in general be longer than the hydraulic development lengths. Accurate predictions of the heat exchanges will thus require a precise understanding of the flow field.

To reduce the experimental uncertainties as far as possible and to obtain significant results, a detailed definition of the design of the SWATH experiments is done in coordination with numerical simulation work. Accordingly, different experimental channel geometries were investigated in both a water model experiment and a salt experiment. This allows the performance of particle image velocimetry measurements in the water model experiment to determine precisely the flow conditions inside the experimental channel and thus compare them to the CFD predictions for an isothermal flow [284]. The work detailed in Ref. [285] focused on phenomena that were identified as being a priority for the MSFR concept and the SWATH experiments (i.e. the turbulence, the radiative heat transfer modelling, and the solidification of water and of molten salt).

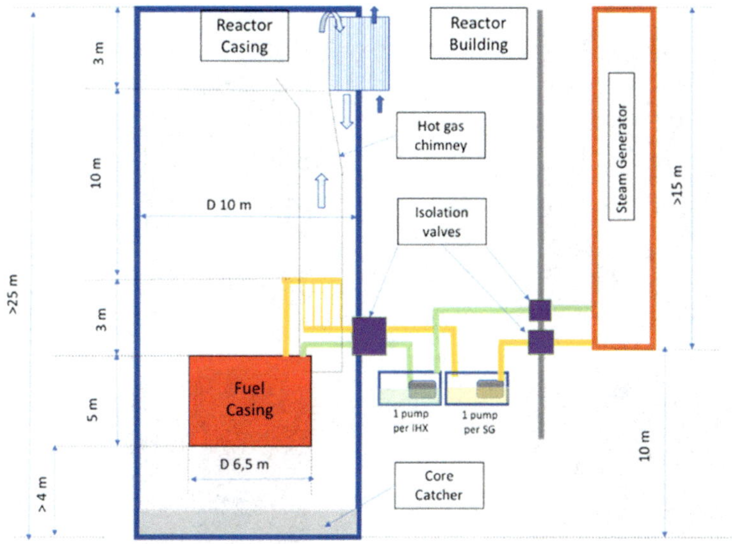

FIG. 54. MSFR plant sizing in the absence of an emergency draining tank (courtesy of E. Merle, Laboratoire de Physique Subatomique et de Cosmologie).

A numerical model for simulating the draining process was introduced and used for analysing different draining scenarios [286].

A multiscale phase change model [287] has been developed to improve the description of the solidification of molten salts and the information available on the properties of the resulting solid phase. Numerical results were obtained using this model, considering natural or forced convection in the liquid phase, to simulate the experimental results obtained with molten salt experiments. A good agreement was observed. Figure 55 shows the temperature and flow fields in the solid and liquid formed during the solidification process. The left and right images show the solidification experiments without rotation and with rotation, respectively.

The SWATH facility [286] is operated on a discontinuous working principle in which the flow is established by regulating the pressure difference between two tanks, rather than using a pump. A glovebox with an argon atmosphere is included between the tanks to protect a special portion of the circuit that can be changed for testing different configurations. Figure 56 presents a diagram and a photograph of this facility.

A closed channel section with different shapes or an open channel section can be placed in the glovebox. Other types of experiments related to solidification and melting processes have been performed to design efficient cold plugs [288] or to validate the solidification model. In the latter case, since it is expected that

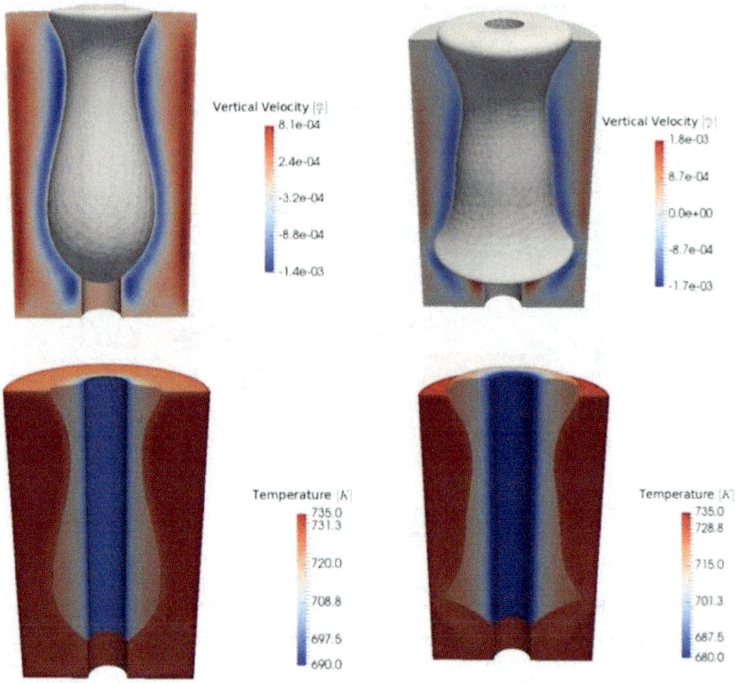

FIG. 55. Temperature (bottom) and flow (top) fields during the solidification process [285], without rotation (left) and with rotation (right) (courtesy of E. Merle, Laboratoire de Physique Subatomique et de Cosmologie).

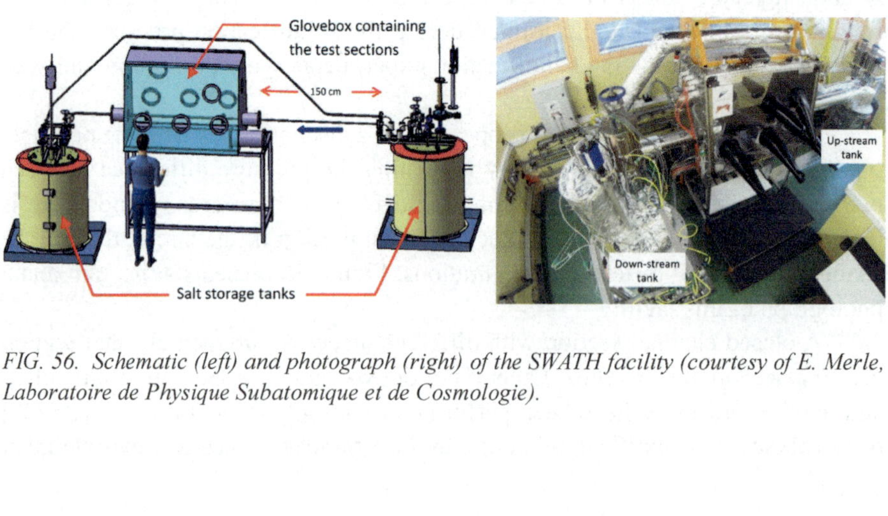

FIG. 56. Schematic (left) and photograph (right) of the SWATH facility (courtesy of E. Merle, Laboratoire de Physique Subatomique et de Cosmologie).

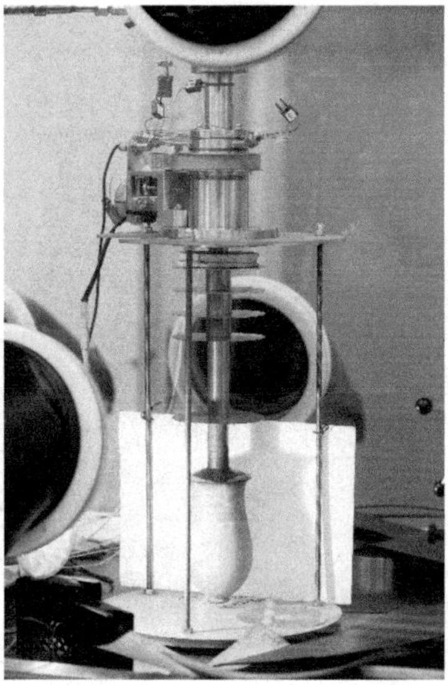

FIG. 57. Photograph of the solidification experiment (courtesy of E. Merle, Laboratoire de Physique Subatomique et de Cosmologie).

the flow by convection in the fluid phase has a significant effect on the shape of the solidification front, two different boundary conditions will be investigated: natural and forced convection. The solidification experiment (see Fig. 57) is based on a tube rotating inside an annular cavity filled with molten FLiNaK salt. The rotating tube contains an inner tube to enable the circulation of argon (cooling gas), which decreases the temperature of the external wall of the outer tube below the FLiNaK melting point, to initiate the solidification process. This figure presents the system after extraction from the salt bath with the solidified salt at its bottom; the salt bath is another part of the experiment.

(c) FFFER facility

The FFFER experiment at LPSC in Grenoble started in 2009, and a cold plug design was first implemented. This facility is a LiF–NaF–KF eutectic (FLiNaK) salt forced convection loop with the following main objectives:

— To study the liquid–gas separation in the case of on-line physical molten salt processing. Since on-line bubbling is an efficient process to capture dispersed non-soluble particles in a liquid and partly absorb dissolved gas, this process was selected for testing with molten salt. It consists in injecting

FIG. 58. View of the FFFER facility with thermal insulation around the pipes (courtesy of E. Merle, Laboratoire de Physique Subatomique et de Cosmologie).

bubbles into the flowing salt in the pipes followed by a liquid–gas separation in another part of the facility.

— To acquire technical experience in operating and designing molten salt experiments at high temperature, including flow control, salt level measurements and safety devices (freeze valve).

A partial description of the facility is given in Ref. [288], and Fig. 58 offers a view of the facility with thermal insulation around the pipes. Tests of helium bubbling and liquid–gas separation were performed in the FFFER facility in 2017, providing satisfactory results in the configuration used (about 1% vol. gas, 1.9 salt L/s); however, the results have not been published. The design of the separator top has to be improved to allow for running of higher flows.

5.8. RESEARCH AND DEVELOPMENT ACTIVITIES IN ITALY

5.8.1. Activities at Politecnico di Milano

The Nuclear Reactors Group at Politecnico di Milano has been involved in the modelling and analysis of MSR concepts with thermal and fast neutron spectra, in the context of several international projects, including the EVOL,

SAMOFAR and SAMOSAFER projects; the IAEA's Coordinated Research Project T12026 [289]; and collaborations with other research institutions such as TU Delft, Paul Scherrer Institute, LPSC in Grenoble, Karlsruhe Institute of Technology, Westinghouse Electric Company and VTT Technical Research Centre of Finland. Simulations and analyses oriented for experiments have been carried out for more than ten years to study different MSR concepts. The need to perform dedicated analyses is due to (i) the scarce experience on these reactor systems compared with the fleet of current nuclear reactors (e.g. Generation II, Generation III, research reactors) and (ii) the uncommon feature of the fuel motion in the case of MSRs with liquid fuel that make the adoption of 'standard' simulation tools for nuclear cores inadequate.

From a modelling point of view, the main peculiarities that characterize the complex environment of such systems are the drift of the delayed neutron precursors, the strong coupling between neutronics and thermohydraulic behaviours, the internal heat generation in the fluid and the possibility to perform on-line reprocessing and on-line fission product removal. These peculiarities have an impact on the dynamics and control of the system and require specific tools for the simulation of the reactor core [290] and burnup analysis, along with tools for the investigation of the heat transfer characteristics and natural circulation capabilities. The activities of the Nuclear Reactors Group focused on reactor physics (advanced modelling on multiphysics calculations), system behaviour and operation (dynamics and control) and component and technology development (experimental thermohydraulic activities).

5.8.1.1. System behaviour, operation, safety and security

The classic control strategies adopted for light water reactors and sodium fast reactors are not suitable for MSRs because of the different characteristics of the MSR systems. To support the development, testing and validation of the control system design, simulation tools specifically conceived for control purposes of MSRs are necessary. As a first step, to evaluate the effects of the fuel circulation on the system kinetics, simplified models [291, 292] were preliminarily set up to allow for the decay of delayed neutron precursors in the part of the primary circuit that is out of the core and assessed against the experimental data available from the MSRE (see Fig. 59, top). The results obtained are relevant for the startup and the shutdown phases of the reactor.

Regarding the analysis of MSR dynamics, different numerical models were conceived for studying the entire system behaviour (core, primary, secondary circuits), based on lumped and one dimensional approaches. In this framework, the approximations related to the adoption of simplified geometries were also investigated [293], highlighting the impact of one dimensional modelling to

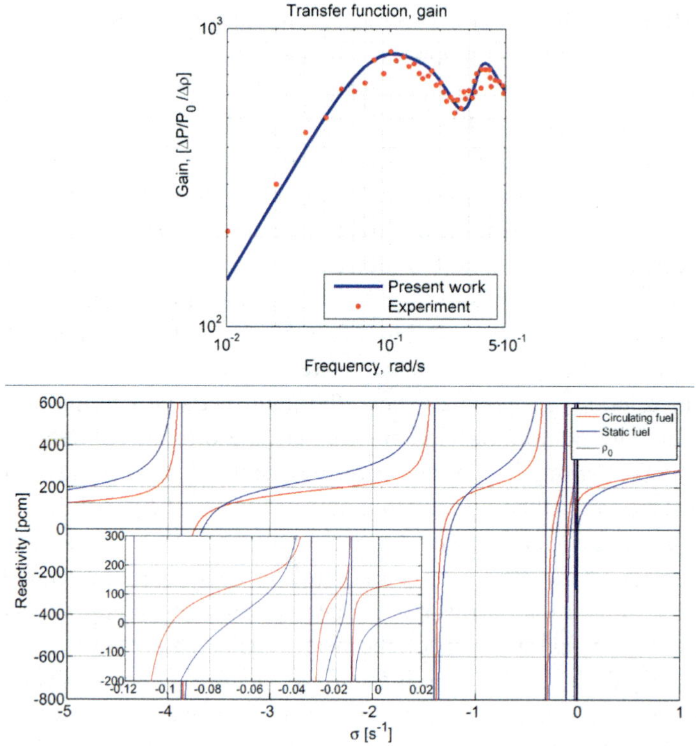

FIG. 59. Top: gain of the transfer function between power variation ($\Delta P/P_0$ where P_0 is the nominal power) and reactivity ($\Delta\rho$) for the MSRE at 8 MW (blue solid line: 'Present work') and comparison with experimental data (red dots: 'Experiment') (reproduced from Ref. [292] with permission). Bottom: inhour equation for the MSBR with circulating and static fuel (ρ_0 being the reactivity compensation for the loss of delayed neutrons and σ being the reciprocal time constant) (reproduced from Ref. [294] with permission).

predict the reactivity loss outside the active region. These models were adopted for linear stability analyses concerning both thermal and fast MSRs [294]. The outcomes indicate that, especially at low power, the dynamics is faster with respect to the non-circulating fuel system and is dependent on the fuel velocity (see Fig. 59, bottom). A reduction in the stability margin is possible when the drift of the delayed neutron precursors dominates over the temperature effects, in particular for moderated systems owing to the possible positive reactivity effect of the moderator.

A simulator that is better designed for the study of the MSFR power plant has been developed in the framework of the SAMOFAR project, based on the Modelica language [295]. The simulator focuses not only on the primary circuit,

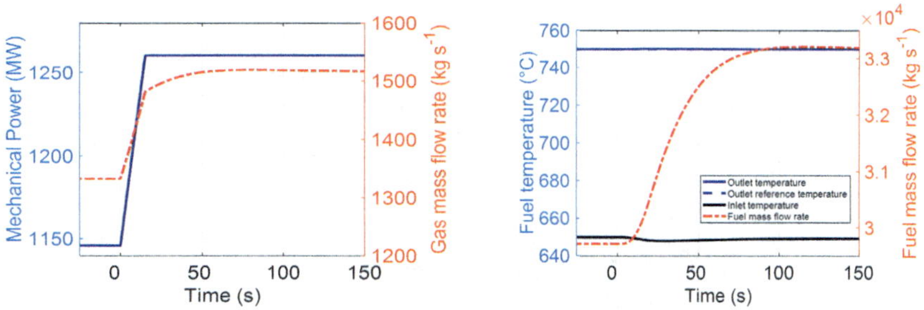

FIG. 60. Transient response of the control system for the MSFR without employing external reactivity (control strategy of mass flow rates). Left: mechanical power and gas flow rate; right: fuel temperatures and fuel mass flow rate (courtesy of S. Lorenzi, Politecnico di Milano).

but also considers the intermediate circuit and the energy conversion circuit. The choice of Modelica allows the use of standard and validated libraries for the modelling of the main components of the power plant, such as pipes, pumps and turbines. A special effort was made in developing an ad hoc MSR library for the neutronics kinetic model, the flow in one dimension of the salt fuel and the motion of the delayed neutron precursors. A preliminary version of a decentralized feedback control scheme for the MSFR was proposed and it was designed using the tools mentioned. Different control schemes were proposed and good results were obtained even if reactivity control by external means is not employed, which can be the case of the MSFR where control rods are not envisaged in the reactor design [296]. The inclusion of the intermediate circuit and the energy conversion circuit in the simulator of the power plant also allow an investigation of a more comprehensive control strategy [297] that confirmed the possibility of controlling the MSFR without such means in the mode of full power. In addition, the dynamics of the plant's electrical power, which is fast controlled, confirms the load following capabilities while keeping the controlled variables in a safe bandwidth (see Fig. 60). The helium Brayton cycle is considered as the energy conversion system in this figure.

5.8.1.2. Reactor physics

Analytic, deterministic and Monte Carlo approaches [298] (see Fig. 61, top) were developed for calculating the effective delayed neutron fraction, β_{eff}, which is an important reactor kinetics parameter for both safety and control. These approaches were used to consider the inhomogeneous spatial importance of neutron distribution and to obtain an accurate estimation of β_{eff} in the MSFR. With reference to the MSFR, the Nuclear Reactors Group at Politecnico di Milano developed also an extended version of the SERPENT-2 code to study fuel burnup

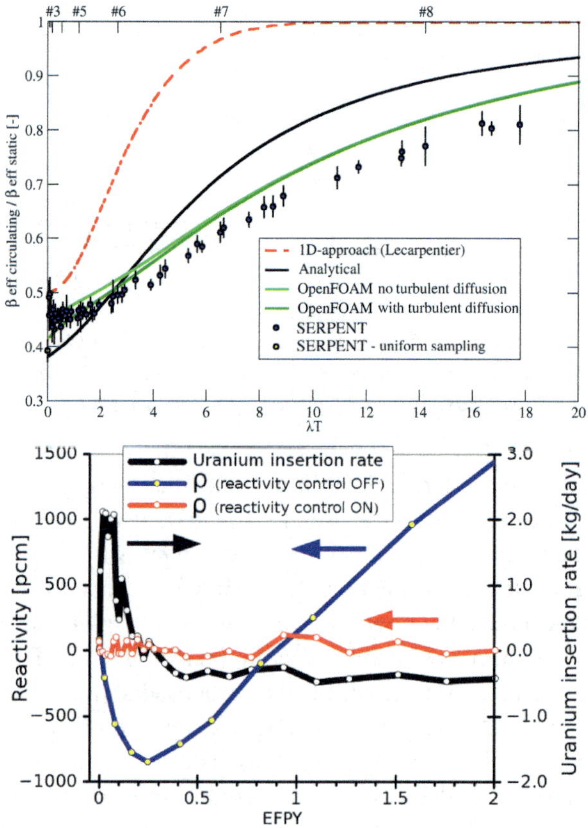

FIG. 61. Top: β_{eff} for the MSFR; comparison between analytical, based on Monte Carlo (SERPENT) and deterministic approaches (OpenFOAM) as a function of the product between the precursor decay constant λ and the fuel salt loop circulation period T (reproduced from Ref. [298] with permission; see Ref. [298] for information reflected in the figure). Bottom: reactivity and uranium insertion rate for the transuranic-started version of the MSFR obtained with the reactivity control algorithm (reproduced from Ref. [299] with permission). EFPY — equivalent full power year.

and core material evolution [299] (see Fig. 61, bottom) and investigated the fuel cycle and the core's physics performance in the long term (in terms of possible initial core loading, actinide burning capabilities, radiotoxicity generation, decay heat and safety features), which also included a systematic comparison with fast reactors with solid fuel [300]. The latter R&D activity was the subject of the IAEA's Coordinated Research Project T12026 on Near Term and Promising Long Term Options for Deployment of Thorium Based Nuclear Energy, in which Politecnico di Milano participated.

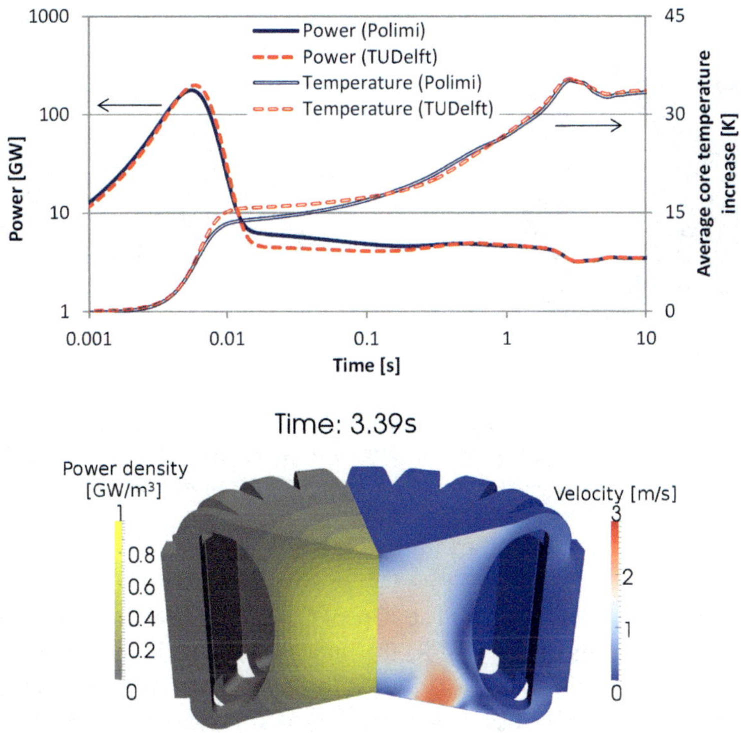

FIG. 62. Top: evolution of power and average core temperature in the MSFR core after a stepwise reactivity insertion equal to 0.2%Δk/k with a two dimensional axial-symmetric model (reproduced from Ref. [302] with permission). Bottom: velocity and power density fields in the MSFR after an accident of a single pump failure with a three dimensional model (reproduced from Ref. [199] with permission). Polimi — Politecnico di Milano.

As a development in the direction of more accurate analyses, a multiphysics modelling approach [290, 301] was undertaken because of the capability of solving different physics in the same simulation environment. With the adoption of this approach, the fluid dynamics (both incompressible and compressible flow with buoyancy effects), the neutronics and the heat transfer can be considered in an inherently coupled way. Both steady state and transient analyses (e.g. reactivity insertion, fuel mass flow rate variations, chilled inlet) can be performed. In this way, a two dimensional axial-symmetric model of a core channel of the MSBR, and two dimensional axial-symmetric [302] (see Fig. 62, top) and three dimensional models of the MSFR [199] (see Fig. 62, bottom) fuel circuit were developed. The multiphysics models were applied using both finite element and finite volume solvers, the latter being also open source and with a high parallelization level. In the framework of the SAMOFAR project, the model

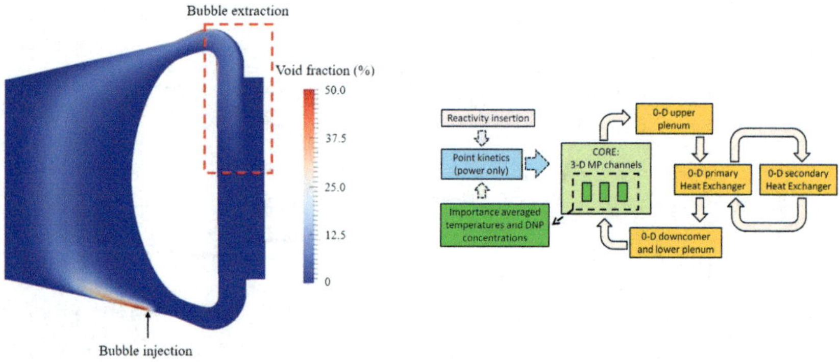

FIG. 63. Left: bubble distribution (void fraction) in the MSFR obtained with the two phase compressible solver (reproduced from Ref. [305] with permission). Right: conceptual scheme of the geometric multiscale model of the MSRE (reproduced from Ref. [292] with permission). DNP — delayed neutron precursors. MP — multiphysics.

based on the finite volume solver (i.e. OpenFOAM) has been improved to study the potential accidents of the MSFR, the impact of the helium bubbling system and the fuel compressibility effects [303]. To this end, the neutronics modelling of the multiphysics solver based on OpenFOAM was extended with multigroup neutron diffusion and the isotropic simplified spherical harmonics (SP3) neutron transport approach [304], and in the thermohydraulic modelling with the inclusion of a compressible thermohydraulic model with two phases, which is based on a Euler–Euler approach [305]. This solver underwent an intense verification phase during the SAMOFAR project, using a benchmark developed ad hoc for this purpose [306].

The fluid dynamics improvement is aimed at studying the effect of the on-line bubbling system for removing fission products that build up in the molten salt (especially the gaseous and the metallic ones). From a neutronics point of view, a correct bubble spatial distribution is essential to obtain a correct evaluation of the void reactivity feedback coefficient because of the importance of the inherent spatial and neutron effects (see Fig. 63, left). This solver will be useful in the design of the on-line bubbling system, in terms of location optimization and in assessing the system's capability for extracting fission products.

To combine the accuracy of the multiphysics modelling approach with acceptable computing power for the analysis of the overall plant dynamics, a geometric multiscale strategy [292] was developed. It consists in modelling the components of the plant according to their dimensionality. In particular, a three dimensional core channel of the MSRE was modelled with a multiphysics modelling approach, whereas the components located out of the core were modelled with a zero dimensional approach (see Fig. 63, right).

The thermohydraulic behaviour of internally heated molten salts flowing in straight circular channels was analysed by applying analytical and numerical

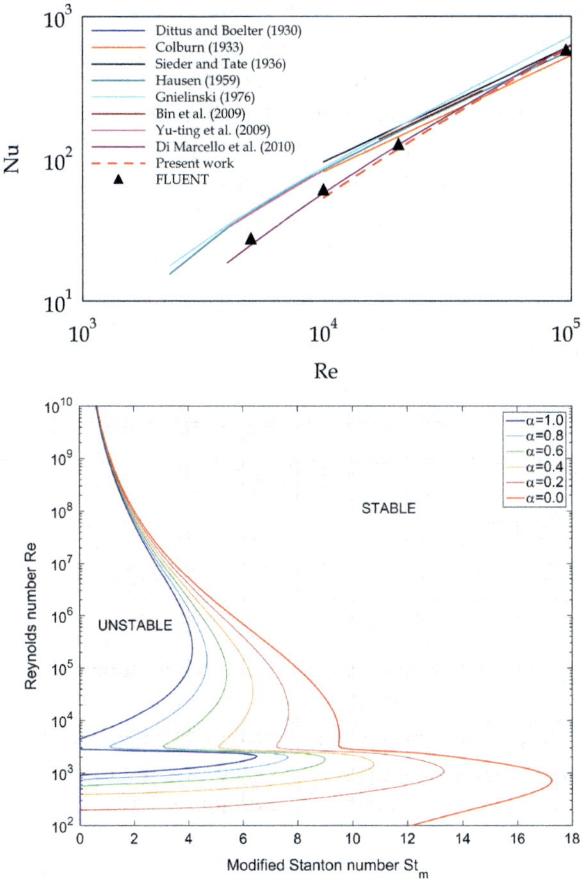

FIG. 64. Top: comparison among heat transfer correlations and CFD results obtained by means of FLUENT software, in the Reynolds range of interest for MSRs (Pr = 11) (reproduced from Ref. [290] with permission.) Nu — Nusselt; Re — Reynolds. See Ref. [290] for information reflected in the figure. Bottom: stability map (modal analysis) of a natural circulation loop for various levels of internal heat generation, from α = 1 (conventional, localized heat flux only) to α = 0 (internal heat generation only) (reproduced from Ref. [313] with permission).

tools (e.g. FLUENT software[15]) [290, 307] to MSRs moderated by graphite and to the MSFR (see Fig. 64, top). In particular, it was shown that the effect of internal heat generation can affect the wall-fluid bulk heat transfer characteristics and that it can be described by means of a corrective factor to be applied to traditional correlations (e.g. Dittus–Boelter or Gnielinski correlations) for the Nusselt number. The developed correlation was used to carry out a parametric investigation of the effect of decay heat on the components that are out of the core of the MSFR. The volumetric power causes higher temperatures at the channel wall than in the channel's inner parts, but the effect is significant only in the case of large diameters, low velocities or both. This effect can be of interest in the case of the channels of the salt reprocessing system [308, 309]. As for the dynamics of natural circulation in a molten salt with internal heat generation, linear analysis methods were developed to investigate the natural circulation features of internally heated fluids for MSR applications [310, 311], along with an advanced approach based on CFD for studying spatial effects [312]. These methods allow the stability of the operational conditions in terms of both asymptotic behaviour (modal analysis; see Fig. 64, bottom) and initial transient energy growth (non-modal analysis) to be studied. They were compared with non-linear models including a system code approach (one dimensional discretization with use of pressure drop and heat transfer correlations) and an approach based on CFD [313].

5.8.1.3. Component and technology development

The DYNASTY testing facility was built at Politecnico di Milano [314] (see Fig. 65, left), with the goals of (i) the experimental investigation of the natural circulation dynamics in the presence of distributed heat generation, with a specific focus on the analysis of thermohydraulic instabilities (stability maps); and (ii) the validation of the theoretical models and simulation tools developed to study the natural circulation capability of a loop with a homogeneously heated molten salt, useful for MSFR design optimization. DYNASTY is a natural circulation loop that can operate in circulation that is driven by buoyancy or with forced flow since it is equipped with a pump and electric heating elements. The latter allow the possibility of creating different conditions in terms of natural circulation configurations (conventional and distributed cases). The facility can run with a solar molten salt (KNO_3, $NaNO_2$, $NaNO_3$), but also can be operated with water and water–glycol mixtures, the latter being a simulant of the thermohydraulic behaviour of a molten salt. The facility has been extended (see Fig. 65, right)

[15] FLUENT is a general purpose CFD software used to model fluid flow, heat and mass transfer, chemical reactions, and more.

FIG. 65. Left and centre: the DYNASTY facility built at Politecnico di Milano; right: in orange, the extension of the facility (e-DYNASTY) to study coupled natural circulation loops (courtesy of S. Lorenzi, Politecnico di Milano).

within the SAMOFAR project to also investigate the dynamics of two coupled loops with natural circulation.

The facility has been characterized in detail through a strong modelling effort using various modelling approaches. This includes the development of models based on a system code approach, CFD models based on the large eddy simulation approach that can unveil flow reversal features during unstable transients and data assimilation techniques.

5.8.2. Activities at Politecnico di Torino

The faculty staff of the Energy Department of Politecnico di Torino, as members of the Nuclear Engineering Modelling research group (known as NEMO), has been involved in research activities on the neutronics of MSRs for more than two decades, in the framework of European projects, the IAEA's Coordinated Research Projects and collaboration with researchers at the international level.

The research activities initially focused on the study of the neutronics of reactors with circulating fuel and aimed to provide proper physicomathematical models for analysis carried out by the research group. The study of the dynamics of circulating fuel first focused on the neutronics part and then moved to a multiphysics framework, including coupling to thermohydraulic phenomena. In more recent years, starting with the SAMOFAR project, the different expertise available within the research group has been applied to carry out research activities with a stronger focus on safety.

5.8.2.1. Reactor physics: neutronics of circulating fuel systems

The former Reactor Physics Research Group at Politecnico di Torino was involved at the beginning of the 1990s in the European project on the impact of the accelerator based technologies on the safety of nuclear fission (the IABAT project). One task within this very broad and ambitious project was devoted to studying the design of a subcritical system with liquid circulating fuel. The neutronics analysis of this kind of system posed questions regarding the correct way to describe the behaviour of the neutron population within the reactor, especially regarding the movement of delayed neutron precursors that are dragged by the fuel motion before emitting delayed neutrons. Taking advantage of the available expertise in the typical methods for reactor physics (e.g. reactor dynamics [315]), various approaches to the neutronics analysis of these kinds of systems were addressed, such as the application of perturbation theory [316] and the definition of a point model suitable for the description of the peculiar physical effect appearing when the fissile material is moving [317, 318]. In particular, the issue related to the reduction of the role of delayed neutrons and the consequent impact on the dynamic response of the system was already addressed in these early works [319].

The involvement in other European projects that were fully focused on the MSR concept allowed further development in the neutronics analysis of systems with circulating fuel. The MOST project (2001–2004) gathered various European institutions interested in analysing the state of advancement of the MSR concept. In this framework, the staff at Politecnico di Torino carried out a research activity on the development of kinetic models for MSRs, leading to a consistent definition of the factorization–projection process adopted in point kinetics and quasi-statics [320, 321]. This process was then applied to transients of interest for these kinds of systems, such as modifying the fuel flow with consequent oscillation of the reactivity and, thus, the power (see Fig. 66 [321]).

During the MOST project, various institutions developed computational tools for the simulation of MSRs as a function of time, with a specific focus on the MSRE, which was carried out in the 1960s at ORNL. The performance of these tools was compared in a benchmark exercise using experimental measurements obtained on the MSRE (see Fig. 67 [322]).

The MOST project was followed by the ALISIA project (2007–2008), in which the activities at Politecnico di Torino on MSR dynamics were further pursued. In particular, the aspects related to the influence of the fluid dynamic aspects on the neutronic behaviour were studied, with an approach initially based

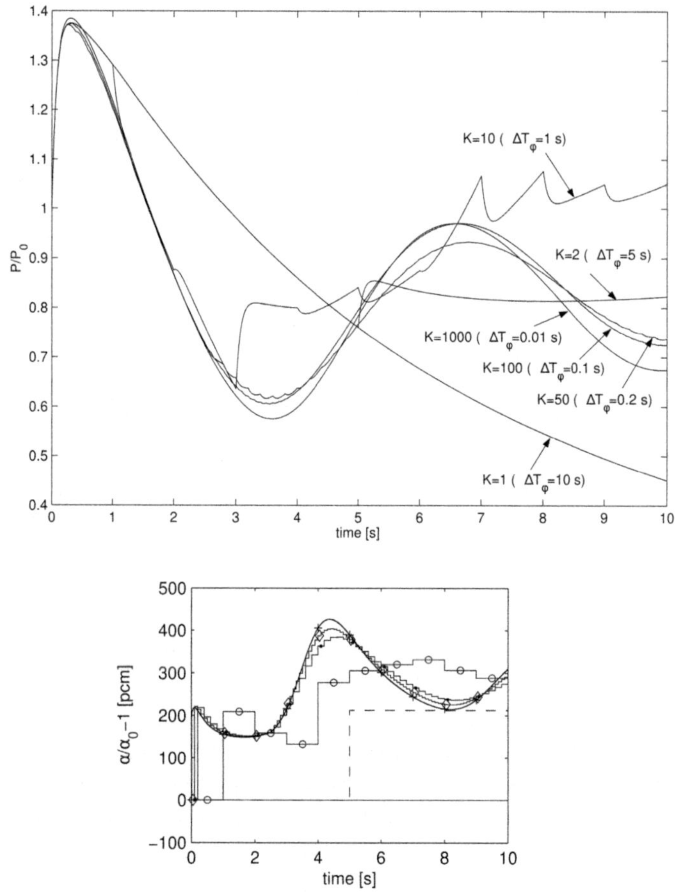

FIG. 66. Top: power evolution, normalized to initial value, for different numbers of shape recalculations (K) for a transient of fluid velocity in a critical system (ΔT_ϕ is the time interval adopted for the shape recalculation); bottom: time behaviour of α kinetic parameter of point kinetics related to reactivity evolution, see Ref. [321]), normalized to its initial value α_0, for different numbers of shape recalculations (solid line: K = 1; dashed line: K = 2; circles: K = 10; dots: K = 50; diamonds: K = 100; plus sign: K = 1000) (reproduced from Ref. [321] with permission).

on single-physics simulations: the fuel velocity field was modified parametrically to demonstrate its neutronic effects [323] and the effect of periodic oscillations of reactivity was analysed [324]. The neutronic and thermohydraulic tight coupling in an MSR then required a more complete simulation of MSR dynamics, leading to a series of activities aimed at the development of a model for the coupled neutronic–thermohydraulic analysis of an MSR, thanks to the collaboration with

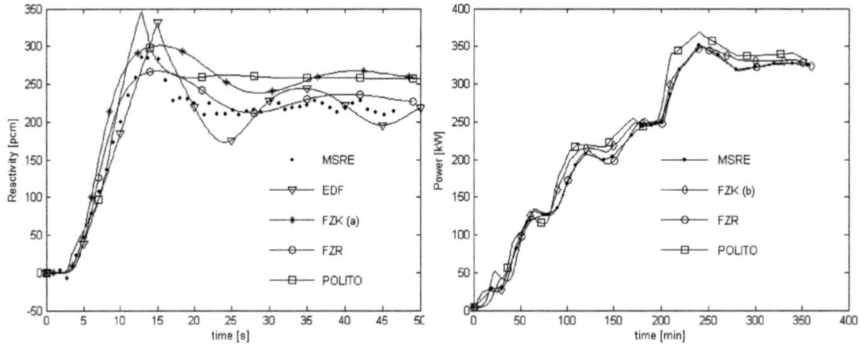

FIG. 67. Pump startup transient in the MSRE (Joint Evaluated File, JEF, decay data) (left) and natural convection transient in the MSRE (right) (reproduced from Ref. [322] with permission). EDF — Électricité de France. FZKa — Forschungszentrum Karlsruhe with SIMMER codes. FZKb — Forschungszentrum Karlsruhe with SimADS codes. FZR — Forschungszentrum Rossendorf. POLITO — Politecnico di Torino.

researchers from Los Alamos National Laboratory in the implementation of suitable numerical schemes for the solution of the coupled system of equations [325–328].

The MSR activities were continued in the framework of another European project, the EVOL project (2010–2013), in which Politecnico di Torino was involved in another benchmark focused on the design of the MSFR. The definition of a reactor design with a fast spectrum, with a profound difference in terms of geometry of the core and fluid patterns, was the motivation for the activities carried out. The activities focused on the neutronic behaviour of the reactor in steady state and relevant parameters for safety analysis and potential breeding of new fissile material, as the possibility of a fuel cycle based on thorium was also foreseen [329]. The result of the comparison among the project partners was summarized in Ref. [254].

In parallel to these European projects, the IAEA launched a Coordinated Research Project on Studies of Innovative Reactor Technology Options for Effective Incineration of Radioactive Waste, where a task was devoted to the analysis of molten salt systems for the purpose of their application in the framework of partitioning and transmutation of nuclear waste. The members of the scientific community involved in the research project had also participated in the previous European projects and the results of such joint efforts are summarized in Ref. [330].

The activities carried out in the framework of this variety of projects also stimulated the interest in more fundamental aspects of the physicomathematical modelling of these topics, leading to further research collaboration. The mathematical characteristics of the neutronics problem for a system with fluid fuel has been studied, comparing neutron transport and its diffusive approximation [331, 332], while the problem of reactivity oscillations due to localized modification of the core composition was approached by assessing the core composition's properties [333].

More recently, in the framework of the SAMOFAR project (2015–2019), an uncertainty quantification study concerning the influence of the uncertainty in the nuclear data on the main neutronic parameters was conducted. Using both a well established technique like the generalized perturbation theory and a reduced order approach known as XGPT, both available in the Monte Carlo code SERPENT-2, the uncertainty on the effective multiplication parameter was estimated for the main nuclides of interest (i.e. ^{232}Th and ^{233}U). To accomplish this task, the nuclear data generation and processing procedure has been used, through the application of codes[16] such as NJOY, SANDY and the T6 package, to generate perturbed cross-section libraries as input for SERPENT. The uncertainty quantification study proved that the uncertainty related to ^{232}Th and ^{233}U seriously impacts the final uncertainty on the multiplication parameter. This suggests the need to extend the study to macroscopic cross-sections, employed in deterministic multiphysics calculations, in order to propagate the uncertainty also to the main thermohydraulic parameters [334].

5.8.2.2. Safety assessment and uncertainty quantification for the MSFR

Politecnico di Torino enlarged the field of research activities pertaining to MSRs in the SAMOFAR project, by launching a new activity that focused on risk analysis and safety assessments. It participated in the project as part of the Interuniversity Consortium for Technological Nuclear Research (Consorzio Interuniversitario per la Ricerca Tecnologica Nucleare, CIRTEN), together with Politecnico di Milano. With the contribution of the various faculties of the Nuclear Engineering Modelling research group, this new field of research could be approached with specific attention given to the safety characteristics of the MSFR design.

The Integrated Safety Assessment Methodology proposed by the Risk and Safety Working Group of the Generation IV International Forum in 2011 was selected as the basic methodology [267]. This methodology was reviewed to better reflect the international standards and rules and to suit the particular

[16] The codes involved are (i) NJOY: nuclear data processing code developed at Los Alamos National Laboratory; (ii) SANDY: nuclear data sampling code compatible with nuclear data files in ENDF-6 format; and (iii) T6 package. The T6 software system consists of the six core codes that are needed to produce a complete nuclear data library: TALYS (code for the simulation of nuclear reactions), TEFAL (code for producing nuclear data libraries in ENDF format), TASMAN (code for covariances, optimization, sensitivities and other statistical information for TALYS) TARES (code for resonance parameters), TAFIS (code for fission neutron quantities), and TANES (code for prompt fission neutron spectra).

case of the MSFR. The approach described in IEC EN 61508 [335] constitutes a milestone for design that is safety driven in the process industry and inspired this analysis. According to the standard, the safety of the systems needs to be studied and pursued from the early design by risk analysis tools. This is done through the definition and analysis of safety instrumented functions in order to understand the effective risk reduction needed in terms of safety systems and additional safety requirements.

In order to select the most comprehensive list of hazards and to improve the efficiency of the risk analysis and detailed design definition, the methodology was enriched with the following risk analysis tools: functional failure modes and effects analysis, master logic diagram and lines of defence method. This work was summarized in deliverable D1.5 of the SAMOFAR project.

The implementation of the methodology started with the identification of deviations that can compromise system safety (in terms of postulated initiating events, the most challenging conditions for plant safety), through two approaches applied at the same time: the functional failure modes and effects analysis, a bottom-up approach, focused on the identification of the functions of the system and the analysis of the consequences of the loss of each of them, and the master logic diagram, a top-down approach that after the selection of a top event identifies its possible elementary causes. In addition to the identification of postulated initiating events, the functional failure modes and effects analysis and the master logic diagram highlighted the lack of information on some systems, procedures or phenomena indicated potential limitations of the design; and allowed for suggestions to enhance the safety of the concept [269]. The lines of defence method was then applied to selected accident scenarios to ensure that every accident evolution of the reactor was always prevented by a minimum set of homogeneous (in number and quality) safety features before a situation with potentially unacceptable consequences might arise. Each event was briefly characterized, identifying also plausible prevention measures. During the application of the lines of defence method, some input data regarding natural behaviour of the plant following the initiating events, with a preliminary evaluation of expected radiological consequences, were fundamental to be able to define the number of safety provisions. The lines of defence method indicated that additional provisions could be necessary to ensure the complete management of the accident (e.g. the addition of a core catcher or equivalent) or recognized the importance of ensuring the availability of some existing components.

The results of this analysis were included in deliverable D1.6 of the SAMOFAR project and presented in Refs [336] and [337]. The functional approach successfully applied to the fuel circuit in the SAMOFAR project has been chosen for the analysis of the fuel treatment unit in the SAMOSAFER project (2019–2023). This was chosen in order to contribute to the development of the design of this unit with a risk oriented approach.

In the framework of the SAMOSAFER project, Politecnico di Torino is also contributing to the analysis of the reactor system by developing methods and models for the early detection of deviations from the operational status of the systems, which will reduce the need to actuate safety provisions.

5.9. RESEARCH AND DEVELOPMENT ACTIVITIES IN JAPAN

As described in Section 2.1, ORNL in the USA accomplished a range of achievements for the development of MSRs between the 1950s and 1970s. Using those results, K. Furukawa and his group in Japan have been investigating the FUJI MSR, a graphite moderated reactor, and related accelerator technology since the 1980s [338, 339]. FUJI is based primarily on the designs for the MSBR and for the denatured MSR, with some improvements. The group has provided comprehensive design results, such as those demonstrating the FUJI-U3 self-sustaining 1 conversion ratio using Th–^{233}U fuel salt. The FUJI concept is described in Section V.2.2 of Appendix V. In addition, the group investigated the following areas: the FUJI-Pu, which uses plutonium as the startup fissile material; the transmutation capability of minor actinides; the super-FUJI of 1000 MW(e); and a mini-FUJI as a pilot plant.

With regard to the initial fissile feeding for an MSR, Furukawa proposed the concept of an accelerator molten salt breeder, or AMSB, in 1981 with a numerical study [340]. The breeder is similar to an accelerator driven system, which utilizes a proton accelerator and molten salt target containing thorium to produce a large amount of ^{233}U for feeding fissile material to an MSR. Inspired by the book by A.A. Harms [341], Furukawa combined an MSR and the accelerator molten salt breeder to establish the thorium molten salt nuclear energy synergetic system (known as THORIMS-NES) in 1990 [342]. A dry reprocessing facility for molten salt fuel is placed between these two facilities, as shown in Fig. 68 [343].

In order to promote the development of MSR technology, Furukawa and his group established the International Thorium Molten-Salt Forum, or ITMSF, in 2008, and in 2010, Thorium Tech Solution Inc. The forum has been an observer member of the Generation IV International Forum MSR provisional System Steering Committee from this committee's beginning.

Under the Atomic Energy Society of Japan, a special committee was started in 2013, named the Nuclear Application of Molten Salt. Within this committee, a working group with members of the International Thorium Molten-Salt Forum proposed general design criteria for an MSR [339], which are based on the

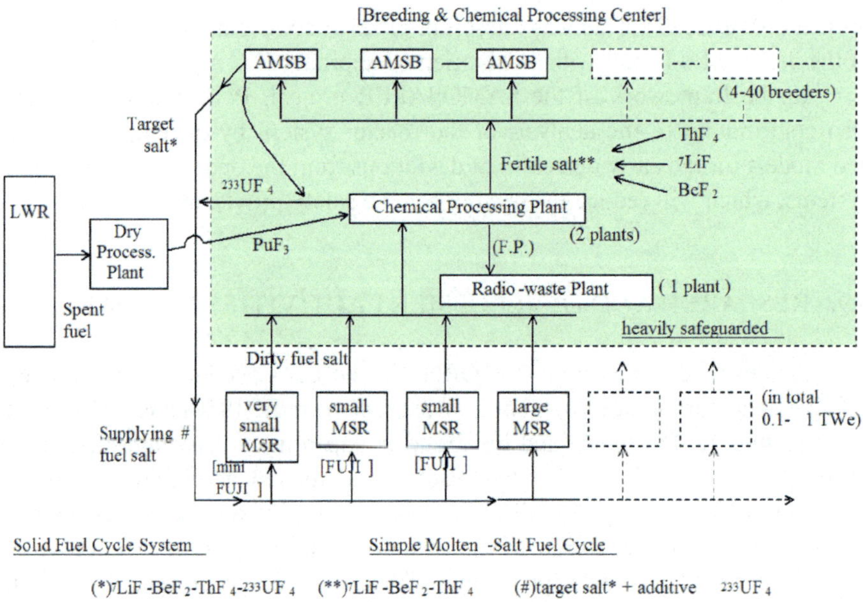

FIG. 68. Thorium molten salt nuclear energy synergetic system (THORIMS-NES) reproduced from Ref. [343]). AMSB — accelerator molten-salt breeder. LWR — light water reactor.

draft general design criteria from the ANS-20.1 working group of the American Nuclear Society, where the criteria for FHRs were discussed in the USA.

A comprehensive summary of MSR safety analysis is provided by the International Thorium Molten-Salt Forum, in which the philosophy for MSR accident analysis with safety criteria and 40 possible accidents with numerical results are described [339].

In addition to these activities, basic studies related to MSRs were conducted in the 2010s at Japanese universities and research institutes, such as the University of Fukui, the University of Tokyo, the Central Research Institute of Electric Power Industry and the National Institute for Fusion Science.

The Japanese Government started to support the development of MSR technology in 2019. Three MSR venture companies were selected: two are promoting an MSR with fluoride salt that is moderated by graphite and one is promoting a fast MSR with chloride salt.

5.10. RESEARCH AND DEVELOPMENT ACTIVITIES IN THE NETHERLANDS

5.10.1. Activities at the Nuclear Research and Consultancy Group

5.10.1.1. Introduction

In the Netherlands, NRG, as the Dutch national nuclear laboratory, operates the High Flux Reactor and the related Hot Cell Laboratories in Petten and carries out a nuclear R&D programme sponsored by the Ministry of Economic Affairs and Climate. An important part of this R&D programme is the study of innovative nuclear systems and, in particular, MSRs. The overall aim of the Dutch molten salt programme is to gain experience with the handling, irradiation, post-irradiation research and waste treatment of molten salts, as well as the modelling of a reactor in which the fuel is liquid. The programme has the following short and medium term goals:

- To support the qualification of materials and components for use in an MSR;
- To improve insight into the behaviour of fission products in molten salt in relation to accident scenarios and decommissioning;
- To develop a technique for the (partial) on-line removal of fission products ('helium bubbling');
- To test and qualify measurement and control techniques required to operate MSRs;
- To design, build and operate a facility for the High Flux Reactor that can serve as a prototype for future first-of-a-kind reactors (a so-called molten salt loop).

The programme has matured in the past few years, and currently consists of several irradiation, modelling and laboratory efforts. Important collaborations are those with the JRC, which provide in-depth knowledge of salt thermochemistry and thermophysical properties, with TU Delft where PhD students work on salt chemistry and helium bubbling, and with the Research Centre Řež in the Czech Republic. Part of the programme is carried out as part of the European SAMOSAFER project, coordinated by TU Delft, which started in September 2019. The JRC in Karlsruhe has provided and continues to provide well characterized fuel salt samples for the salt irradiations carried out in the High Flux Reactor, and the Research Centre Řež provided the salt samples needed for gamma irradiation tests.

5.10.1.2. Reactor physics

NRG upgraded the capability of the Sophisticated Plant Evaluation Code for Thermal-Hydraulic Response Assessment (SPECTRA) [344–346], which is a thermohydraulic system code originated by NRG and developed for the thermohydraulic analysis of nuclear power plants. The purpose of the upgrade is to enable the simulation of systems fuelled with molten salt. Originally developed for light water reactors, the code has a flexible set-up that allows application to high temperature gas cooled reactors, liquid metal cooled fast reactors, or LMFRs, and with the possible extension to MSRs. The computer code is intended for transient and accident analyses that could entail loss of coolant accident scenarios, operational occurrences and other safety event scenarios in nuclear power plants. The code features numerical models and correlations for multidimensional two-phase flow, non-equilibrium thermodynamics, transient heat conduction in solid structures and a general package for heat and mass transfer with built-in models for steam, water, non-condensable gases, under natural and forced convection, condensation and boiling. For molten salt and liquid metal reactor applications, users are given the freedom to define the fluid properties and heat transfer correlations. A point reactor kinetics model can be included, with an isotope transformation model to compute concentrations of important isotopes (e.g. ^{135}Xe). The package for radioactive particle transport in the code deals with radioactive fission product chains, release of fission products, aerosol transport, deposition and resuspension.

With all the extensions made to SPECTRA, the code can simulate the following relevant phenomena for systems fuelled with molten salt:

- Delayed neutron precursor drift;
- Fission product transport in MSRs (e.g. ^{135}Xe);
- Noble gas and noble metal behaviour (based and validated with Refs [75, 347, 348]);
- Noble metal extraction (data comparison with Ref. [349]);
- Chromium leaching and deposition (based on Ref. [350]).

Reference [351] offers a more elaborate description of the code and its application to the MSRE. An example output of the code is shown in Fig. 69. The figure shows the noble gas ^{135}Xe distribution in the MSRE at steady state.

5.10.1.3. Fuel and coolant chemistry and supporting technology

Work at NRG is focused on irradiation and modelling efforts, while the JRC in Karlsruhe and the Research Centre Řež have provided the fuel salt

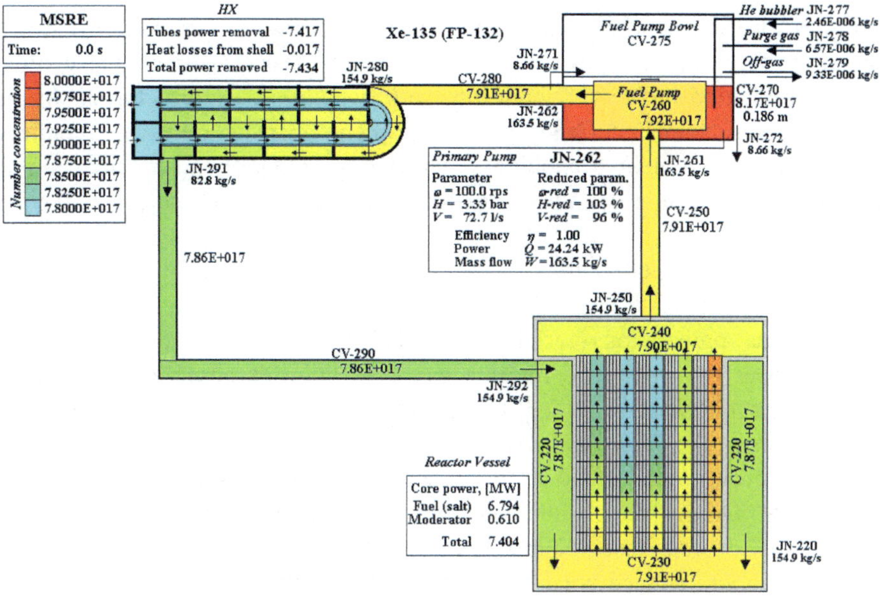

FIG. 69. Steady state distribution of noble gases in the MSRE using the SPECTRA code (courtesy of F. Roelofs, NRG).

FIG. 70. Overview of the High Flux Reactor (courtesy of R. Hania, NRG).

samples for the salt irradiations in the High Flux Reactor (see Fig. 70). Work on supporting technologies is therefore generally carried out in the context of the irradiation projects. Supporting work, which is described further in the following subsections, currently consists of gamma irradiation tests, laboratory testing to optimize helium bubbling conducted at TU Delft and electrochemical measurements on fluoride salts in support of the SALIENT-03 project.

(a) The SAGA experiment

During preparations for the SALIENT-01 fuel salt irradiation tests, it became clear that radiolytic fluorine gas production as observed and studied previously at ORNL [246, 247, 351] is an issue affecting the quality of irradiation experiments in the High Flux Reactor. A capsule irradiation in the reactor runs for many cycles, each of 30–31 days, with intermediate shutdown periods of 4–30 days. During these shutdown periods, samples cool down quickly to below 65°C. It was observed at ORNL that at temperatures below 150°C, the rate of radiation induced bond cleavage in MSRE salt exceeds the rate of bond re-formation through thermal motion, resulting (with varying delays) in the production of fluorine gas. The corrosive gas F_2 is itself a manageable safety hazard, but perhaps more important to the quality of the experiment is that this leaves the salt sample in a reduced (and non-representative) state. It is not yet clear whether the same issue occurs for chloride salts.

The salt gamma, or SAGA, experiment was designed with the objective of measuring the efficiency and amount of F_2 gas released from salt samples in a gamma field, and studying the mechanism of radiolytic gas production for several salt species. The idea of the experiment is to continuously measure pressure in sealed salt filled capsules during high dose gamma irradiation using spent fuel from the High Flux Reactor (at an average gamma dose rate of ~30 kGy/h in air filled ionization chambers). Absorption of radiation is registered by ionization chambers and converted to sample dose using a Monte Carlo N-particle model, and temperature in the system is monitored by thermocouples (45–60°C). The SAGA irradiation facility is reloadable. For the first experiments, it was loaded with powder samples of LiF, BeF_2, UF_4, ThF_4 and a LiF–BeF_2–UF_4 mixture, with one empty reference capsule. A collaboration with the Nuclear Research Institute Řež in the Czech Republic was established for the fabrication of the fuel salts and fluorination of the capsules.

The construction of the facility and loading with salt samples was finalized in the last quarter of 2019 (see Fig. 71). The first results of the gamma irradiation were recorded in early 2020. The observed pressure buildup at least confirms the expected trend that heavier (higher *Z*) salts absorb more energy and therefore generate more fluorine gas. In 2020, the fluorine gas was recombined with the salt

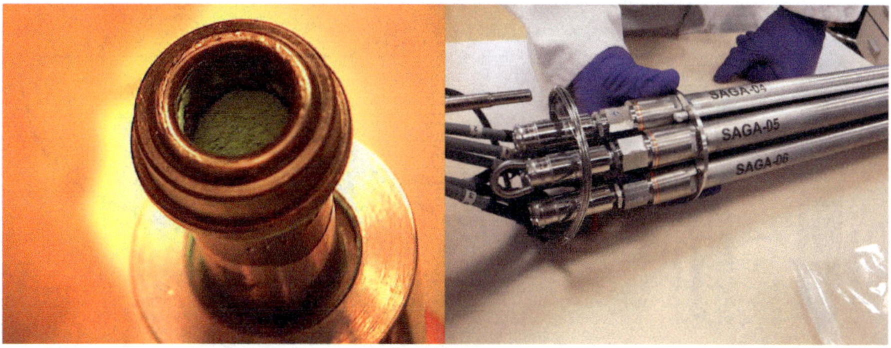

FIG. 71. Left: one of the salt filled SAGA capsules before sealing; right: the completed assembly of six capsules before loading into the SAGA facility (courtesy of R. Hania, NRG).

samples by heating the system to up to 400°C. As of early 2023, a repetition of the experiment was being assembled, with the goal of checking for consistency of results. This experiment contains an NaCl sample instead of the empty reference, to test for chloride release.

(b) Helium bubbling

In a collaboration between NRG and TU Delft on the chemistry of molten salts, research has been conducted with a focus on the chemical speciation of the main fission products in fluoride salts and on the on-line noble metal extraction via helium bubbling. A CFD model for two-phase flow has been used by NRG to model the flotation of noble metals; this model is to be validated by comparison with laboratory tests on the flotation efficiency. Preliminary CFD calculations have been carried out by NRG to establish the influence of helium bubbles on the temperature profile under irradiation (see Fig. 72). The simulations show that bubbling promotes heat transport to the edges of the salt column, resulting in a more uniform temperature distribution, but also that the influence of bubbling on the temperature profile is small except for high flow rate and small bubbles. Figure 72 gives results for the smallest bubble size tested of 0.2 mm and flow rates in the range of 5–50 μL/s. Results for larger bubbles, which are not shown, reveal that the temperature profiles with and without bubbles are practically the same. However, the temperature profiles are clearly different in the case of bubbles with a size of 0.2 mm and a flow rate of 50 μL/s. The simulations therefore confirm that the new SALIENT test can be conservatively designed based on 'no bubbling' conditions.

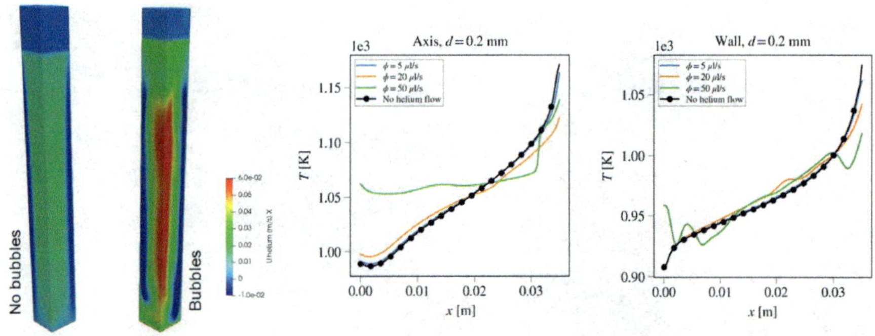

FIG. 72. Left: salt velocity profile; centre and right: axial temperature profiles for different flow rates of bubbles of diameter 0.2 mm (centre: profile at the centre of the salt column; right: profile at the salt–capsule interface) (courtesy of F. Roelofs, NRG).

5.10.1.4. Performance of materials

A number of experiments were carried out to evaluate the performance of materials, as described in this subsection.

(a) The ENICKMA project

Nickel based alloys are foreseen to be used in MSR systems for their high corrosion resistance. Nickel is, however, sensitive to thermal neutron irradiation, producing helium that consequently may cause embrittlement of the material. The ENICKMA irradiation project was carried out to study this embrittlement in nickel based materials. For this scoping study, the irradiation targets were defined and material suppliers contacted. Table 15 presents an overview of the various materials: 316 L(N) is SS316 with low carbon content and added nitrogen developed by CEA in the framework of the Advanced Sodium Technological Reactor for Industrial Demonstration (known as ASTRID), which is a sodium cooled fast reactor; GH3535 is a Chinese analogue of the well known nickel–molybdenum–chromium alloy Hastelloy N; MONICR (molybdenum–nickel–chromium) is a closely related alloy developed in the Czech Republic; and Hastelloy 242 has a relatively high molybdenum content for better high temperature strength. Figure 73 shows MONICR samples as obtained from the Research Centre Řež.

The effect of helium embrittlement was investigated using different techniques. After irradiation, tensile and low cycle fatigue samples were loaded to study irradiation effects. Small punch test samples were also loaded to study the plastic deformation behaviour, and foils for transmission electron microscopy

TABLE 15. OVERVIEW OF THE VARIOUS MATERIALS FOR THE ENICKMA IRRADIATION EXPERIMENT

Material	Supplier
316 L(N)	CEA (France)
Hastelloy N	Haynes (USA)
GH3535	SINAP (China)
HN80MTY	COMTES FHT (Czech Republic)
MONICR	COMTES FHT (Czech Republic)
Hastelloy 242	Haynes (USA)

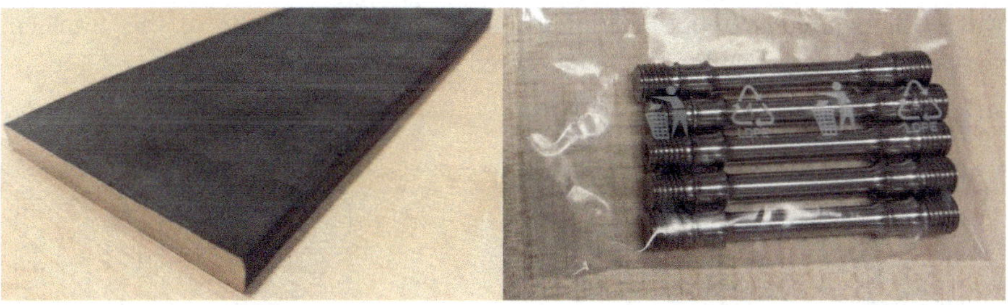

FIG. 73. Left: a segment of MONICR plate as supplied by the Research Centre Řež; right: tensile samples fabricated at the Research Centre Řež from the same material (courtesy of R. Hania, NRG).

were included to investigate the microstructural behaviour after irradiation. The concept design was completed in 2018; in 2019 safety assessments were conducted with nuclear and thermomechanical analysis, and the experiment was conducted in 2021–2022. Post-irradiation mechanical tests and microstructure analysis were ongoing in early 2023.

(b) SALIENT-03

Following SALIENT-01 and considering the lessons learned from that experiment (see Section 5.10.1.6), SALIENT-03 is a molten salt capsule irradiation focused on the in-pile corrosion of alloy N. This experiment is performed in collaboration with the JRC and is described in Section 5.6.3.

5.10.1.5. System behaviour, operation, safety and security

Transient simulations for molten salt systems have been performed using the SPECTRA code. For a system cooled by molten salt (e.g. Mk1 PB-FHR), simulation results of a loss of primary coolant flow and natural convection cooling through a passive direct reactor auxiliary cooling system loop were compared with results obtained using the RELAP5 and SAM codes developed in the USA [352]. Figure 74 presents a snapshot of the simulation. In Ref. [353], the results with the SPECTRA code are shown to be in good agreement with the RELAP5 and SAM codes. The differences are mainly related to modelling assumptions for the pumps.

Simulations for the thermal convection heat removal test of the MSRE were performed recently by F. Alcaro [354]. This required a precursor simulation to model the state at the start of the transient. The precursor simulation performed represented the 'history' of the system prior to the start of the transient. Based on the available data from ORNL reports dating back to the MSRE era, the model was implemented in the SPECTRA code. Simulation results of this transient showed good agreement with the measured data from the MSRE.

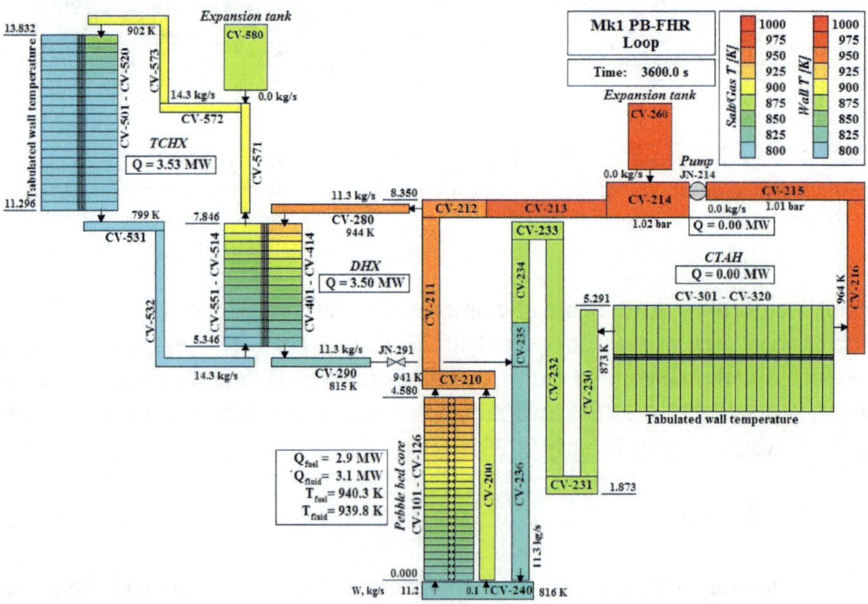

FIG. 74. *Transient results at a particular time for the Mk1 PB-FHR using the SPECTRA code (courtesy of F. Roelofs, NRG).*

5.10.1.6. *Component and technology development*

SALIENT-01 (see Fig. 75 and Ref. [244]) is the first molten salt fuel irradiation conducted since the 1960s (when salt capsule irradiations, salt loops and the MSRE were operated at ORNL). The primary goal for the irradiation was to obtain hands-on experience. Graphite was selected as the crucible material because molten salt corrosion is not considered an issue for this material. The salt composition, 78LiF–22ThF_4 mol% (99.9% ^{7}Li, natural ^{232}Th), was selected based on the possibilities for synthesis and purification at the JRC in Karlsruhe in Germany at the start of the project in 2015 (these possibilities have significantly expanded since then). The salt LiF–ThF_4 is also the basis for the European MSFR concept.

The irradiation rig consists of a stack of four samples of ~1.5 cm^3 of 78LiF–22ThF_4 mol% salt in open graphite crucibles, allowing passage of fission gases into the first containment space. A fifth graphite crucible at the top of the stack is added as a reference. The five crucibles are enclosed in a sample holder made of SS316 providing double containment. The first (inner) containment is sealed by welding.

In the design phase, it was found that the fuel samples were susceptible to radiolysis upon cooling to below 150°C, whereby fluorine gas is released [246, 247]. These concerns prompted the start of the SAGA gamma irradiation project (see Section 5.10.1.3), and were the reason for the termination of the twin experiment SALIENT-02. Following an evaluation of the consequences of radiolytic F_2 production, the SALIENT-01 irradiation started on 10 August 2017 and finished on 17 August 2019, after 17 cycles (508 full power days) of irradiation.

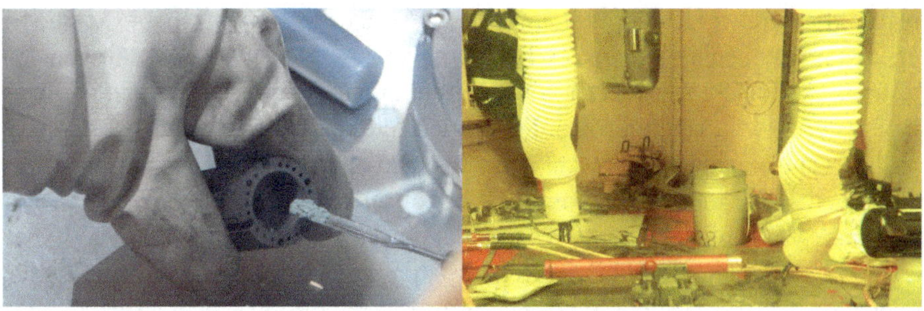

FIG. 75. Left: loading of the LiF–ThF_4 salt samples in an inert glovebox environment at the JRC in Karlsruhe; right: the SALIENT-01 sample holder after irradiation. Following cutting of the instrument lines in the DM-cell of the High Flux Reactor, the sample holder is inserted into a gastight holder for transport to the Hot Cell Laboratories (reproduced from Ref. [244] with permission).

The following specific questions were to be addressed:

- How far does the fuel salt penetrate into the graphite, and what does fission product penetration into the graphite look like?
- Is fission gas release indeed ~100% as expected based on the low solubility of xenon and krypton in molten salts?
- What is the ultimate size distribution of noble metal fission product particles observed on nickel foil and on graphite, and what is the relative deposition rate ('sticking factor')?
- Does prolonged contact with salt have an influence on the surface quality of the graphite?
- Can the effects of radiolytic fluorine gas production (uranium carbide formation, interstitial fluorine in the graphite and CF_4 gas in the plenum space) be observed?

A post-irradiation examination campaign in the Hot Cell Laboratories in Petten aims to answer these questions. Planned examinations include gamma spectrometry, puncturing of the first containment for fission gas analysis and electron microscopy with chemical analysis.

5.10.2. Activities at TU Delft

5.10.2.1. Introduction

The Delft University of Technology, referred to here as TU Delft, has a long tradition with homogeneous nuclear reactors, starting with the vision of J. Went, the first professor of nuclear reactor physics at TU Delft and director at the KEMA research centre, to develop a homogeneous reactor in Europe. In 1957, KEMA opened a new nuclear laboratory for the development of the KEMA suspension test reactor, or KSTR, a reactor with small uranium–thorium–oxide fuel particles in a water flow pumped through the primary circuit. By continuously refuelling particles and cleaning the coolant, uninterrupted operation over a long period was envisaged. After construction and studies for ten years in a zero-power reactor, the construction of a reactor with a thermal power of 1 MW started in 1963. After a few years of operation, this reactor was shut down in 1977. Meanwhile, in the 1960s, D.G.H. Latzko worked on the use of molten salts as heat transfer fluids for power generation.

The interest in homogeneous reactors in TU Delft has intensified as part of a research programme on Generation IV reactors. The research programme concluded that the breeding of new fuel and complete destruction of long lived actinides could best be accomplished in a homogeneous reactor. Research on the thermal MSR started in 2005 and focused on the thorough understanding of the MSRE, and on improving the thermal MSBR design, both originally developed by ORNL. Various alternatives were investigated, including one-fluid, one-and-a-half fluid and two-fluid reactor core designs. Studies were also carried out in close collaboration with the JRC in Karlsruhe, Germany, that focused on the chemical and physical properties of various fuel salts. Although the thermal MSR design has viability, research at TU Delft has focused on the MSFR design originally proposed by the CNRS and adopted as the reference design for Generation IV reactors in Europe. This culminated in two projects funded by the European Horizon 2020 programme: SAMOFAR in 2015–2019 and SAMOSAFER, which started in 2019.

The university coordinated the SAMOFAR project and is currently coordinating the follow-up SAMOSAFER project. It contributes with fundamental research regarding the computational challenges in the MSFR (coupled neutronics–CFD, melting and solidification of salts), numerical and experimental work to measure viscosity and other parameters of the (fuel) salt, as well as various applications such as assessing safety by analysing transients and the design of freeze plug devices and passive mechanisms for decay heat removal.

Furthermore, TU Delft investigates the phase equilibria and thermodynamic properties of several types of fluoride and chloride salt systems (fuel relevant compositions with inclusion of fission products), their chemical interaction with structural materials, and the chemical speciation of fission products generated during irradiation using a combination of experimental studies and thermodynamic modelling assessments using the CALPHAD method (see Section 5.6.1.1).

Numerical and experimental work on the extraction of fission products via gas bubbling and part of the thermodynamic modelling studies are carried out in the framework of a scientific cooperation between TU Delft and NRG.

Many of the results provided below are described also in the PhD theses of Refs [355–362].

5.10.2.2. Numerical code development and transient analysis

During the SAMOFAR project, completed in 2019, the TU Delft multiphysics simulation package, originally based on the neutron diffusion code DALTON [363], has been redeveloped and extended, combining a state of the art discrete ordinates neutron transport code (PHANTOM-SN) with a newly developed simulation code for fluid dynamics (DG-Flow). This multiphysics code package [364] was verified with code to code comparisons with other SAMOFAR partners and separate benchmarks from the literature [365]. The package was utilized to investigate the behaviour of the MSFR under both steady

FIG. 76. Evolution of the fuel salt's temperature distribution during a total loss of power transient [362] (courtesy of J.L. Kloosterman, Delft University of Technology).

state and transient conditions. The transients studied were unprotected loss of heat sink, unprotected loss of fuel flow, total loss of power, unprotected pump overspeed and salt overcooling. The main conclusion from these simulations is that the reactor response to these scenarios is excellent. No major weak points were identified regarding the safety of the MSFR design: modest temperature variations and power variations are strongly damped by negative reactivity feedback. An example of the fuel salt's temperature distribution obtained by multiphysics analysis is shown in Fig. 76.

5.10.2.3. Development of a reduced order model

Multiphysics analyses of the three dimensional MSFR are inherently expensive from a computational point of view. A strong effort was made at TU Delft to develop reduced order modelling to partially counteract this cost. The focus has been on methods that are non-intrusive to be able to work with complex models without requiring access to the underlying model equation (i.e. a black box approach [366]). The newly developed methods are based on adaptive sampling of the parameter space combined with interpolation for the amplitudes of a set of modes that adequately describe the system, the latter determined by proper orthogonal decomposition. The set of modes and their amplitudes are adaptively enriched by sampling the parameter space where the errors are largest until specified error criteria are met. The methodology was also extended to transient scenarios where time is treated as any other input parameter, which enables efficient handling of parametric dynamic models. The reduced order model was subsequently used for various purposes, such as uncertainty quantification [367]. Steady state analysis has been performed for 30 independent input parameters originating from cross-sections, neutron precursors and thermodynamics parameters. The influence of various parameters on the observables, such as k_{eff} and the maximum temperature, has been ranked and probability distributions of these various key parameters were made. An example of such distributions is shown in Fig. 77.

5.10.2.4. Generalized polynomial chaos development and application

The university TU Delft has also made significant progress in developing and applying methods of generalized polynomial chaos (GPC) expansion to perform fast and accurate uncertainty quantification and sensitivity analysis of the MSFR despite the computationally expensive multiphysics modelling required. Just like reduced order modelling, GPC also represents a black box approach that builds efficient meta-models of selected important system responses as a function of uncertain physical and chemical properties and undefined design parameters. It has the advantage of using multidimensional polynomial approximations (instead of hierarchical linear interpolation as in the reduced order modelling approach). Therefore, it can sometimes be more efficient in terms of overall computational cost than reduced order modelling methods, especially in cases where strong non-linearity is present and only key performance parameters are of interest [368, 369]. Moreover, it provides a straightforward approach to derive local sensitivities as well as sensitivity metrics based on global variance (Sobol indices). Generalized polynomial chaos was used to perform a detailed analysis of the MSFR steady state (using high fidelity coupled neutron transport–CFD calculations). Figure 78 shows the results for the salt temperature in the reactor (1/16th of the rotationally symmetric design), highlighting that (i) the upper

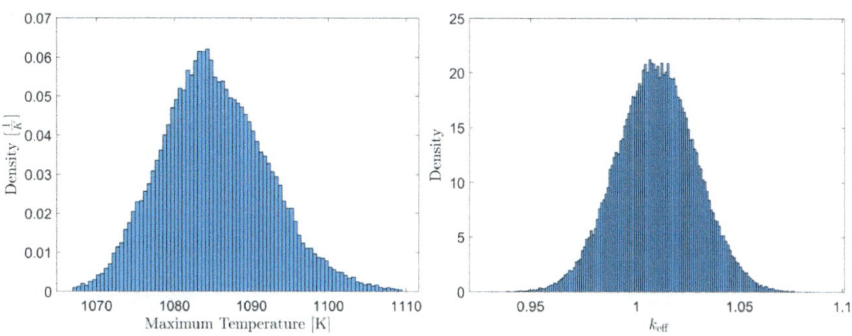

FIG. 77. Probability distributions of the maximum salt temperature in the core and of the multiplication factor k_{eff} under the influence of input parameter variation (courtesy of J.L. Kloosterman, Delft University of Technology).

parts of the core have a non-negligible probability of surpassing the maximum temperature design specification of 1023 K and (ii) the lower parts of the heat exchanger and the vessel have a similarly non-negligible probability of cooling below the minimum temperature specification of 923 K. Thus, GPC allows the type of in depth analysis that can identify parameters with a significant effect on performance metrics, thereby helping to optimize the MSFR.

Generalized polynomial chaos was also used to quantify the uncertainty regarding the salt composition. As the MSFR will use molten salt as fuel, knowing the physicochemical properties of the salt is paramount under all conditions. This is heavily hindered by the lack of data available and the

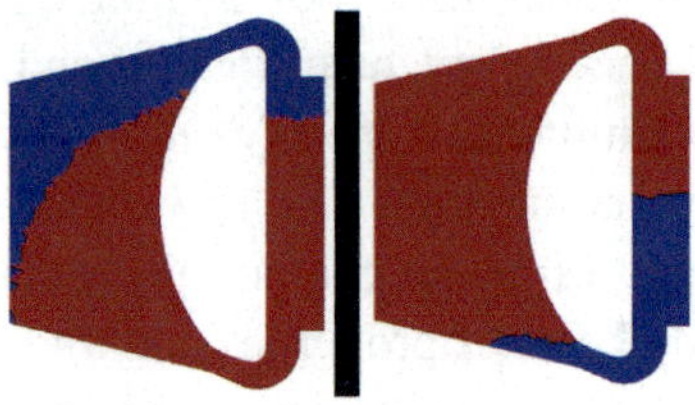

FIG. 78. Left: domain where T<1023 K with probability higher (red) and lower (blue) than 95%; right: domain where T>923 K with probability higher (red) and lower (blue) than 95% (courtesy of J.L. Kloosterman, Delft University of Technology).

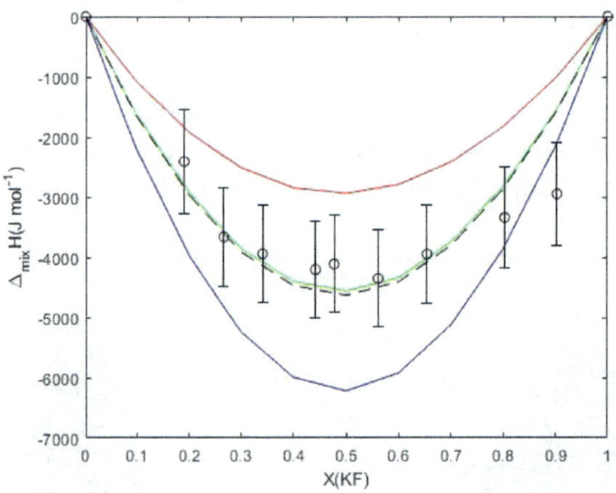

FIG. 79. Comparison of measured and CALPHAD simulated mixing enthalpy values together with derived uncertainties (courtesy of J.L. Kloosterman, Delft University of Technology). Red, green and blue lines represent percentiles 0, 50 and 100 of the calculated phase diagram, while the black circles showcase the measured values together with their experimental uncertainties. The black dotted line corresponds to the mixing enthalpy computed using the thermodynamic model implemented in the JRCMSD thermodynamic database (see Section 5.10.2.12).

inherent difficulty of predicting properties for such multicomponent systems, containing many different elements resulting from the emergence of fission products during operation. Significant first steps have been made by TU Delft in addressing this issue, by combining expertise and GPC tools with the CALPHAD method, allowing a quantification of the uncertainties in the phase diagrams and thermodynamic properties computed using optimized thermodynamic models (see Section 5.10.2.12). Figure 79 shows the result of such a calculation, where the uncertainties on Gibbs energy parameters were propagated to the mixing enthalpy values of a LiF–KF system and compared with experimentally determined values, showing good agreement. The true power of such a coupled GPC–CALPHAD methodology — that is, using CALPHAD (for predicting phase diagrams and other properties) in combination with GPC (providing an efficient connection between the input and output parameters of CALPHAD) — is that by analysing the relation between the uncertain input parameters and the experimentally determined uncertainties, the input parameters can be tuned such that the resulting phase diagrams are in accordance with experiments, and these values can be used to also give statistical information about parameters that can be calculated, but are difficult to measure.

5.10.2.5. Ultrasonic technique for measuring the viscosity of radioactive fluids at high temperature

An innovative method based on the propagation of ultrasonic waves is being developed for the simultaneous determination of the viscosity and density of the molten salt fuel at elevated temperatures. A thin plate is used as a waveguide to transmit shear waves separating the transducer from harsh environments such as elevated temperatures, highly corrosive fluids and/or radioactive fluids. At the solid–fluid interface, the echo signal of the ultrasonic wave depends on the operating frequency, the physical properties of the fluid (viscosity and density) and the properties of the waveguide (density and shear modulus). The research at TU Delft deals with the study of the factors for determining the physical properties of the salt fuel and the solutions required to maximize the accuracy of the proposed technique, especially at very low viscosities. The rheology of non-Newtonian fluids is under investigation as well.

5.10.2.6. Physicochemical effect of mixing molten salt fuel and water

The physical effects of mixing were studied for the hypothetical situation in which molten fuel salt and water come in contact. A series of scoping experiments were performed to understand the interaction of solid fuel salt, which was used at the MSFR, with water in various concentrations to simulate different situations

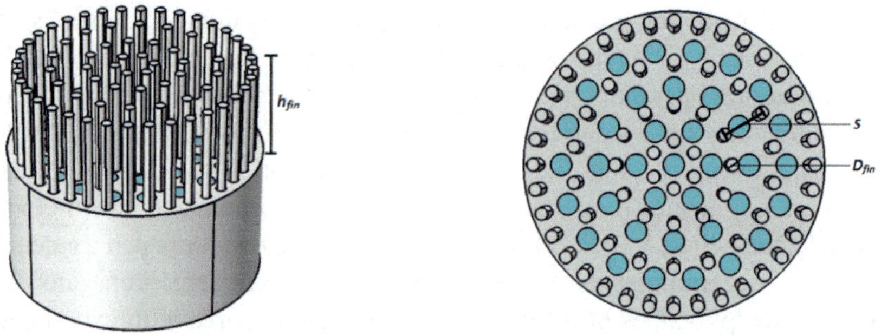

Design including cooling fins.

Top view of design including cooling fins.

FIG. 80. Preliminary design of a freeze plug for the MSFR. Fins are installed to enhance the melting of the plugs that are present in the distribution plate (courtesy of J.L. Kloosterman, Delft University of Technology).

such as fuel leakage in water or vice versa. The analysis included investigating the dissolution in water of fuel salt (LiF–ThF_4 or LiF–ThF_4–UF_4) and FLiNaK (46.5 LiF–11.5 NaF–42 KF (mol%)). The measurements were performed under both irradiating (gamma source) and non-irradiating conditions to understand the effect of irradiation on the salt solubility. The solvent was examined by chemical analysis after the dissolution experiments for evaluating the solubility of the cations and the formation of hydrate compounds. Reference [359] provides additional information on the physicochemical effect of mixing molten salt fuel and water.

5.10.2.7. Development of a fast, passive freeze plug

Passive protection in the MSFR is provided by a so-called freeze plug. The freeze plug is an actively cooled blockage consisting of solidified (frozen) salt. In the case of a postulated accident, such as a station blackout, the plug will melt because the active cooling will stop, enabling the reactor vessel contents to flow into underground tanks. This process needs to be completed within 8 min to prevent damage to the reactor vessel and components in the MSFR. One of the promising designs consists of a thick distribution plate that contains holes of a certain diameter (see Fig. 80). This design increases the melting surface and diminishes the drainage time of the reactor vessel. Plate material and geometry were varied. According to preliminary calculations, the drainage time can be reduced to 6 min, which is below the prescribed 8 min. The use of additional cooling fins is under investigation.

5.10.2.8. Molten salt–structural materials interaction

The suitability of the structural material is defined to a large extent by its resistance against corrosion by the molten salt. The MSRE experience showed that Hastelloy N, a nickel–molybdenum–chromium alloy, could withstand the harsh operational conditions in the reactor (high temperatures, contact with the corrosive salt, high radiation dose), and prospective alloys in current MSR designs include the chemical elements in this alloy. The rate of corrosion is primarily determined by the redox potential of the salt, which is controlled by the UF_4/UF_3 ratio when a molten salt based on fluoride is used. During irradiation with this kind of salt, free fluorine is formed that reacts with UF_3, hence increasing this ratio, and leading to oxidation reactions such as Cr(alloy) + $2UF_4$(salt) → CrF_2(salt) + $2UF_3$(salt) and subsequent formation of voids in the nickel based structural alloy. Among the large variety of MSR designs under investigation, reactors with and without redox control or salt cleanup are under consideration. Because nickel is the main element of the structural alloy and chromium is the most likely element to be dissolved, the phase equilibria of the fuel salt with potential nickel and chromium corrosion products have recently been studied. A thermodynamic model for the AF–NiF_2, AF–CrF_3 (where A is lithium, sodium or potassium) and CrF_2–CrF_3 systems was developed for the first time. The model can feed simulation codes that model the behaviour of the liquid fuel during normal and accident conditions.

5.10.2.9. Chemical speciation of fission products

The importance of chemical speciation of fission products in nuclear fuels is beyond doubt. The fate of these elements and their influence on the fuel properties strongly depend on their chemical state, which in turn depends on the reactor parameters such as temperature and redox potential. For reactors that utilize molten fluoride salt mixtures as nuclear fuel, a thorough thermochemical analysis of the most important fission products is not yet complete. In order to predict the stable phases and the physicochemical properties of the mixtures, thermochemical analysis and studies of phase diagrams are employed. Two thermodynamic databases are under development at TU Delft, one including the main salt-soluble fission products (i.e. caesium and iodine) and the other including the noble metals and their fluoride phases.

The chemistry of caesium and iodine is particularly relevant as their potential release into the environment is a subject of primary concern for the safety of the population and the environment. The isotopes ^{135}Cs, ^{137}Cs and ^{131}I are indeed the main cause for the radiological consequences of an accident and their behaviour in nuclear fuels needs to be carefully evaluated. A comprehensive thermodynamic assessment of the lithium, caesium, Th–F, iodine system was therefore performed at TU Delft in collaboration with the JRC in Karlsruhe and NRG, and combines experimental investigations of phase diagrams, vapour pressure measurements and thermodynamic modelling.

For the fission products consisting of noble metals, molybdenum, ruthenium and niobium are particularly relevant. Molybdenum and ruthenium have the highest fission yield and are expected to be in the metallic state under the reducing conditions that would normally be maintained in the reactor as a result of the redox control of the salt during operation. The behaviour of niobium (dissolved in the salt or precipitating) depends directly on the redox potential conditions. Knowledge of their thermochemistry, and in particular of the formation of noble metal fluoride phases, is essential to understand their fluorination behaviour and establish a proper fission product management strategy. A thermodynamic model of the noble metal elements molybdenum–niobium–ruthenium–fluorine has been developed for this purpose, by combining calculations based on first principles and modelling assessments of the same type as CALPHAD.

5.10.2.10. Influence of melting and solidification on heat transfer

A good understanding of the melting and solidification phenomena of the $LiF–ThF_4–UF_4$ fuel in the MSFR is required for the design of the freeze plug, a key safety component, for an accurate prediction of the possible formation and growth of a solid layer of salt on the reactor walls during normal operation, as well as for the analysis of accident scenarios where the solidification of the fuel salt might pose a risk. As such, the goal of this research, being part of the SAMOSAFER project, is to improve the knowledge and understanding of the physics underlying the process of phase changes. To this end, the effects of phase change will be included in a CFD model based on the discontinuous Galerkin approach through the so-called enthalpy method, where the melting/solidification front is tracked implicitly. Validation of the applied numerical models will be performed through a set of experiments of phase changes, for both laminar and turbulent flow regimes. In addition, the role of forced and mixed convection on the process of these changes will be investigated. Both types of convection can occur in an MSFR during normal operation.

5.10.2.11. Noble particle extraction

One of the advantages of having a liquid fuel is the possibility of controlling and extracting fission products while operating the reactor. At present, helium bubbling is seen as the most mature technique for in-core removal of fission products and it can be employed to extract the gaseous fission products. Moreover, this process has also the potential to remove (via flotation) the insoluble particles, such as noble metals, that might otherwise deposit on relatively cold metallic surfaces. A feasibility and performance study for the removal of particles via flotation is ongoing at TU Delft in collaboration with NRG. It aims to gain a better understanding of the underlying mechanisms involved in the process and evaluate the dependence of the extraction efficiency with the process parameters. An experimental set-up using simulant fluids was developed for this purpose to provide the required data (local, instantaneous particle concentrations) for process optimization and scale-up for molten salt systems. Measurements of the particle concentrations with the help of laser induced fluorescence are ongoing for particles with sizes varying from about 100 nm to 0.1 μm.

5.10.2.12. Development of a thermodynamic database for multicomponent systems

The JRCMSD thermodynamic database has been developed over the years by the JRC in Karlsruhe (see Section 5.6.2.5) for fluoride and chloride salts. It is based on the CALPHAD methodology and quasi-chemical formalism in the quadruplet approximation that allows the computation of thermodynamic equilibrium properties of multicomponent systems. The models in the JRCMSD rely on the great wealth of phase diagram and crystallographic studies produced by the original MSRE programme at ORNL, as well as other miscellaneous sources and more recent measurements performed at the JRC. Since 2016, TU Delft has been contributing to the development of this database with experimental structural and thermodynamic studies and modelling activities, in a joint effort with the JRC. Fission product (caesium, iodine, niobium, molybdenum, ruthenium) systems and corrosion product (nickel, chromium) systems were investigated as part of this effort. Moreover, the LiF–UF_4, NaF–ThF_4 and KF–ThF_4 systems were recently re-examined, as part of an effort to systematically review and model the AF–AnF_4 systems containing actinides (where A is lithium, sodium, potassium, rubidium or caesium and where An is thorium or uranium).

5.10.2.13. Structural studies and advanced structural thermodynamic modelling

Depending on the conditions of composition, temperature and redox potential, the local structure of the molten salt (an ionic liquid) can vary widely with cases where the ions in the melt are completely dissociated, form molecular species or even exhibit some degree of network formation (polymerization). The local structure of the melt can be directly related to the thermodynamic and transport (viscosity, thermal conductivity) properties. A fundamental understanding of the relationship between the local structure of the molten salt and its physicochemical and thermodynamic properties is needed to gain greater predictive capability over the dynamic (far from ideal) behaviour of the fuel salt. Through in situ extended X ray absorption fine structure spectroscopy studies at high temperature in the molten states, interpreted with the aid of molecular dynamics, the structural characteristics of the AF–AnF_4 melts (where A is lithium, sodium, potassium or caesium and where An is thorium or uranium) were explored. A dedicated furnace set-up for molten salts was developed at TU Delft for the extended X ray absorption fine structure measurements of hygroscopic, corrosive and radioactive salts. Moreover, in combination with the quasi-chemical formalism, CALPHAD models that reproduce the structure of the melt simultaneously with the phase diagram and key thermodynamic properties (such as mixing enthalpies, heat capacities and activity coefficients) were developed and optimized. This is ongoing work, but to date, the advanced models have been successfully applied to LiF–BeF_2, LiF–AnF_4 (where An is thorium or uranium), NaF–ThF_4 and AF–ZrF_4 (where A is sodium or potassium).

5.11. RESEARCH AND DEVELOPMENT ACTIVITIES IN THE RUSSIAN FEDERATION

For more than 20 years, R&D on MSRs in the Russian Federation has focused mainly on fast spectrum concepts with or without thorium support [125, 370, 371]. These concepts have been recognized by the Generation IV International Forum as an alternative for the long term to fast neutron reactors with solid fuel, as they have attractive features, namely strong negative reactivity feedback coefficients, smaller fissile inventory and simplified fuel cycle. A negative coolant temperature reactivity coefficient is universally recognized as a desirable safety feature for power reactors. For example, the MOSART concept can operate within technical limits using an alloy with high nickel content as container material and with different fuel loadings and make-up based on TRUs (from used light water reactor fuel with a ratio of minor actinides to TRU that is up to 0.45) as a special actinide transmuter, a self-sustainable system (conversion ratio equal to 1) or even as a breeder (conversion ratio >1).

Basic information concerning the fuel cycle of the MOSART power plant is provided in Ref. [371] and integration with facilities of the Experimental Demonstration Centre (see Fig. 81). The Experimental Demonstration Centre is under construction in the Russian Federation at the site of the Mining and Chemical Combine, and after 2020 will begin the reprocessing of spent nuclear fuel from VVER-1000 reactors on the basis of innovative technology, providing recovered nuclear material (refined products) for recycling in thermal and fast reactors with solid fuel. The highly active raffinate, containing long lived minor

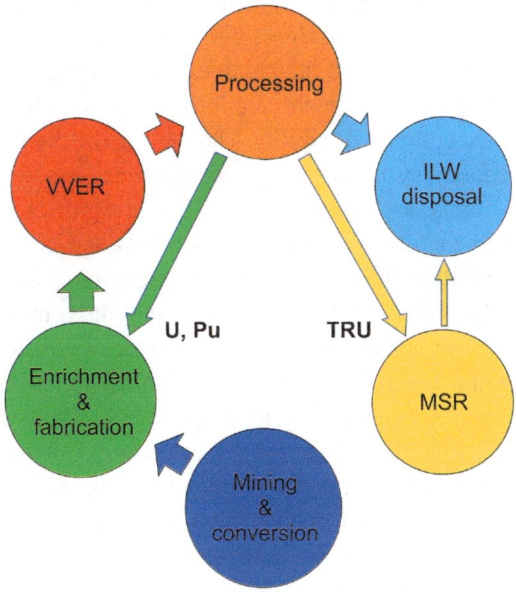

FIG. 81. Nuclear fuel cycle with MOSART at the Mining and Chemical Combine site (courtesy of M. Gurov). ILW — intermediate level waste. TRU — transuranic element.

actinides, is sent for conditioning, as shown by the light blue arrow. Using the MOSART system, most of the uranium and plutonium can be recycled to be used as solid fuel for thermal and fast reactors (green arrow in Fig. 81), and this process may reduce the volume and radiotoxicity of radioactive waste. The main feature of the MSR technology in this application is the flexibility for using pyroprocessing of spent fuel that was cooled for a short time and multiple recycling for better use of resources, reduction of waste and gaining additional profits as compared with the traditional fuel cycle using solid fuels.

It was proposed to use the technological capabilities of the Mining and Chemical Combine site to place a MOSART system in the immediate vicinity of an aqueous reprocessing plant for spent nuclear fuel, linking it to the infrastructure of the Experimental Demonstration Centre. The main design objective of MOSART is to close the nuclear fuel cycle for all actinides, including neptunium, plutonium, americium and curium. It is assumed that the fuel cycle of this complex will be organized as follows (see Fig. 81): the bulk of the removed uranium and plutonium returns to thermal and fast reactors with solid fuel, and the remaining TRU elements are transferred for utilization in the MOSART system. The collocation of the MOSART reactor and the spent nuclear fuel reprocessing plant will provide the Mining and Chemical Combine site and surrounding customers with electricity (7.92 TWh(e) per year) and reduces the problems of transport of nuclear materials and radioactive waste management. As with fluid fuel based on fluorides, the entire fabrication process of solid fuel elements is avoided, providing exceptional flexibility. The fuel can be directly mixed with the reactor's primary coolant system as needed at any time. In addition, long cooling times and interim storage are not necessary. Thus, a significant part of the front end effort (including radioactive doses for workers) and cost is eliminated for MOSART. All fresh fuel of fluoride molten salts containing significant quantities of fissile materials for initial loading and make-up will be manufactured on-site at the Experimental Demonstration Centre through the hydrofluorination process.

The main advantages of MOSART are the ability to vary widely the minor actinides content in the fuel salt without losing the inherent safety, and the absence of fabrication of solid fuel and refabrication. This leads to significant proliferation resistance and safeguards implications related to the fuel make-up and chemical processing in a MOSART plant: (i) continuous variation of isotopic concentrations in the fuel salt from both actinide transmutation and chemical processing; (ii) the refuelling scheme includes the ability to continuously feed the core with fresh fissile material; and (iii) the plate-out of noble metals in the primary circuit could complicate inventory tracking. In addition, the fuel becomes less attractive for fissile material diversion after each recycling. During 50 years of operation, a 2400 MW(th) MOSART can utilize more than 12 t of minor actinides.

To reduce the possibility of diversion of nuclear material, the MOSART plant is integrated at the front end with the VVER spent nuclear fuel aqueous reprocessing plant, and at the back end with the high temperature fuel salt cleanup facility; all facilities are located at the Mining and Chemical Combine site. In a pyroprocessing facility for molten salt, higher actinides would always accompany the plutonium. This operation would therefore never produce a 'clean' material that would be attractive for diversion.

The Russian Federation established an experimental complex in support of future R&D on MOSART. A 10 MW(th) MSR ($LiF–BeF_2–AnF_n$ fuel salt) test reactor facility with a homogeneous core at the Mining and Chemical Combine site is currently under design within the framework of a project supported by Rosatom. Major achievements of recent MOSART R&D activities are summarized below.

5.11.1. ISTC project number 1606 (2001–2008)

Various Russian institutes, including the All-Russia Scientific Research Institute of Technical Physics (Snezhinsk), Kurchatov Institute (Moscow) and

Institute of High Temperature Electrochemistry (Yekaterinburg), initiated ISTC project no. 1606 entitled 'Experimental Study of Molten Salt Technology for Safe and Low Waste Treatment of Plutonium and Minor Actinides in Accelerator Driven and Critical Systems' in 2001. These institutes had already contributed to MSR technology, including through key technical solutions for a promising MOSART concept with a single stream. Experimental and theoretical studies were performed in the following main areas: (i) neutronics and thermohydraulics of the reactor unit, (ii) key physical and chemical properties of fuel salt and (iii) container materials for the fuel circuit. The main results achieved in this project are summarized below.

The 2400 MW(th) MOSART system developed within this project has a homogeneous core with an epithermal to fast spectrum of neutrons. The fuel salt is a molten $15LiF–58NaF–27BeF_2$ (mol%) mixture with the addition of plutonium and minor actinide fluorides. The specific power of the fuel salt is approximately 43 W/cm^3. To minimize actinide losses in reprocessing, a removal time of about 300 days for soluble fission products (rare-earth trifluorides) is considered. Lithium is enriched to 99.99% ^7Li. There is no strong requirement on gas permeability (10^{-8} cm^2/s) for the graphite reflector of the MOSART core, but molten salt should be excluded from the open pore volume (pore structure less than 10^{-6} m). The MOSART core can satisfy some important neutronic and thermohydraulic considerations. These include (i) $AnF_3 + LnF_3$ (where An refers to actinides and Ln refers to lanthanides) concentration in the fuel salt is truly within the solubility limit of $AnF_3 + LnF_3$ for molten $15LiF–58NaF–27BeF_2$ (mol%) at a minimum fuel salt temperature in the primary circuit of 600°C for the fuel cycle scenarios under consideration; (ii) regions of reverse, stagnant or laminar flow are avoided; and (iii) the maximum temperature of solid reflectors is low enough to allow their use for the suitable time of four years.

Several nuclear data libraries, codes and computational models were employed to compute safety related neutronics parameters for the MOSART system. The results show that the parameters are favourable for reactor safety, mainly because of the strong density and fuel Doppler effect. A preliminary study of transients has demonstrated that the MOSART design is an inherently stable reactor design on account of its large, negative fuel temperature coefficient in combination with its negative graphite reflector reactivity coefficient. The MOSART reactor is not expected to be seriously challenged by the major, unprotected transients such as loss of flow without scram, ultimate loss of heat sink and overcooling. A preliminary evaluation indicates that the MOSART system has attractive features in terms of performance and the efficient transmutation of actinides, while having lower amounts of total materials and of waste than prior MSR designs, including the MSBR.

Experimental studies for the molten LiF–BeF_2–NaF system found quite a wide range of ternary compositions, with minimal LiF (15–17 mol%) and BeF_2 (25–27 mol%) content, which achieve the required PuF_3 solubility at 600°C and maintain an adequate melting point (<500°C). In the design of MOSART, a Russian modified commercial Hastelloy-N alloy (e.g. HN80M-VI, HN80MTY) is the specified material for nearly all metal surfaces in contact with the fuel and coolant salts. The results of a loop corrosion experiment with on-line redox potential measurement demonstrated that operations at high temperature with a LiF–BeF_2–NaF salt are feasible using carefully purified molten salts and loop internals. In an established interval of salt redox potential of 1.25–1.33 V relative to a beryllium reference electrode, corrosion is characterized by a uniform loss of material from the surface of samples with a very low rate of corrosion. No significant change in the corrosion behaviour of samples of HN80M-VI and HN80MTY alloys was observed due to the additions of plutonium trifluoride and tellurium to the LiF–BeF_2–NaF salt.

5.11.2. ISTC project number 3749 (2008–2016)

The purpose of ISTC project no. 3749 was to develop the Th–U MOSART concept, focusing mainly on improvements in the areas of fuel circuit configuration and system safety and on the conduct of experimental studies on the basic properties of fuel salt and the fuel salt's compatibility with structural materials.

The main goal of this project consisted in the selection and experimental study of the properties of molten salt compositions and metallic structural materials for prospective nuclear power systems and fuel cycle facilities. Eutectic mixtures, containing lithium, calcium and thorium trifluorides, were selected for physicochemical studies. The following characteristics were experimentally obtained for selected compositions: melting temperature and enthalpy, temperature dependences of density, isobar thermal capacity, viscosity, thermal conductivity and plutonium trifluoride solubility. The electrochemical properties of two metals, neodymium and plutonium, and their fluorides, dissolved in the selected molten salts, were investigated. The processes of reductive extraction of neodymium, lanthanum and thorium trifluorides from molten fluoride salts into liquid bismuth, as well as a process of re-extraction of neodymium, lanthanum and thorium from a system consisting of liquid bismuth/molten lithium chloride, were studied experimentally. The compatibility of fuel salt containing tellurium additions with nickel based alloys, at controlled redox potential and different mechanical loadings on the alloy samples, was tested. Tests on the compatibility of nickel based alloys with the selected melt were carried out in a thermal convection loop under controlled redox potential. A detailed investigation into the change of the structure and strength properties of the alloys after around 1000 h of exposure in circulating molten salt was completed. Investigations on hydrogen permeability of molten fuel salts and nickel alloys were also carried out.

The concept of a Th–U MSR with a homogeneous core and a fast neutron spectrum was developed. Several scenarios for fuel loading of initial and feeding compositions of TRU elements from the spent fuel of light water reactors as well as thorium, as a fertile material, were considered. It was found that an MSR with a homogeneous core offers flexibility in the fuel cycle, as the MSR plant is able to operate with a wide spectrum of fuel and fertile material loadings without reactor shutdown and with no need to change the core design. Such a plant can quickly be re-oriented and included in any strategy of a nuclear power programme.

Therefore, the development of an MSR with a homogeneous core both as a Th–U breeder and as a new element in a nuclear power programme for the burning of TRU elements from spent nuclear fuel is possible. Thermohydraulic calculations were carried out for the most promising variants of this system. Safety and non-proliferation characteristics were also considered.

5.12. RESEARCH AND DEVELOPMENT ACTIVITIES IN SWITZERLAND

5.12.1. Motivation

Activities in Switzerland for the R&D of MSRs were launched in 2013 and concentrated within the domain of the Swiss Federal Institutes of Technology. The activities are carried out predominantly at the Paul Scherrer Institute and at the universities Swiss Federal Institute of Technology Zürich (Eidgenössische Technische Hochschule, ETH Zürich) and the Swiss Federal Institute of Technology Lausanne (École Polytechnique Fédérale de Lausanne, EPFL) through students' projects. Historically, a dedicated MSR research project was conducted in the 1970s and 1980s at EIR, the predecessor to the Paul Scherrer Institute (see Section 2.2.1). The work focused on chloride fast MSRs, and the last concept developed was entitled SOFT [40]. The SOFT concept was a reactor with 3 GW(th) power and a four loop layout, which was based on natural chlorides without isotopic enrichment (see Fig. 82).

Switzerland is a member of the Generation IV International Forum, and in November 2015 signed a memorandum of understanding to join the forum's MSR project. The Paul Scherrer Institute acts as the Swiss implementing agent in the Generation IV International Forum. The institute's Nuclear Energy and Safety Research Division is also a national competence centre for nuclear reactors. The division's missions include the two major topics, safety of current reactors and

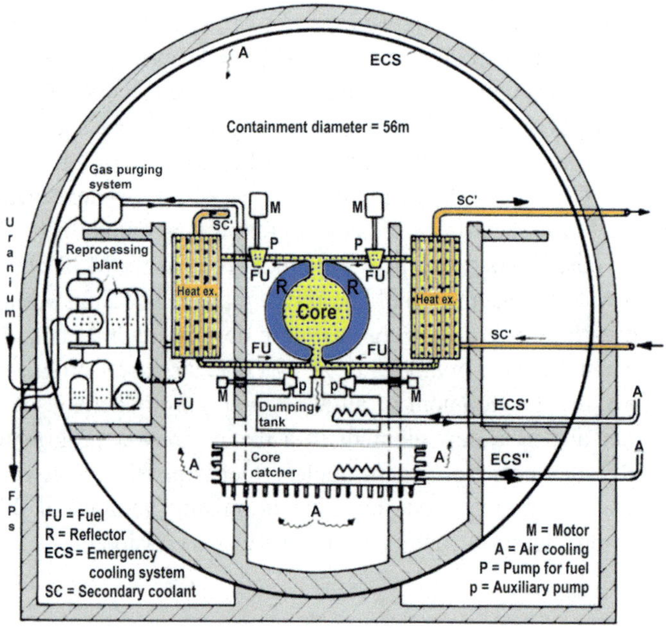

FIG. 82. Layout of the SOFT reactor containment (reproduced from Ref. [40] with permission). FPs — fission products.

waste management, as well as advanced reactor research. The motivation for advanced reactor research in Switzerland, including on MSRs, is driven by the following factors:

(a) The potential of these reactors to combine unique safety features with high fuel utilization and waste minimization;
(b) The many novel and multidisciplinary topics that this research offers;
(c) The attractiveness to students of this field of research and its potential for funding from grant agencies such as the Swiss National Science Foundation.

The research activity on innovative MSR technology also represents an important asset to maintain excellence in nuclear engineering, to ensure successful recruitment of a highly motivated workforce in the coming years and to provide an attractive workplace for future generations of scientists and engineers. At the same time, the unparalleled combination of high fuel utilization with unique safety features of advanced reactors can promote a shift of public acceptance towards nuclear energy in the future. Molten salt reactors have the potential for both risk reduction and radioactive waste minimization. Moreover,

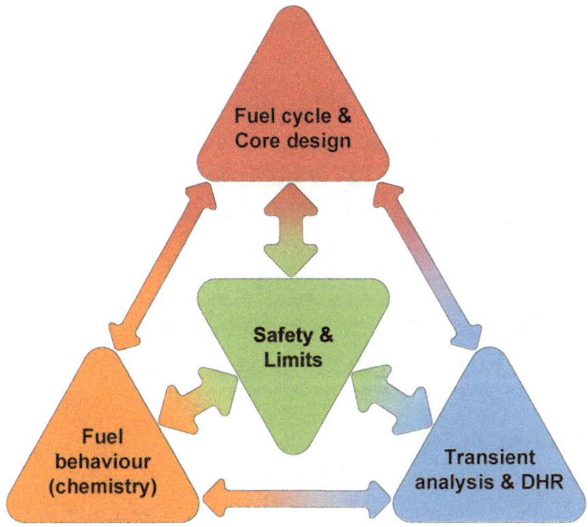

FIG. 83. Illustration of the four research areas at the Paul Scherrer Institute and their interdependencies (courtesy of J. Křepel, Paul Scherrer Institute). DHR — decay heat removal.

once the technology is mastered and advanced, the inherent safety and possible simplicity of MSRs might allow for a reduction of capital costs.

The research on MSR technology at the Paul Scherrer Institute integrates several past and ongoing national and international projects. The most important project is the participation in the European SAMOFAR and SAMOSAFER projects. National projects relate predominantly to sustainability and safety. In 2014, the Swiss National Science Foundation financed a PhD project dedicated to designing a modular MSR for low waste production. The major outcome of this PhD project was the proposal of the breed-and-burn fuel cycle [87, 372] in a molten chloride fast reactor [373]. Another project, entitled 'Feasibility and Plausibility of Innovative Reactor Concepts in a European Electricity Supply Environment', was financed in 2015 by ETH Zürich and the Fund for Projects and Studies of the Swiss Electric Utilities. Finally, swissnuclear, the nuclear energy section of swisselectric, financed a project in 2016 entitled 'Chemical thermodynamic aspects of LWR Pu and MA burning in MSR'.

Research on MSR technology in Switzerland is based predominantly on research by students. Research at the Paul Scherrer Institute focuses on MSR safety and sustainability. However, it covers a wide range of different topics, which can be structured into four research areas (see Fig. 83). The assessment of MSR safety is the ultimate objective. The other three areas address MSR sustainability, fuel salt chemistry for nominal and accidental conditions, and transient analysis and decay heat removal. Knowledge from these last three areas is necessary to evaluate MSR safety.

5.12.2. Major achievements

In three of the four research areas illustrated in Fig. 83, the focus is on modifying existing tools or developing dedicated tools, to ensure that the peculiarities of MSRs can be fully addressed by the simulations carried out at the Paul Scherrer Institute. The following major codes have been developed or modified:

— EQL0D and EQL3D for equilibrium fuel cycle procedures;
— GEMS (Gibbs Energy Minimisation Software) for thermodynamics modelling;
— AMoDy for molecular dynamics;
— GeN-Foam, a multiphysics tool for MSR core analysis;
— TRACE/PARCS, a system code for MSR transient analysis.

Of these codes, TRACE/PARCS was not developed at the Paul Scherrer Institute and was only modified for applicability to MSRs. The EQL0D and EQL3D procedures are dedicated to the analysis of the fuel cycle at equilibrium. EQL3D was originally developed for the simulation of fast reactors with solid fuel and later modified for MSR application. EQL0D was developed from the beginning as a dedicated MSR tool. The multiphysics solver GeN-Foam, based on OpenFOAM, was originally developed for reactors with solid fuel. It now includes several necessary models for MSR simulation and some additional models are being implemented. GEMS has been developed at the Paul Scherrer Institute since 2000 for several different purposes. In the past, it was applied to pyrochemical reprocessing of molten salt, hence some of the necessary compounds are already included in its thermodynamics database, which nonetheless needs further extension and modification for MSR application. The AMoDy code is a molecular dynamics tool being developed at the Paul Scherrer Institute with some unique features. Each of these achievements is described in more detail below.

5.12.2.1. Fuel cycle simulation (EQL0D)

An MSR with liquid fuel can be designed as a breeder reactor in both U–Pu and Th–U cycles, and it can also burn the legacy waste from spent fuel or utilize enriched uranium. In all cases, the equilibrium fuel composition will determine the neutronic performance. Therefore, the equilibrium procedures EQL0D and EQL3D were developed at the Paul Scherrer Institute. The

EQL3D procedure [374, 375] is based on ERANOS[17] [376] and includes modifications [300] to simulate the continuous removal of fission products in MSRs. Later, another zero dimension procedure named EQL0D, specifically dedicated to MSR simulation, was developed [377–379]. The latest version (v3) of the procedure couples MATLAB (matrix laboratory) with the Monte Carlo code SERPENT-2 [207] through the burnup matrix and solves the Bateman equation by the same Chebyshev rational approximation method [380] as SERPENT. The previous versions (v1 and v2) are MATLAB scripts coupled through the reaction rates of actinides with the cell transport code ECCO[18] [376] or SERPENT, respectively. Accordingly, they did not account for fission products.

Three basic MSR designs were compared using EQL0D v2 without a fission product model: (i) graphite moderated MSR, (ii) homogeneous fluoride fast MSR and (iii) homogeneous chloride fast MSR. For each of these three options, two different fuel salts were considered. The six resulting options were compared with the performance from ten reactors with solid fuel [381].

The EQL0D v3 procedure, including the fission products model, was applied for a scoping study of graphite and non-graphite moderated MSRs. Five different fluoride salts and six different moderators were evaluated. For each of these 30 combinations, the fuel composition at closed Th–U cycle equilibrium was simulated for many salt to moderator ratios and salt channel diameters. Figure 4 in Section 4.2 presents the results of thousands of EQL0D simulations. For better readability, the reactivity is plotted only for values above one. The values on the right side of each square correspond to the performance of the respective salt in an infinite homogeneous fluoride fast MSR without moderator.

According to Fig. 4, the LiF salt is the best performing carrier salt in the thermal spectrum, followed by the LiF–BeF_2 salt. Moderators based on beryllium and deuterium are the best and outperform graphite. These two types of moderator would, however, need coating or cladding, which was not accounted for in Fig. 4. Accordingly, the resulting performance of the core may differ. Additional simulation showed that only SiC as coating or cladding can preserve this performance [87].

The EQL0D v3 procedure was also applied to homogeneous fluoride and chloride fast MSRs. Five potential fluoride and three potential chloride carrier

---

[17] ERANOS (European Reactor Analysis Optimized calculation System) aims to provide a suitable basis for reliable neutronic calculations of current as well as advanced fast reactor cores. It consists of data libraries, deterministic codes and calculation procedures developed within the European Collaboration on Fast Reactors.

[18] The ECCO code prepares self-shielded cross-sections and matrices by combining a slowing-down treatment in many groups (1968 groups) with the subgroup method within each fine group.

salts have been analysed in the closed Th–U and U–Pu cycle. Figure 84 shows the breakdown of the reactivity excess. Reference [381] gives a description of the breakdown method. The highest reactivity excess, and thus the best performance in both fuel cycles, is provided by the $Na^{37}Cl$ and $Ac^{37}Cl_4$ carrier salts. Their tremendous reactivity excess in the U–Pu cycle can even enable the utilization of a breed-and-burn fuel strategy [135]. The best performance of the fluoride carrier salts contains ^{7}LiF. This is the same reference carrier salt used in the European SAMOFAR and SAMOSAFER projects. A parametric spectral study at equilibrium cycle was accomplished with the EQL3D procedure [375]. A hybrid spectrum MSR core, with thermal and fast spectrum zones, was evaluated [168], the simplified reprocessing scheme assessed [382] and the continuous and batchwise reprocessing schemes compared [383].

The EQL0D procedure has also been applied to breed-and-burn fuel cycle analysis since 2015 [87, 135, 372]. At the beginning of the analysis, graphite moderated and homogeneous fast MSRs were considered for both fluoride and chloride salts and in both Th–U and U–Pu fuel cycles. Accordingly, all possible combinations were analysed (eight combinations from 2 × 2 × 2: moderated or unmoderated, chloride or fluoride salt, Th–U or U–Pu fuel cycle). As expected, the graphite moderated MSRs have insufficient performance for this type of fuel cycle in both Th–U and U–Pu cases and for both fluoride and chloride salts (dashed lines in Fig. 85). Nonetheless, the obtained minimal subcriticality was surprisingly good for the fluoride salt and the Th–U cycle (red dashed line in Fig. 85, top). The homogeneous fluoride fast MSR showed a poor performance and cannot be used for the breed-and-burn fuel cycle (full lines in Fig. 85, top). The homogeneous chloride fast MSR performs well in the U–Pu cycle (blue full lines in Fig. 85, bottom). The reactivity is on the edge and the breed-and-burn cycle is not possible in the Th–U fuel cycle (red full line in Fig. 85, bottom).

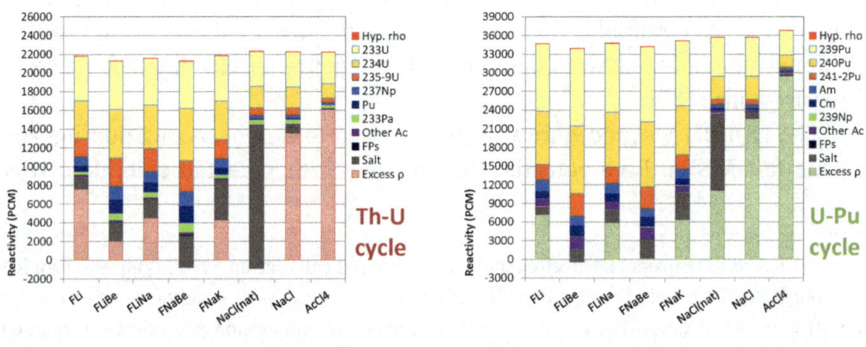

FIG. 84. Comparison of eight fast molten salt reactor designs using excess and hypothetical (for zero parasitic captures) reactivities with Th–U cycle (left) and U–Pu cycle (right) (courtesy of J. Křepel, Paul Scherrer Institute).

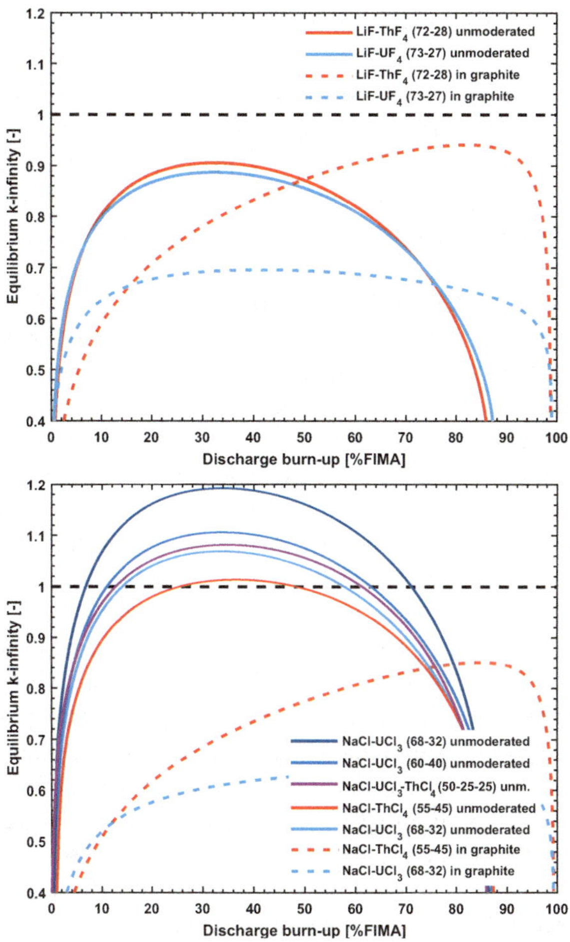

FIG. 85. Evolution of infinite multiplication factor in a breed-and-burn reactor as a function of fuel discharge burnup [372] (courtesy of J. Křepel, Paul Scherrer Institute).

Since the U–Pu cycle provides sufficient excess reactivity, several additional cases were considered with higher actinides share in the salt or with a mixed fuel cycle of U–Pu and Th–U. For all chloride salt cases, the enriched salt with ^{37}Cl was applied. The mixed Th–U and U–Pu cycle seems to be possible (purple full line in Fig. 85, bottom).

Figure 85 shows the simulation results for an infinite lattice. The major difficulty of designing a breed-and-burn core is the minimization of neutron leakage. The chloride salts are transparent for neutrons and the leakage minimization results in bulky cores, which may not be acceptable [87, 139, 372]. One possible option can be the multizone or multiliquid core layout [138].

5.12.2.2. Thermodynamic simulation (GEMS)

The GEMS[19] package is an open source thermodynamic modelling package that plays a vital role in MSR research. It has been developed at the Nuclear Energy and Safety Research Division of the Paul Scherrer Institute since 2000 and was and is applied in numerous European Union and international projects[20], as well as in a number of national projects financed by a consortium of Swiss utilities (i.e. swissnuclear)[21].

The GEMS package is applied for thermodynamic modelling in different areas, such as light water reactor related systems (e.g. crud formation, water chemistry), waste management, modelling the release of fission products during normal and accident conditions and to MSR related systems (simulation of phase diagrams and speciation) with the extended HERACLES database [384]. GEMS has a modular structure and flexible database with an interface built around an efficient open source GEMS3K numerical kernel code [385] for solving phase equilibria in complex non-ideal systems. Moreover, the extent of GEMS applicability can be significantly broadened by coupling this code with other transport or multiphysics codes. GEMS3K also contains the TSolMod code library [386], which has more than 25 state of the art mixing models for aqueous, gaseous, solid solutions and melts. This gives GEMS a broad applicability to various types of materials, including molten salts, if extended with the necessary datasets, and mixing models for binary interactions between the compounds.

For applications to nuclear materials and reprocessing of spent fuel from light water reactors, the HERACLES [384] thermodynamic database for low pressure and high temperature non-aqueous systems has been developed and maintained at the Paul Scherrer Institute since 2010. The database contains thermodynamic properties of pure compounds and binary interaction parameters for various systems. Currently, the HERACLES database covers data for over 650 condensed compounds, including melts and liquid condensates, over

---

[19] Gibbs Energy Minimisation Software developed by the Paul Scherrer Institute for Geochemical Modelling.

[20] Projects include ACSEPT (on actinide recycling by separation and transmutation), SACSESS (on the safety of actinide separation processes), the SAMOFAR and SAMOSAFER projects, and TCOFF (on thermodynamic characterization of fuel debris and fission products).

[21] Projects include 'Chemical thermodynamic aspects of LWR Pu and MA burning in MSR', ATHESC (advanced characterization and thermodynamic study of CRUD), ASTAM (advanced source term analysis with MELCOR, a fully integrated, engineering-level computer code developed by Sandia National Laboratories for the US Nuclear Regulatory Commission to model the progression of severe accidents in nuclear power plants) and MOCH ATF (modelling and characterization of ATF).

350 gaseous compounds, and is valid up to a temperature of at least 3000 K, covering most of the elements of interest.

The HERACLES database is used to study the behaviour of complex systems, speciation in selected molten salts and Gibbs energies of formation of different species in selected melts. Figure 86 compares vapour pressures for different species of the LiF–ThF_4–CsF mixture calculated using the HERACLES database with experimental and calculated results from literature [242]. Figure 87 shows vapour pressure evolution over time during salt heat-up for a more complex LiF–ThF_4–UF_4–Cs–I system evaluated with GEMS.

Furthermore, this approach has been used to assess the standard potentials for molybdenum or MgO species in the LiCl–KCl and LiF–AlF_3 melts to build the E–pO^{2-} stability diagrams and to provide a broad and thorough picture of their speciation [387].

By coupling GEMS with the severe accident code MELCOR [388], the cGEMS code was created [389]. It was applied on a generic pool filled by MSFR fuel salt with dissolved iodine and caesium fission products. The salt was slowly heated up by the decay heat. Figure 88 shows the evaporation behaviour during this heat-up process. The cGEMS code can be used for a more comprehensive accident analysis of MSRs once a detailed description of the reactor confinement and accident sequences are available and more fission product elements have been added to the HERACLES database.

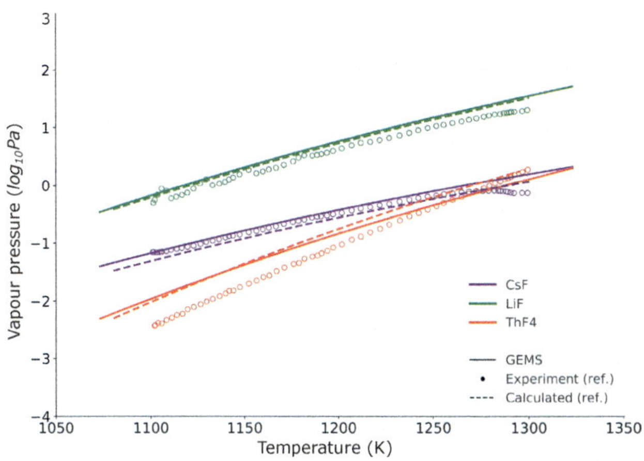

FIG. 86. Vapour pressure data for the LiF–ThF_4–CsF system calculated at the Paul Scherrer Institute with GEMS and compared with Ref. [242] (courtesy of J. Křepel, Paul Scherrer Institute).

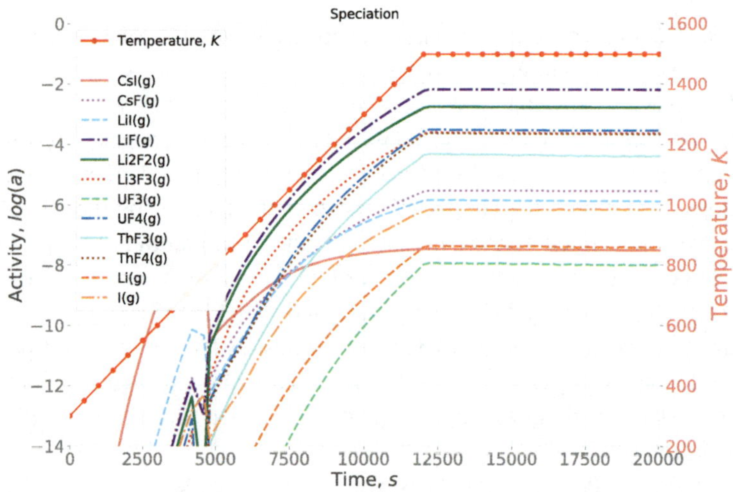

FIG. 87. Vapour pressures for the LiF–ThF$_4$–UF$_4$–Cs–I system calculated with GEMS (courtesy of J. Křepel, Paul Scherrer Institute).

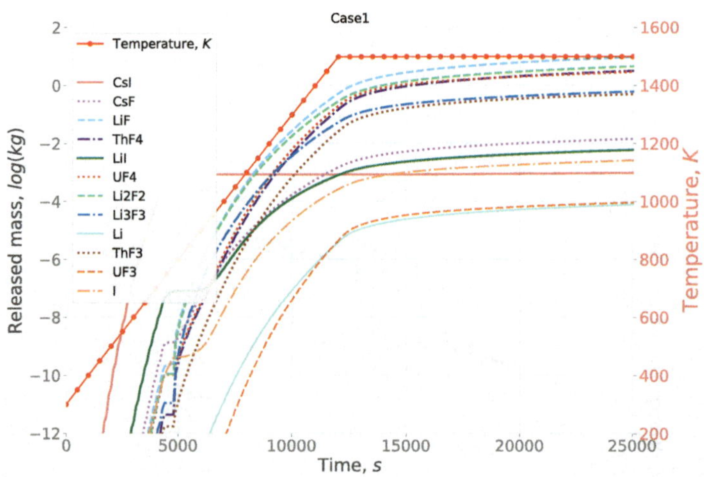

FIG. 88. Fission product release from a molten salt pool during the heating process (courtesy of J. Křepel, Paul Scherrer Institute).

5.12.2.3. Molecular dynamics simulation (AMoDy)

The AMoDy[22] molecular dynamics tool used at the Paul Scherrer Institute is another important asset for MSR analysis. In cases when the information on binary interactions between different compounds in binary systems is missing, the so-called atomistic modelling is used to fill the information gap. In particular, the molecular dynamics approach is used to study mixing enthalpies, excess volumes for different binary systems, polar as well as non-polar [390, 391] structures, thermal conductivity [392, 393] and diffusion in liquid as well as in solid state. As an example, the information on the excess properties can be evaluated (see Fig. 89). In this case, AMoDy allows for direct estimation of mixing properties and serves as a reliable tool in predicting these properties for mixing models in thermodynamic modelling, while at the same time providing information at the atomistic level, thus improving the understanding of the underlying processes.

AMoDy also can be potentially used for the direct estimation of phase diagrams for the systems of interest. One such modelling experiment is presented in Fig. 90, where the phase diagram for LiF–KF is shown together with six snapshots of molecular dynamics results. The snapshots refer to 24% (left) and 75% (right) of LiF share in the mixture for three different temperatures. Unstructured regions in the top left and right snapshots represent the molten region, the structured zones in the middle and bottom snapshots represent the solid region with a regular crystal structure. The molten region is extending with increasing temperature, leading to a complete melting of the two selected LiF–KF mixtures at about 1000 K. In the simulated case, AMoDy reproduces experimental observations with a high degree of accuracy.

AMoDy was recently extended with a new force field model. This model is based on the combination of a pair potential (in the form of a Morse potential) and a multibody interaction potential (based on the modified embedded atom model, or MEAM). It allowed for significant improvement of the modelling capabilities of this code. In particular, it can now model systems that have a very complex crystal structure in solid state, for example ThF_4. The behaviour of these systems cannot be reproduced with a simple force field model based on the pairwise interaction between ions in a system.

With the developed force field model, the AMoDy code can be applied not only to the structural properties of complex systems, but also to the dynamic properties of molten salt mixtures used in MSRs. Figure 91 shows AMoDy results

[22] AMoDy molecular dynamics tool used at the Paul Scherrer Institute. It calculates thermal conductivity and viscosity using an improved Green–Kubo algorithm, providing quite accurate values of the studied properties without the need to perform multiple simulations to gather sufficient statistics.

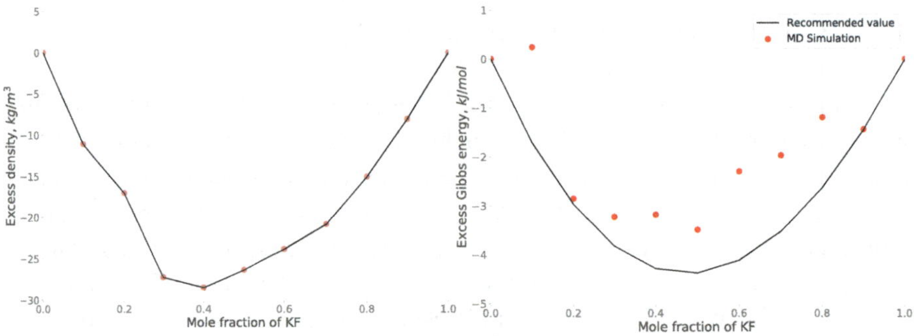

FIG. 89. Excess density (left) and excess Gibbs energy (right) of the Li–K–F system at a temperature of 1100 K; left: the dots show the calculated values and the line is a linear interpolation between two consecutive dots (courtesy of J. Křepel, Paul Scherrer Institute). MD — molecular dynamics.

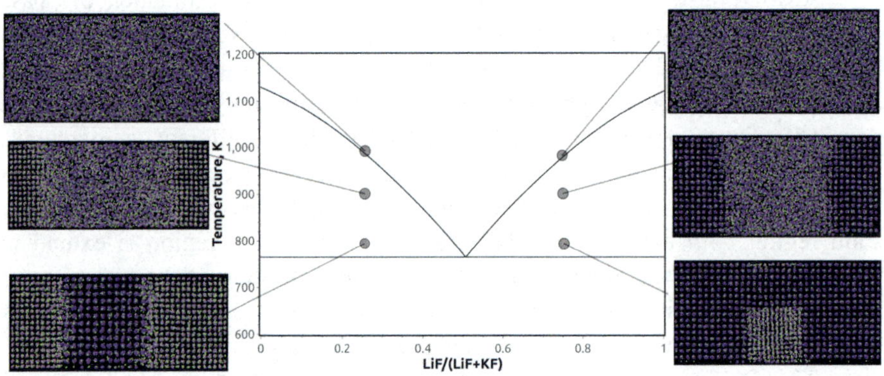

FIG. 90. AMoDy modelling of the LiF–KF system phase equilibrium behaviour (courtesy of J. Křepel, Paul Scherrer Institute).

for thermal conductivities of a LiF–ThF_4 system as a function of composition (top) and temperature (bottom). The results are compared with available recommended values [394]. These calculations provide detailed information about the thermal conductivity dependence on the composition and temperature of a LiF–ThF_4 system. They also complement the experimental information and extend the data to a wider temperature range and the whole range of ThF_4 composition in the salt.

The viscosity of a LiF–ThF_4 system calculated by AMoDy is presented in Fig. 92 (top) for several ThF_4 molar fractions as a function of temperature, and in Fig. 92 (bottom) for several temperatures as a function of ThF_4 molar fraction. The top figure also includes the available experimental [394] and molecular dynamics calculated [395] values and shows good agreement of the AMoDy

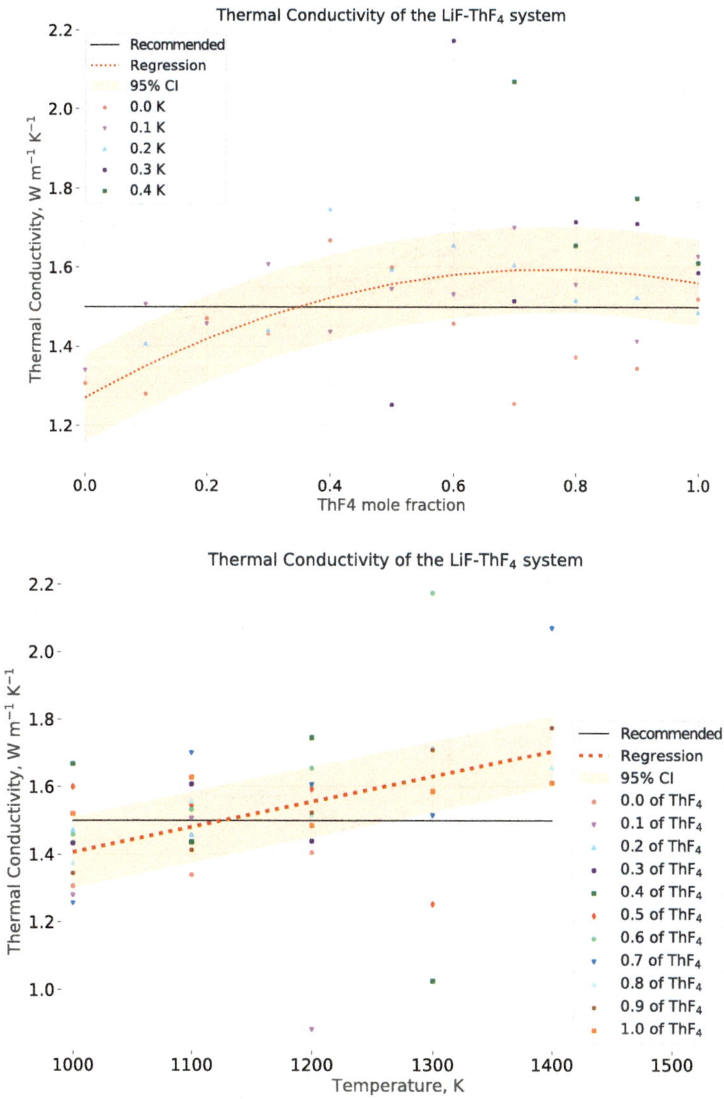

FIG. 91. Thermal conductivity of the LiF–ThF$_4$ system as a function of composition (top) and temperature (bottom); values calculated by AMoDy and recommended in Ref. [394] are compared (courtesy of J. Křepel, Paul Scherrer Institute).

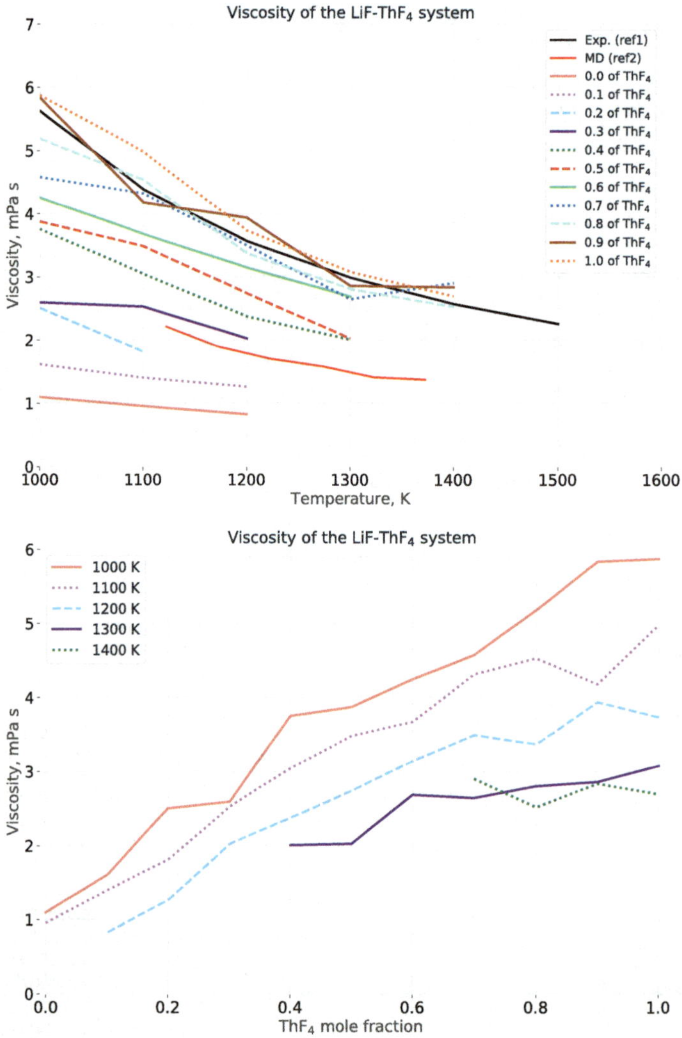

FIG. 92. Calculated viscosity of the LiF–ThF$_4$ system as a function of temperature (top) and composition (bottom); the temperature dependency calculated by AMoDy is compared with values from Ref. [394], labelled as (ref1), and Ref. [395], labelled as (ref2) (courtesy of J. Křepel, Paul Scherrer Institute).

results. Compared with sparse experiments, AMoDy allows for the performance of more simulations at different temperatures and composition covering a wider range of parameters and extending available theoretical data.

In general, the results of the AMoDy calculations are in good agreement with other published results but provide much more detailed information on the dependencies of molten salts on temperature and on the concentration of specific constituents of the molten salt. This feature is of great importance and serves as a source of information on the physical and dynamic properties of salt mixtures applicable in MSRs.

5.12.2.4. Transient behaviour simulation

The activities grouped in this area are dedicated to analysis of system behaviour and dynamics in nominal and accident conditions. They are driven by the internal MSR project at the Nuclear Energy and Safety Research Division of the Paul Scherrer Institute as well as European projects such as SAMOFAR and SAMOSAFER. Two different approaches for MSR transient analysis were developed. One is a high fidelity approach based on the OpenFOAM solver GeN-Foam [201, 396], and the other is based on the coupled system code TRACE/PARCS [397–399]. Not all codes can simulate MSR transients because of MSR peculiarities, such as the drift of delayed neutron precursors. Also, the heat exchange between volumetrically heated liquids and solid surfaces is a scientifically challenging issue, which can have a strong relevance to safety. The following tasks have been accomplished in this area:

- — Development of the GeN-Foam multiphysics solver [201, 396].
- — Inclusion of salt properties in the TRACE code and pre-evaluation of the heat exchangers [397].
- — Modification of point kinetics in the TRACE code for delayed neutron precursor drift [398].
- — Modification of TRACE/PARCS for delayed neutron precursor drift and its application to MSRE transients [399].
- — Assessment of decay heat distribution in the fuel stream [400].
- — Completion of two MSc studies in 2016. The GeN-Foam solver [401] and the TRACE/PARCS coupled code [402] were preliminary applied to the MSRE and the MSFR, respectively.
- — Completion of one PhD study related to GeN-Foam application to MSR. It relates to salt freezing phenomena, but covers also the design of an MSR concept [139].

As an example of the application of TRACE/PARCS, Fig. 93 shows results [402] for the steady state distribution of two selected groups of delayed neutron precursors in the axial cut of the MSFR benchmark geometry. Figure 94 shows another illustrative result [139] from modified GeN-Foam code on a breed-and-burn molten chloride fast reactor design entitled 'Tap'. The figure illustrates the impact of baffles in the core on turbulence, where eddy viscosity, as a selected measure, is reduced by at least an order of magnitude in the core with baffles.

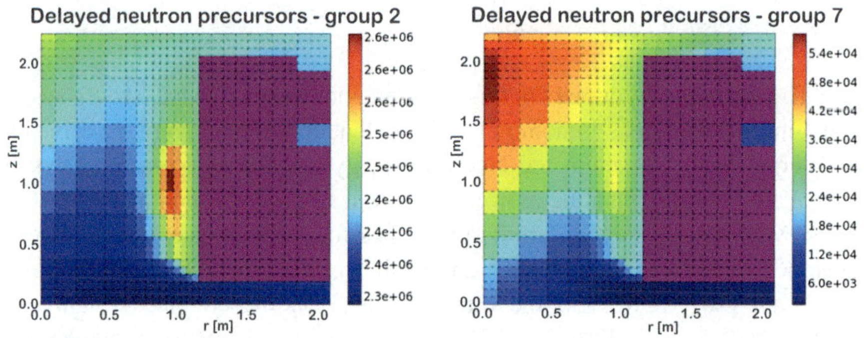

FIG. 93. Distribution of group 2 (left) and group 7 (right) of delayed neutron precursors in the axial cut of MSFR core benchmark geometry [402] (courtesy of J. Křepel, Paul Scherrer Institute).

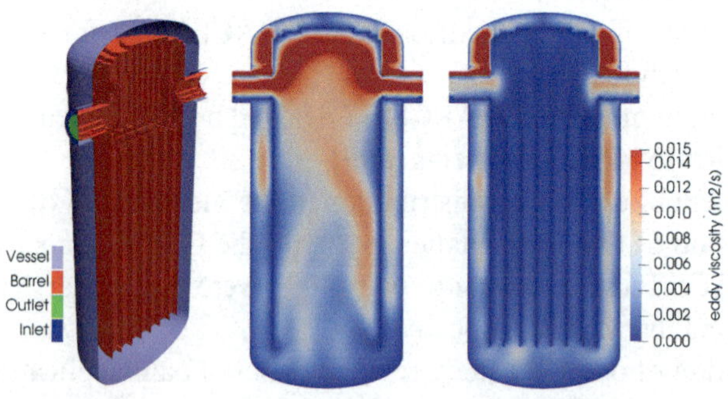

FIG. 94. Impact of baffles on eddy viscosity in the MCFR core. Diagram of baffles (left), eddy viscosity without baffles (centre) and eddy viscosity with baffles (right) [139] (courtesy of J. Křepel, Paul Scherrer Institute).

5.13. RESEARCH AND DEVELOPMENT ACTIVITIES IN THE UNITED STATES OF AMERICA

The nuclear industry has the responsibility to design, construct and operate commercial nuclear power plants in the USA [403]. However, the US Department of Energy has statutory authority under the Atomic Energy Act to promote and support nuclear energy technologies for commercial applications. The role of the Department of Energy falls more in the area of R&D.

Most support by the US Government for advanced reactors is provided through technology independent, competitive award processes. The overall focus of activities supporting MSR technology by the US Government is to facilitate industry success in the deployment of commercial MSRs. Activities by national laboratories focus on developing the understanding and the technology needed to design, operate, maintain and evaluate MSRs, as well as providing technical support for safety evaluation tools and methods for commercial development in the near term.

Current activities supporting MSR technology sponsored by the US Government include the following:

(a) Developing an efficient and effective regulatory process that is technology independent.
(b) Providing direct support to national laboratories. ORNL is conducting the Department of Energy's MSR campaign and has addressed MSR reactor design, instrumentation and control, and sensors for harsh environments.
(c) Providing direct support to advanced reactor developers through competitive award processes.
(d) Providing developer support for access to national laboratories.
(e) Providing small business innovative research awards to develop MSR technologies.
(f) Providing support for university research and educational activities.
(g) Providing innovative, advanced reactor technology development support through grants from the Advanced Research Projects Agency – Energy.
(h) Developing technology at national laboratories in support of capability demonstration in the fields of safety and safeguards.
(i) Measuring (and providing open access to) fundamental data on the thermophysical and thermochemical properties of fuel salt.
(j) Engaging the international community through the Generation IV International Forum and IAEA initiatives. The forum runs working groups on the risk and safety and proliferation resistance and physical protection for MSRs.

(k) Supporting the development of industry consensus standards. For example, the American Nuclear Society has a working group for the development of a design safety standard for liquid fuelled MSRs (ANS 20.2).
(l) Developing advanced modelling and simulation tools.

6. CURRENT CHALLENGES TO DEPLOYING MSRs

This section focuses on the challenges associated with deploying MSRs and possible solutions to these challenges. The methods or processes for overcoming any specific challenge are mainly dependent on decision making by the technology development organization facing the challenge, as well as governments and regulatory bodies. Other stakeholders in an MSR project may also support the development of these methods or processes and this decision making. Furthermore, many of the issues are specific to the region in which the MSR is to be deployed.

The following categories of challenges are considered to be the most significant:

(a) Supply chain;
(b) Fuel supply;
(c) Regulation;
(d) Fuel salt waste disposal;
(e) Safeguards and security;
(f) Maintenance and operations;
(g) Programme documentation.

The following subsections discuss each category. This section does not include the detailed technical challenges associated with MSRs because they are described and discussed in other sections of this publication. Furthermore, this section does not discuss existing policy challenges related to the deployment of MSRs because of the significant differences in policy positions between Member States. It remains incumbent upon each Member State and organization developing MSRs to assess the policy positions relative to its region of interest to understand the challenges associated with the development in that region.

Potential solutions to specific challenges are discussed at the end of each subsection.

6.1. SUPPLY CHAIN CHALLENGES

Many of the components and structures required by various MSR designs do not have an existing supplier base or a full set of technical specifications necessary for their specialized materials and components. Depending on the regulatory framework, MSRs might be able to rely on commercial grade high temperature technology for industrial processes. The lack of other components and the scarcity of manufacturers lead to the challenge of sourcing new suppliers. This challenge will also include difficulties in developing, controlling and maintaining a nuclear safety culture; obtaining technical expertise; and ensuring knowledge transfer across the global supply chain. Additionally, the qualification of materials and manufacturing methods is critical, whether nuclear or non-nuclear standards for high temperature are being applied.

The solution to the supply chain challenge depends on the market for MSRs. Since basic manufacturing technology was established for the MSRE in the 1960s [404], supply chains for current MSRs might be easily established. That is, if there is a demand, a supplier will appear. This rule does not necessarily work the other way around.

6.1.1. Qualification of suppliers

Inspecting and certifying manufacturing facilities for MSRs is currently difficult if the manufacturer follows an established, conventional nuclear standard such as, for example, the Boiler and Pressure Vessel Code of the American Society of Mechanical Engineers [81], which is not directly applicable to molten salt media. In addition, attempts to use limited manufacturing facilities may cause long lead times. The risk of incurring long lead times is increased if the required components come from a link within the supply chain that has limitations on capacity. The difficulty associated with qualifying facilities and suppliers is common among advanced high temperature reactor developers, especially those that have not chosen to use materials qualified by existing standards.

Because many developers are new entrants to the industry, they have limited experience or capability to qualify a supplier. Developers need to be diligent and oversee the suppliers to ensure that correct design and quality assurance standards are followed. Without fully vetting and qualifying suppliers, the risk of substandard components being supplied increases.

The solution for the qualification of suppliers also depends on the market for MSRs. Taking lessons identified from the MSRE and with the participation of current nuclear plant equipment suppliers in the deployment stage, this issue can be resolved. For the suppliers to sell their products, they need to be able to be qualified for this purpose.

6.1.2. Infrastructure considerations

Although infrastructure is generally considered to be regionally specific, the status of existing infrastructure to support MSR deployment presents a challenge. Roads, ports, railways, shipping lanes, heavy haul routes, heavy transportation and heavy lifting, most of which are project and technology specific, need to be adequate. The unavailability of satisfactory infrastructure for a specific MSR deployment project can result in significant challenges to deployment and to meeting schedules associated with critical paths.

The solution for establishing infrastructure could be to utilize the current systems for nuclear plant deployment. The deployment of MSRs as SMRs might be facilitated through the smaller and lighter structures and components of SMRs, which might not need as sophisticated an infrastructure as large nuclear projects.

6.1.3. Reactor developer and supplier engagement

Early engagement between reactor developers and suppliers has long been viewed as critical to success. However, a successful engagement is a delicate balancing act between timing, need, availability, cost, price, competition and risk. Accomplishing this requires a strategic and cooperative fit between the supply chain strategy of developers and the competitive strategy of the suppliers — an arrangement that is very difficult to achieve. Developers have historically chosen to strategically delay the selection of suppliers, whereas suppliers tend to drive towards securing early relationships with developers. Although the timing, engagement and decisions made by any particular developer of an MSR will be different for almost every component or service required to deploy its particular design, the need for both the developer and the supplier to be ready for engagement is paramount to success.

The solution to the lack of engagement between developers and suppliers depends on the progress of the deployment of MSRs in the market. Since developers and suppliers are likely to be based in different countries and probably on different continents, international cooperation is required.

6.2. FUEL SUPPLY CHALLENGES

Current developers of MSRs are contemplating a multitude of fuel types and associated suppliers. Fuels include enriched uranium, plutonium from used nuclear fuels, TRU elements from used fuels and ^{233}U produced from fertile thorium. Each MSR using liquid fuel will need to have a fuel salt specification to support purchasing. Fuel salt composition and tolerance standards are still being developed.

Most MSRs with fast spectra will require uranium enrichment greater than 5% for their initial cores. Hence, for Member States that currently impose a regulatory limit on the production of low enriched uranium to less than 5%, regulators and government agencies need to consider how to authorize production, transportation and use of this material. More generally, a strategic plan for MSR fuel cycles showing where and when materials will be generated, transferred, used, stored and disposed of needs to be developed — just like for any other nuclear fuel.

Since the current fuel supply for low enriched uranium and mixed oxide fuels is based on the solid fuel industry, producing molten salt with uranium, plutonium or thorium requires new facilities and equipment. Nevertheless, these facilities and equipment are similar to those used in the existing chemical industry. Some MSRs are often compared with chemical reprocessing plants more than with conventional nuclear power plants. The same is assumed here too: if demand exists, supply will appear.

6.2.1. Supply challenges for high assay low enriched uranium fuel

Data available from Ref. [405] provide evidence that many MSRs under development are planning to utilize fuel enrichment levels that require high assay low enriched uranium, which has a uranium enrichment between 5% and 20% and poses a supply challenge. This challenge will not be encountered by MSRs planning to utilize an enrichment below 5% or located in a Member State in which high assay low enriched uranium is available.

The use of high assay low enriched uranium is potentially problematic in many Member States because of the lack of regulatory requirements for the fabrication, transportation and use of this type of fuel. Its introduction for advanced reactors, including MSRs, is being promoted by a nuclear organization in the USA [406].

6.2.2. Supply challenges for uranium–plutonium and transuranic fuel

A significant investment may be required to develop the infrastructure necessary for the reprocessing of used nuclear fuel from currently operating reactors, which would make available the feedstock material required for the fabrication of the fuel for an MSR. On the other hand, reprocessing used nuclear fuel is expected to reduce the amount of radioactive waste for final disposal. This reduction has benefits in several areas, including financial and environmental.

The solution for utilizing plutonium or minor actinide fuels can be limited to the following two scenarios:

(1) For those Member States with commercial reprocessing plants, the solution might be straightforward as it involves adding only a conversion facility to produce the fuel salt with uranium or plutonium;
(2) For other Member States, construction of an entirely new reprocessing plant would probably be needed.

6.2.3. Supply challenges for thorium fuel

The supply of thorium fuel is characterized by different types of challenge than the supply of high assay low enriched uranium or plutonium. The primary challenges remain acceptance by industry; the widespread adaptation of existing uranium based fuel fabrication technology for manufacturing thorium based fuel; and the accompanying development of the significant supporting infrastructure for thorium based technology. Furthermore, fuels that are based on uranium already have an infrastructure that is well established; these fuels are therefore a natural preference to less popular thorium. For these reasons, there may be significant challenges in the foreseeable future for the deployment of reactor technologies that are based on thorium. Nonetheless, if reactor developers increasingly pursue a design that utilizes thorium, the market for thorium fuel will grow.

6.3. REGULATORY CHALLENGES

This publication recognizes the existence of multiple regulatory agencies worldwide and makes no attempt at evaluating their effectiveness in regulating any reactor technology. Furthermore, it does not identify the challenges faced by any of these agencies in carrying out their individual responsibility to their respective governments with regard to the licensing and obtaining of permits for nuclear activities involving MSRs. However, in meeting regulatory demands, there are unique, technology specific challenges associated with the licensing of MSR designs irrespective of the regulatory jurisdiction. It is these, more generally applicable challenges from a regulatory perspective that are included here. It remains incumbent upon each individual applicant seeking regulatory approval to evaluate the specific challenges for licensing or permitting of an MSR within its chosen regulatory jurisdiction. Providing a comprehensive listing of regulatory challenges associated with each of the many regulatory agencies is beyond the scope of this publication.

The solution to regulatory challenges for the MSR technology might be different depending on each domestic market, but it is desirable to have criteria or guidelines that are applicable worldwide.

6.3.1. Acceptance of evaluation tools by regulators

The scarcity or lack of available technical information to allow developers to create successful designs is a challenge. Hence, providing meaningful and credible evidence to enable the regulator to align or develop its regulatory requirements according to the safety characteristics of MSRs is a challenge as well.

Many of the issues impeding industry interest in MSRs are in the areas of safety evaluation and regulatory licensing tailored to the characteristics of MSRs. Consequently, common safety and licensing challenges exist from the lack of specific safety evaluation tools and capabilities. The outstanding issues are, among others, lack of a consensus on a set of potential MSR accidents, the form and content of a regulatory licence application, and experimental validation or benchmarks of safety evaluation tools for MSRs. Moreover, current regulatory practices mean that the acceptance of the newly developed tools is expensive and time consuming.

6.3.2. Evaluation of accident progression

Depending on the regulatory framework, the licensing of an MSR may require evaluating the progression of potential accidents for MSRs. No comprehensive consensus set of potential MSR accidents or safety models of these accidents is currently available. This lack of documented and credible MSR accident scenarios, in which significant quantities of radionuclides are dispersed, might be due, at least partly, to the good safety characteristics of MSRs. A robust justification of these safety characteristics is expected to completely, or at least partially, resolve this issue.

However, it is unlikely that tools for modelling MSR accidents will need to be highly precise; this type of nuclear power plant may be able to trade its large safety margins against modelling precision. Overall, a comprehensive set of design basis accidents could be developed for MSRs based on their common characteristics in a similar way to those initially developed for light water reactors, well before they had accumulated considerable operating experience.

The technology of MSRs offers considerable flexibility to reactor developers, which leads to substantial design variations between prospective developers and increases the difficulty of validating modelling tools for general purposes. Moreover, MSRs will vary substantially by type and by vulnerability to accidents that occur beyond the reactor system. However, these accidents may also lack the energetic potential to disperse radionuclides beyond the plant site, owing to the inherent properties of molten salts and the low operating pressure of the primary system.

One means to provide assurance that a proposed list of design basis accidents for MSRs adequately spans the potential set of accidents is to employ maximum hypothetical accident evaluations such as those used by the US Nuclear Regulatory Commission [407]. This approach led to historic (and current non-power) reactor licensing that relied on maximum hypothetical accidents to define the list of potential accidents. Molten salt reactors might be able to follow a similar strategy by showing that even credible beyond design basis accidents, such as prompt criticality or massive failure of the boundary wetted by the fuel salt, or both, do not result in significant radionuclide release while crediting only barriers and passive structures, systems and components for removing decay heat.

Measurements of fuel salt properties are required to provide data to determine, in subsequent testing, which radionuclides are released from the fuel salt under normal and accident conditions. The combination of fuel salt property measurements and performance models will form the foundation for fuel salt qualification.

Fuel accidents can also be bounded by maximum hypothetical accidents. If even prompt criticality or unintended criticality outside the core region of the fuel boundary does not result in a release of radionuclides from the plant, the fuel performance specification and consequently the testing required to validate fuel properties could be relaxed substantially. In essence, even if fuel salt that is well beyond design specifications is loaded into the reactor, the probability of radionuclide release into the environment remains low, provided that the fuel continues to be cooled; some MSR designs provide this cooling by passive means. Thus, a fuel specification would be limited to validating the thermophysical properties of the fuel salt rather than its nuclear properties. These thermophysical properties are not anticipated to change substantially even after many years of use, which substantially decreases the rigour required in the measurement of these properties [408]. These are the benefits of using a fuel salt that may simplify the licensing process.

Any accidents that might occur during the entire fuel cycle or after the disposal of the used nuclear fuel also need to be evaluated. Regulators expect safety cases to be supported by the Member State. A waste disposal plan needs to be prepared and its safety reasonably demonstrated by the licence applicant.

6.3.3. Consensus on industry standards

The development of MSRs encounters issues and challenges before the MSRs are ready for deployment. One of the issues is a lack of consensus in the use of industry standards. Some of the required standards for development include the following:

- — MSR design safety standards;
- — Salt composition standards;
- — Fuel salt transportation standards;
- — Standard methods for performing the accounting of MSR fissile materials;
- — Standard methods for evaluating vital areas of MSRs related to physical protection.

The development of licensing standards has started in the USA (e.g. ANS-20.2 by the American Nuclear Society standards working group[23]). However, the above list of recommended standards is not exhaustive and further development is needed.

6.4. FUEL SALT WASTE DISPOSAL CHALLENGES

Liquid salt does not accumulate physical damage like solid fuel. Also, to the extent that its composition can be controlled to remain within a specification, it does not suffer from ageing. Fission products can potentially build up to equilibrium concentrations and most designs include on-line or batch continual chemical processing (polishing) of the molten salt. The extent of this phenomenon will therefore be dependent on the developers' choice of fission product and fission gas management techniques and design choices. Consequently, used MSR fuel salt can be considered for use in future generations of reactors. However, further advances in technology development and demonstration are required for an MSR to operate on a closed fuel cycle [117]. Nevertheless, MSRs still require a disposal path for fuel salt waste and defining this path could significantly support the licensing of MSRs.

The Office of Nuclear Energy of the United States Department of Energy has recently provided overviews of the potential waste processing and waste form options for MSRs [409] and has evaluated the status of molten salt waste technologies, focusing on fast spectrum systems [410]. Thermal spectrum reactors still remain to be evaluated, and challenges in the development of disposal techniques of different fluoride and chloride salts need to be overcome.

---

[23] Information on the scope of the working group on ANS-20.2, Nuclear Safety Design Criteria and Functional Performance Requirements for Liquid-Fuel Molten-Salt Reactor Nuclear Power Plants, can be found at https://ans.org/standards/involved/voloppor/rarcc/advanced/

The basic technology for creating stable salt waste forms is generally known [411, 412]. However, no implementation details are available for the technology or licensing of waste disposal systems. A validated MSR waste stream roadmap, which would help to minimize the technical risks, is yet to be completed.

Some components wetted by fuel salt may have significant amounts of fission products plated onto and possibly embedded into their surfaces. Moreover, container alloys will have been activated while in use and might include surface contamination with actinide materials. The activation of nickel within nickel based alloys might (depending on the shielding design) result in the structural alloy being classified as a radioactive waste and would require proper disposal. Several Member States are working towards developing nuclear waste disposal strategies that, in principle, would be the same as approaches used for MSR and light water reactor waste. The recycling of used nuclear material is also possible.

The knowledge of the waste streams and viability of solutions to treat the used fuel and dispose of the waste are important aspects in developing and licensing MSRs. Technical options exist for reusing actinides indefinitely and for stably packaging the remainder of the fuel salt. The remainder of the solid waste can largely be left to decay sufficiently to be acceptable in surface storage sites.

6.5. SAFEGUARDS AND SECURITY CHALLENGES

Molten salt reactors incorporate features that are fundamentally different from conventional reactors. At the same time, MSRs have limited operational experience, restricted to research reactors from a time that precedes modern best practices for safeguards and security. A modern safeguards and security framework for MSRs is largely untested and reflects important uncertainties regarding the deployment of MSRs as power reactors at large scale. Consequently, reactor developers need to establish, verify and validate adequate safeguards and security provisions for MSRs. In parallel, regulatory stakeholders — including legislators, international organizations and national regulators — will need foresight, imagination and initiative to proactively accommodate an expansion of nuclear power at large scale, which includes SMRs and MSRs, as prescribed by many scenarios for reducing greenhouse gas emissions [412].

Designs for MSRs vary considerably, which has a significant impact on safeguards and, to a lesser extent, security. Some MSR designs circulate liquid fuel salts while actively controlling the chemistry of the fuel salt, which are features resembling facilities for reprocessing spent nuclear fuel. This is particularly true for MSRs that incorporate on-line refuelling and fuel salt let-down as well as on-line separation of fission products and actinides. Other MSR designs do not exchange considerable amounts of radioactive material with other systems within the nuclear power plant and are more similar to conventional solid fuel reactors. Nuclear material accountancy is expected to be easier to implement for these MSR designs.

Historically, safeguards and security measures were 'added on' rather than developed concurrently with the first nuclear power plants. Despite this, the safeguards and security measures implemented in conventional nuclear power plants have been effective. This is evident from the lack of severe incidents related to external attacks, sabotage, theft or other diversions of fissile material from commercial nuclear power plants. In designing and developing commercial MSRs, the proactive consideration of and discourse about safeguards and security from an early design stage will ensure the effective implementation of efficient measures that will also retain overall economic competitiveness.

Even though the current legal framework for IAEA safeguards [413] focuses primarily on existing nuclear power plants and processing facilities, already established methods and practices are expected to be applicable to some extent to MSRs, although adaptation might be necessary. Nonetheless, since to date no power generating MSR has been licensed for commercial operation in any regulatory jurisdiction, the implementation of necessary safeguards and security protocols remains a challenge to deployment.

6.5.1. Material control and accountancy

Fissile material control and accountancy will be a key differentiating element for the licensing of reactors with liquid fuel. Some Member States plan to adopt material accountancy equivalent to what is currently required for agreements between them and the IAEA, as outlined in INFCIRC/153 [414]. Other approaches, such as including safeguards early in the design process, can be implemented as well.

Depending on the specific reactor design, the application of safeguards to MSRs with liquid fuel might need to consider (i) the mixture of fuel, coolant, fission products and actinides (along with high radiation fields); (ii) the variation in isotopic concentrations in the fuel salt including the removal of fission products, rare earth elements and noble metals; (iii) high operating temperatures; (iv) on-line processing, where some fraction of the inventory of fuel salt can be removed while the reactor is operating; (v) unique refuelling schemes, including the ability to continuously feed the core with fresh fissile and fertile material; and (vi) the presence of fuel outside the reactor vessel [415]. Approaches to accomplish the required material control and accountancy at MSRs are still under development and no general consensus has been reached to date.

Nevertheless, some of the safeguards techniques and equipment described by the IAEA in Ref. [416] could be directly implemented at MSR facilities, such as the weight or volume counting of the fuel salt within a single material balance area, among other non-destructive analysis techniques. With only one or a limited number of material balance areas encompassing the MSR nuclear power plant, material control and accountancy can be achieved in a simplified manner, which would correspond to trends in advanced reactor design where a simplified approach to safety and safeguards is preferred to costly and redundant engineered solutions. When looking at reactor design and operation, the accounting for fuel moving across material balance areas is complicated, as even fuel for a pebble bed reactor could be treated as a bulk item versus a batch item [417]. Measuring and accounting for fuel in continuous pebble or liquid flow, or even the solid salt forms, may require a change in the framework for verification and accountancy measures in order to facilitate the appropriate safeguarding of such facilities [418]. The safeguards framework will have to acknowledge that liquid material with a modified isotopic form cannot be drawn off from the circulating liquid fuel of an MSR and needs to be controlled as a batch item. A mechanism for setting a limit on the material measurement accuracies and discrepancies, as with powders and liquids in fuel, will need to be developed for MSRs and pebble bed reactors.

Additionally, containment and surveillance techniques, which complement measurements, do not necessarily have to differ from the existing ones. If safeguards are considered early in the process of designing an MSR, the complexity of accountancy processes can be significantly reduced. Solid fuel salt delivered to an MSR nuclear power plant prior to melting can be treated as fuel pellets currently used in light water reactors and counted as batch items. Transport and storage accountancy will not differ either. On the other hand, the transport and handling of the nuclear material will increase because of the rapid development and deployment of MSRs and closed nuclear fuel cycles. This will require infrastructure and more personnel capable of dealing with sensitive technology and a more sophisticated approach to handling bulk nuclear fuels with acceptable uncertainties in verification activities.

6.5.2. Security challenges unique to MSRs

Molten salt reactors have various structures, systems and components that are different to those used by the current fleets of light water reactors. The structures, systems and components that are specific to MSRs, such as those used for on-line fuel reprocessing, could be targets of malicious attacks. Accordingly, measures for their physical protection and cybersecurity might have to be developed and then reviewed for regulatory approval. The development and review are related to the regulatory challenges mentioned above and are to be addressed.

Alternative physical security requirements for advanced reactor technologies with enhanced engineered safety and security features are needed, such as those that the Nuclear Energy Institute in the USA has proposed developing [419]. The alternative requirements would have the key result of shifting the responsibility for protecting facilities from the plant owners to local law enforcement, substantially reducing the ongoing expense related to security by the plant owners. The structures, systems and components that are specific to MSRs, such as those used for on-line fuel reprocessing and the interface for transferring decay heat to the environment, will need to be robust against assaults so that local law enforcement can be relied on for their physical protection.

6.6. MAINTENANCE AND OPERATIONS CHALLENGES

The MSRE, which operated and produced energy just below 10 MW(th) for around four years, is a major source of knowledge on the maintenance and operation of an MSR. Detailed reports on maintenance equipment and procedures are available, as well as accounts of the successful implementation of totally remote and long handled tools for the maintenance of the MSRE, including a mock-up facility of an MSR specifically designed for the testing of remote handling technology for the replacement of major components [112, 420]. Remote maintenance is required because of the high temperature and radioactivity of the fuel. Harsh conditions inside an MSR facility can damage the remote handling equipment. As a result, the irradiated equipment might need to be serviced, thereby adding another level of complexity to maintenance and operations.

Moreover, the operational challenges of the decontamination and cleanup of spillages of active salts needs to be addressed by Member States. For example, should any in-core structures and components be replaced, either because of a break or end of life caused by irradiation, specific maintenance actions will be required. Measures used in conventional light water reactors can be adopted in MSRs to some extent, but more advanced and remote solutions might be needed depending on the strategy for dealing with the decontamination and replacement.

Improved modelling tools for process operations would facilitate the scale-up and modelling of MSR operation for an extended duration. Extensive, automated and remote maintenance and component replacement may be required for a commercial plant. Developing and demonstrating this maintenance and replacement, as well as adequate modelling and simulation of maintenance and operation processes, are key elements and challenges to deploying MSR designs. Both the maintenance and operation of an MSR could be significantly simplified with the use of modern technology, such as virtual reality, where a digital twin of a facility could be used for training on and planning of the activities. However, challenges that may be faced during the maintenance and replacement of the remote handling equipment will require solutions that might not yet be commercially viable.

6.7. PROGRAMME DOCUMENTATION CHALLENGES

The MSRE was viewed as a very successful test reactor and the overall MSR R&D programme in the USA was executed over decades, which would suggest that no major technology development is needed in order to construct the first commercial MSR. However, this understanding is not widespread and the technology status and the unique features and benefits of MSRs are not well known beyond the people who have been involved in the various MSR development programmes.

For example, the corrosion of pipes and other components in contact with salts has been extensively researched and documented. Few people outside of a very narrow specialist community, including regulatory agencies, are aware that, as early as 1967, ORNL had concluded that the technology for materials resistant to corrosion was in an advanced stage of development [421]. ORNL came to this conclusion following an extensive corrosion research programme, utilizing several families of fluoride mixtures involving commercial and high temperature alloys that were under development at that time. These same studies also concluded that container materials were available that demonstrated extremely low corrosion rates at temperatures expected to be encountered during anticipated MSR operations [422].

Throughout the 1970s, salt chemistry control was further developed, substantially decreasing the significance of corrosion as a limitation to MSR deployment. Currently, multiple companies are carrying out their own corrosion experiments.

Irrespective of these examples of R&D results, which have been available for several decades, the status of MSR technology developments and the unique features and benefits of MSRs are generally not well known. Developing programme documentation of a high quality that is readily accessible to the non-specialist technical community is more important for the development of MSRs than for other advanced reactor classes.

6.8. SUMMARY OF IDENTIFIED ISSUES FOR MSR DEPLOYMENT

The near term deployment of MSRs is potentially viable, as extensive R&D activities have been undertaken in several Member States. The knowledge gained from the MSRE has been used and progress has been achieved with modern research and technology. As a result, various advanced designs have been proposed and their development continues to be pursued. However, several challenges to commercial deployment remain unresolved.

Addressing the challenges mentioned in this section could be facilitated by the joint effort of the international community involved in the development of MSR technology. In particular, the following issues have been identified: (i) reactor developer and supplier engagement; (ii) supply of high assay low enriched uranium fuel; (iii) supply of U–Pu and TRU fuel; (iv) safety, including the preparation of guidelines; (v) fuel salt waste disposal; (vi) material control and accountancy; (vii) security; (viii) maintenance and operation; and (ix) programme documentation.

7. SUMMARY

Molten salt reactors are a broad and diverse category of reactors, in which molten salt plays one or more significant functions in the reactor core (e.g. fuel chemical form, liquid fuel carrier, coolant, moderator). The fundamentals of MSR technology were developed by the US Government through the ANP programme and later the MSBR programme in the 1940–1970s. Several other Member States developed complementary programmes after publication of the work at the second United Nations 'Atoms for Peace' conference in 1958. The MSRE at ORNL demonstrated the overall soundness of the MSR concept. The engineering challenges associated with MSR technical viability were largely solved by the mid-1970s. However, MSRs did not receive the level of funding necessary to proceed into large scale development and deployment.

A few major technical and non-technical challenges in the development of MSRs are still to be solved prior to the scale-up of the technology and its deployment. The successful operation of the MSRE in the 1960s was adopted as a key reference by the current developers to facilitate the development of the technology. It also supports efforts to establish MSRs within the modern energy supply and regulatory frameworks to reduce the risk of failure in commercial deployment. There are numerous intrinsic advantages of liquid fuels and near-ambient pressure coolants. The primary technological challenges encountered are in the qualification of materials because of the extreme radiation environment and the high temperatures. The following major technical issues need to be addressed: (i) the need for modelling software and databases of salt thermophysical properties; (ii) the mitigation of fast degradation of the structural materials induced by the harsh corrosiveness in molten salts and the extreme radiation environment of the core; and (iii) the design and manufacture of components (e.g. heat exchangers, pumps, valves) capable of operating in such extreme environments. Collaboration among universities, national laboratories and commercial developers is needed to attract venture capital and expedite the deployment of this technology.

There are several approaches to developing a taxonomy for MSRs. One such approach is to classify according to the location of the primary heat exchange between the fuel and the coolant. In this publication, the MSRs are divided into three major classes according to the type of materials present in the core. Accordingly, the active core of the first class includes graphite and salt, the second class relies solely on salt and the third class includes structural material that separates the fuel from the coolant or moderator.

Member States and international organizations have been conducting R&D activities on MSRs under various project schemes. In Canada, the Canadian Nuclear Safety Commission adopts a risk-informed regulatory framework for licensing advanced reactors. Two small modular MSRs, the IMSR by Terrestrial Energy and the SSR-W300 by Moltex Energy, are among a range of SMR designs under the pre-licensing vendor design review. In parallel, experienced Canadian electric utilities are working with SMR developers to advance design engineering towards potential demonstration projects. Several MSR research projects were launched in 2016 at CNL's Chalk River Laboratories site. MSR R&D activities are also performed in several Canadian universities, including the University of New Brunswick.

In 2011, SINAP of the Chinese Academy of Sciences restarted a research programme, known as the Thorium Molten Salt Reactor (TMSR) programme, to achieve the safe and economic use of Generation IV MSRs and the highly efficient use of the thorium resource. For more than ten years, key TMSR technologies have been developed, such as molten salt production and purification. In 2019, SINAP completed the construction of the TMSR-0 simulator (mock-up) reactor for testing, verifying and licensing technologies and the training of reactor operators. For future application, a 168 MW(e) SMR named smTMSR-400 is being designed as a thorium convertor and in situ burner driven by low enriched uranium.

In the Czech Republic, MSR technology has been developed since 1999 with a focus on experimental research and verification in the areas of MSR physics and fluoride salt neutronics, the development of the Th–U fuel cycle, materials and the development of components. The Research Centre Řež is the main research organization responsible for the domestic development of MSR technology.

In Denmark, work on MSR technology is conducted at DTU, Copenhagen Atomics and Seaborg Technologies. Contributions to the MSR community include R&D activities such as coupling schemes of neutronic–thermohydraulic codes, the design of molten salt loops and freeze valves and the development of alternative MSR designs.

The JRC of the European Commission has a wide spectrum of capabilities, including the synthesis of actinide fluorides and chlorides; purification techniques and purity characterization methods for molten salts; development of density measurement of liquid halide salts; post-irradiation examination of nuclear fuels, including MSR irradiated fuels; and the development of structural components for irradiation purposes. The JRC also determines melting points of MSR fuel and coolant salts; solubility limits for noble gases (e.g. helium, krypton and xenon) in molten salt fuels; and mixing enthalpy of multicomponent halide systems.

Research in the European Union is carried out alongside the European R&D Framework Programmes and projects evolved from research on historical ORNL concepts to the MSFR proposed by the CNRS. The research aims to understand, model and design all physics and chemical processes in the fuel circuit of the reactor and in the fuel salt processing stages, and to assess the sustainability and safety of the reactor design. The research is based on experimental work and on advanced numerical modelling combined with experimental validation.

In France, the CNRS has been involved in MSRs since 1997. The MSFR concept was proposed as a result of extensive parametric studies. Various configurations of reactor core, reprocessing performances and salt compositions were first investigated to adapt the reactor such that it can be deployed worldwide as part of a reactor fleet based on thorium. The 'reference MSFR', which was a large power reactor operated in the thorium fuel cycle, was selected for further studies by the Generation IV International Forum in 2008. A safety analysis methodology for such a reactor with circulating liquid fuel has been developed in collaboration with IRSN, Framatome and Politecnico di Torino in the framework of the SAMOFAR project. The SAMOFAR project later became the SAMOSAFER project that included also CEA and Politecnico di Milano. Research activities are also carried out in collaboration with an increasing number of French partners, which reflects a growing national interest in MSR technology in France.

The R&D activities in Italy started in the context of the first European projects on the development of MSR technology (e.g. ALISIA, EVOL) and have continued since 2010. Politecnico di Milano and Politecnico di Torino are the two main institutions involved. The main focus is on neutronics, thermohydraulics, controls, and modelling and simulation. Politecnico di Milano built the DYNASTY testing facility for investigation and model validation of natural circulation.

Using the results from ORNL's MSR activities in the 1970s, Japan has since the 1980s been developing the fluoride fuelled graphite moderated FUJI MSR. The following MSR designs have been developed: FUJI-U3, FUJI-Pu, the super-FUJI of 1000 MW(e), and a mini-FUJI as a pilot plant. In 2019, the Japanese Government started to support the development of MSR technology and three MSR venture companies were selected; two are promoting an MSR with fluoride salt that is moderated by graphite, and one is promoting a fast MSR with chloride salt.

In the Netherlands, NRG in Petten and TU Delft have conducted research on MSRs. At NRG, fuel salt samples have been irradiated by the SALIENT experiments in the High Flux Reactor, and the first samples were unloaded and prepared for post-irradiation analysis at laboratories in Petten and at the JRC in Karlsruhe. The SPECTRA thermohydraulic code originally designed for light water reactors was extended to be used for MSRs. Experimental and calculation studies on flotation and the extraction by gas bubbling of insoluble particles like noble metals are carried out jointly by NRG and TU Delft. Research at TU Delft focuses on the development of an advanced multiphysics numerical code system for transient analysis and fluid dynamics codes. More fundamental structural studies of molten salts using extended X ray absorption fine structure spectroscopy provide an insight into the structural characteristics of the molten fuel salt.

For more than 20 years, the Russian Federation has been conducting R&D activities on MSRs that focus on fast spectrum concepts with or without thorium support. These concepts have been recognized by the Generation IV International Forum as an alternative to solid fuelled fast neutron reactors, as they have advantages such as strong negative reactivity feedback, smaller fissile inventory and simplified fuel cycle. The MOSART concept, for example, can operate using high nickel alloy for container material, with different fuel loadings and make-up based on TRUs as a special actinide transmuter, self-sustainable system or even as a breeder. An experimental centre was established to support future R&D on MOSART. A 10 MW(th) homogeneous core test reactor using LiF–BeF2–AnFn fuel salt is under development.

In Switzerland, MSR R&D activities have been carried out at the Paul Scherrer Institute in cooperation with the universities ETH Zürich and EPFL in Lausanne. In the 1970s and 1980s, a dedicated MSR research project was also conducted at EIR. It focused on the possibility of using $AlCl_3$ salt as a direct boiling coolant for a homogeneous chloride fast reactor and for a solid fuel fast reactor. Later, several options for a fast heterogeneous MSR cooled by salt were considered, including the option to use blanket salt as the primary coolant. The latest studies were dedicated to homogeneous chloride fast MSRs with natural chorine isotopic composition. Nowadays, MSR research at the Paul Scherrer Institute focuses on safety and sustainability. The fuel cycle performance of the fluoride salt cooled reactor, the graphite moderated MSR, the homogeneous fluoride and chloride fast MSR and the non-graphite moderated MSR has been assessed.

The US Government has sponsored a range of activities supporting MSR technology. These activities include providing direct support to national laboratories, developers of advanced reactors and universities for work on MSR technologies. Research awards and grants are provided to support the development of technologies for MSR and innovative advanced reactors. Other activities include the development of industry consensus standards and advanced modelling and simulation tools, and the measurement of fundamental data on thermophysical and thermochemical properties of fuel salt. The USA also participates in international engagement on MSRs through the Generation IV International Forum and IAEA initiatives.

Although a wide range of MSR technologies are under development, a number of issues and obstacles need to be overcome prior to commercialization. The challenges to MSR deployment include supply chain, fuel supply, regulatory framework, disposal of fuel salt waste, safeguards and security, and operation and maintenance. Globally, many companies, universities and regulatory bodies are working on solutions, as well as within collaborative projects. International collaborations need to be enhanced to facilitate the deployment of MSRs.

Appendix I

HISTORY OF MSR TECHNOLOGY IN POLAND AND SWITZERLAND

The history of MSR technology development in Poland and Switzerland is associated with one person: Mieczyslaw Taube. Taube worked in the Department of Radiochemistry of the Institute of Nuclear Research in Warszawa in the 1960s. At that time, he proposed the chloride fast reactor concept [35, 36]. It was cooled by boiling aluminium trichloride, which was in direct contact with the fuel salt. By the end of the 1960s, Taube had joined EIR, which was the predecessor of the Paul Scherrer Institute, in Switzerland. In one of the first EIR publications, aluminium trichloride vapour was analysed as a coolant of a fast reactor with solid fuel [37]. However, the boiling salt and the directly cooled concepts were abandoned because of many technical difficulties. In the early 1970s, a series of heterogeneous chloride fast reactor concepts were proposed by Taube [38, 152, 423], in which the chloride fissile fuel salt was typically in pins being cooled by the fertile blanket salt. These concepts aimed at breeding in the U–Pu fuel cycle and at transmutation of legacy waste. In the case of transmuting waste, beryllium chloride was used as a diluted moderator to increase the burning rates.

When the EIR-215 report, Ref. [38], was published in 1972, Taube sent with it a letter to several other States and institutes that were interested in the MSR technology, including the UK Atomic Energy Authority, CEA in France, Euratom in the Netherlands, and the University of Tennessee and ORNL, both in the USA. He received many answers, such as the one from M.W. Rosenthal (ORNL) dated December 1972, which included the ORNL-4812 report; however, the MSR project at ORNL was stopped soon after. At that time, three EIR laboratories were partly involved in MSR neutronics, chemistry and materials research.

The heterogeneous core layout was replaced by homogeneous concepts in the mid-1970s to avoid technical difficulties with the separating structural material. In 1974, the chloride fast reactor concept was proposed, which combined breeding in Th–U and U–Pu fuel cycles [39]. An EIR internal technical memorandum from 1975 also exists, in which the molten chloride salts were considered as an energy storing medium [424]. In addition, a dedicated waste burner concept was proposed at that time [425]. It had a central subcritical burning zone, which was separated by a moderator from the outer driving zone, which provided neutrons to the central, or actually driven zone. This concept is presented also in a report from 1975 [134], a summary report from 1978 [85] and in the Swiss contribution to the International Nuclear Fuel Cycle Evaluation — Working Group 8: Advanced Fuel Cycle and Reactor Concepts from 1978 [85]. In the last three named publications, E. Ottewitte is listed as the EIR co-author. He published a concept operating in the Th–U cycle in 1978 [426] and defended his thesis at the University of California, Los Angeles, on this topic in 1982 [38]. This thesis is one of the first documents where the abbreviation MCFR (which stands for molten chloride fast reactor) is used for fast chloride based reactors. Taube used the term molten chloride fast breeder reactor, or MCFBR [423], and J. Smith, in Ref. [45], used molten salt fast reactor, or MSFR, in 1974. The abbreviation MSFR is, nonetheless, nowadays used rather for a fluoride fast reactor. The EIR report [85] and the thesis of Ottewitte [41] include Fig. 95, which is quite iconic for a fast MSR.

At the end of the 1970s, the SOFT reactor concept was proposed by Taube. It was a homogeneous U–Pu breeder [40]. The spherical core was reflected with another chloride salt. The MSR research at EIR was almost exclusively relying on simulation at that time. One of the few small experiments was the chloride salt irradiation near the core of the EIR's SAPHIR reactor (a swimming pool reactor designed by ORNL as an exhibition reactor for the first 'Atoms for Peace' conference in 1955) and chemical speciation of the produced sulphur [427]. In the 1980s, MSR research faded out at EIR. It restarted at the Paul Scherrer Institute in 2013 with a scoping study [374] and in 2015 the breed-and-burn fuel cycle option became a major research direction [135, 372].

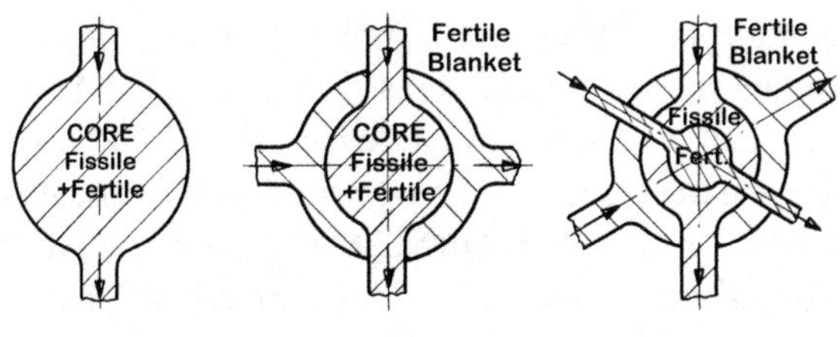

FIG. 95. Basic layouts for homogeneous fast reactors for single-fluid (left), two-fluid (middle) and three-fluid (right) designs and placement of fissile and fertile materials (reproduced from Ref. [78] with permission).

Appendix II

HISTORY OF MSR TECHNOLOGY IN CHINA

In February 1970, Premier Zhou Enlai stated that Shanghai should develop a nuclear power plant to solve the disparity between Shanghai's economic development and its energy shortage. An MSR for high temperature was initially selected for the nuclear power plant project, which would generate 25 000 kW(th) [49].

In the 1970s, a critical experiment device was established for research on the physics characteristics of MSRs at SINAP (then Shanghai Institute of Nuclear Research), and a series of zero power experiments and related results were obtained. The device was later transformed into a uranium–water lattice experiment device to allow a change in the research orientation. A major advantage of this device is the flexibility in the arrangements of the fuel and moderator elements in the core, allowing different moderating ratios. The purpose of the device was to verify experimentally the results from theoretical calculations; determine important characteristics, such as the static and dynamic features, the unit inventory gram reactivity and the temperature effect of the molten salt used; and obtain the relevant design data such as the control rod calibration and its temperature effect and the fertile fuel conversion ratio [50–53].

The device is an experimental, zero power reactor moderated by graphite, fuelled with BeF_2 powder and UF_4–ThF_4 powder mixed in proportion. The core, located in the centre, is a graphite cylinder with 497 fuel or moderator channels arranged in the shape of equilateral triangles, as shown in Fig. 96 [428, 429]. As the name implies, the fuel channels are used for the insertion of fuel elements and the moderator channels are for the insertion of graphite elements for various cell structures with different moderating ratios. The diameter of the channels is approximately 30 mm. The diameter and the height of the cylindrical core are 120 cm and 129.6 cm, respectively. The graphite reflector surrounding the core includes an upper reflector (that can be lifted flexibly for cell replacement in the core), a lower reflector and a side reflector that is octagonal in shape. The thickness of the upper and lower reflectors and the minimum thickness of the side reflector are 63 cm, 61.5 cm and 63 cm, respectively. The channels in the side reflector are used for various detectors, control rods and neutron source, with a channel depth that depends on its function. The air cavity between the reflector and the concrete shield can play the role of thermal insulation to some extent. The cylindrical concrete shield is on the periphery of the reactor and its thickness is 70 cm. In addition, the heavy shielding concrete plate above the upper reflector can be removed to satisfy the experimental requirements. A control rod was made of cadmium and manufactured by simply wrapping up an aluminium

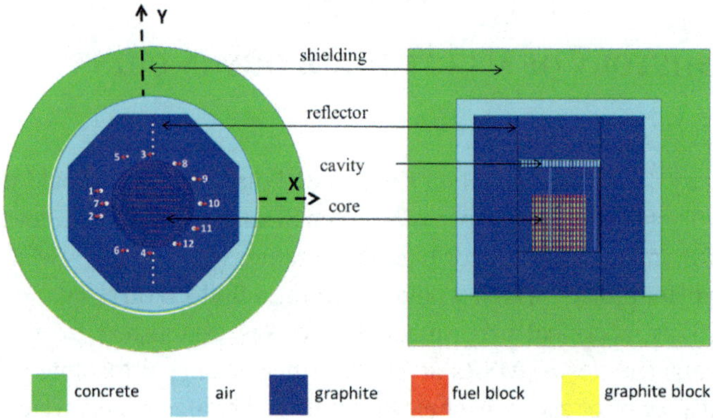

FIG. 96. Schematic diagrams of the critical experiment device; left and right: horizontal and vertical cross-sections of the device, respectively. Control rod channels are marked 1 to 6; experimental detector and test sample location are marked 7 to 12 (courtesy of Y. Zou, Shanghai Institute of Applied Physics).

bar with cadmium [429]. The diameter of the cadmium rods changed with the inner aluminium bar diameter. The length of the part of the control rod wrapped with cadmium sheet was 120 cm, which was the same as the height of the fuel elements. For the sake of safety, there were six cadmium rod channels marked Nos 1–6 in the side reflector, as shown in Fig. 96. Channels 1 and 2 were for shutdown rod insertion, channels 3 and 4 were spare shutdown rod channels and channels 5 and 6 were for shim rod insertion and regulating rod insertion, respectively. A fuel element is made up of fuel blocks, or fuel blocks and graphite blocks alternately loaded in a graphite annular tube, as shown in Fig. 96, right, with fuel block number and graphite block number ratios for various active region heights and fuel volumetric shares. Each fuel block contains 1.6 g of UF_4 and 8.1 g of BeF_2 and the average fuel powder density is 0.979 g/cm^3. The height and the diameter of a graphite block is about 39.1 mm and 20 mm, respectively. Furthermore, several graphite rods are inserted in the outer channels of the core with their heights almost equal to the height of the core, and they are used for adjusting the active region equivalent diameter and the graphite volume ratio [429]. The reactor could achieve a maximum power of 200 W(th), but all experiments mentioned were performed without coolant at zero power state and room temperature.

The results of the experiments performed on this device are summarized as follows:

— The critical mass calculation is consistent with the experimental value within the error range of 5%, and the physics theoretical calculation model and method are accurate.
— There is a large deviation between the measurement results of neutron flux and the calculation results.
— Molten salt at high temperature has a significant effect on the corrosion of structural materials in the reactor, which impacts the safety of the reactor. The material and fuel compatibility issues have been solved, and the adoption of Hastelloy alloy will effectively reduce corrosion, improve safety and extend the life of the reactor.

In July 1973, after two years of research and tests, many problems remained to be solved, such as dealing with the high radioactivity of molten salt fuel and finding a suitable intermediate heat carrier. Steel alloy that is resistant at high temperature and corrosion resistant did not satisfy the requirements. A chemical reprocessing process had not been found. The test results demonstrated that further scientific research of the MSR technology was required. The amount of time that would be needed for this research was not consistent with the original intention to build a nuclear power plant of medium size as soon as possible. The MSR project was therefore replaced by a project for a pressurized water reactor test and design.

Appendix III

HISTORY OF MSR TECHNOLOGY IN FRANCE

This appendix provides detailed technical information about the French MSR designs developed between 1998 and 2008.

III.1. WORK ON GRAPHITE MODERATED MSRs: THE AMSTER CONCEPT, 1998–2005

The AMSTER concept was first proposed in 1999 [430] and was based on the MSBR [108] developed by ORNL. The MSBR was a 1 GW(e) reactor design based on the 8 MW(th) 'reference' MSRE that was successfully operated between 1965 and 1969. Enriched uranium was used in this first version of AMSTER, but subsequent studies used thorium fuel, like the MSBR, because it can be near the breeder in the neutron spectrum when using a graphite moderator [109, 431–433]. All technical characteristics of the MSBR were applied in AMSTER: fluoride salt, graphite moderator, Hastelloy N and on-line reprocessing of the salt (removal of the fission products). However, the MSBR required intense reprocessing activities (whole core reprocessed in ten days to extract ^{233}Pa as quickly as possible and achieve a 1.06 fissile conversion ratio). Since the aim of AMSTER was to burn plutonium and minor actinides and not to maximize breeding rate, the time for reprocessing the whole core was increased to 300 days for the soluble fission products. Two reactors were optimized in Ref. [434]: the AMSTER TRU burner, which is an actinide burner loaded with TRU elements coming from the used fuel from a pressurized water reactor; and the AMSTER iso-breeder, which is a simplified version of the MSBR, with a conversion ratio equal to one. A brief description is provided below, and Table 16 presents information on their fuel.

- *AMSTER TRU burner.* The initial loading of 82.7 t/GW(e) of thorium and 1.16 t/GW(e) of TRU elements coming from the used fuel from a pressurized water reactor represents 1.4% of the heavy nuclei. The soluble fission products are removed in 300 days and the TRU elements are fully recycled within the reactor. It was found that 22 kg/TWh(e) of TRU elements and 76 kg/TWh(e) of ^{232}Th are consumed at equilibrium (without a need for the addition of ^{235}U). The ^{233}U and fission products rapidly tend to an equilibrium.
- *AMSTER iso-breeder.* The reactor consumes only 98 kg/TWh(e) of ^{232}Th. To achieve this performance, a fertile zone around the periphery of the core was

designed, representing half of the salt volume. This zone is undermoderated by increasing the diameter of the salt and thus reducing the graphite available to thermalize neutrons (see Fig. 97).

TABLE 16. FUEL OF THE TWO AMSTER REACTOR TYPES WITH 300 EFPD REPROCESSING TIME[a]

	AMSTER TRU burner			AMSTER iso-breeder		
	Initial loading (kg/ GW(e))	Equilibrium loading (kg/GW(e))	Equilibrium feeding (kg/ TWh(e))	Initial loading (kg/ GW(e))	Equilibrium loading (kg/GW(e))	Equilibrium feeding (kg/ TWh(e))
Thorium	82 700	80 000	76	74 000	80 000	98
TRU	1 160	1 940	22	n.a.[b]	170	n.a.
U-233	n.a.	1 240	n.a.	n.a.	1 460	n.a.
U-235	n.a.	n.a.	n.a.	2 000	n.a.	n.a.
U-238	n.a.	n.a.	n.a.	8 050	n.a.	n.a.
Fission products	n.a.	750	n.a.	n.a.	750	n.a.

[a] EFPD — equivalent full power day.

[b] n.a.: not applicable.

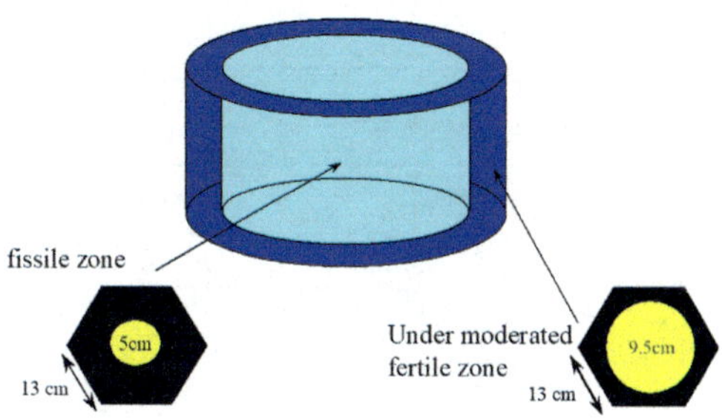

FIG. 97. Representation of the two moderation zones of the AMSTER iso-breeder (black: graphite; yellow: salt) (courtesy of D. Lecarpentier, Électricité de France).

This work has been extended to the main R&D challenges of MSR feasibility, including materials and reprocessing simulation and safety at the French [435–441] and European levels. Reference [442] states:

> "The state-of-the-art review of MSR technology, performed in the MOST... project supported by Euratom in 2002–2004 (5th Framework Program), confirmed the potential of MSR as breeders or burners.
>
> ....
>
> "The main achievement of the project consists of the 9 review reports issued by the end of 2003 and covering all the main aspects of MSR (reactor physics, design and operation, safety, systems and components, structural materials, fuel salt chemistry, fuel processing, economics, proliferation resistance)."

Regarding MSR safety, codes for the simulation of transients have been developed by EDF [443] and by the partners of the MOST project [322] and applied to transients measured in the MSRE. The temperature feedback coefficients of the MSRE [444] and the MSBR [108] were re-evaluated [445]. It was shown that the Apollo II [446] neutron transport code used by EDF was able to compute precisely the feedback coefficients measured by ORNL on the MSRE. It was concluded that the feedback coefficient computed by ORNL was underestimated and that the MSBR feedback coefficient was slightly positive.

The MOST project concluded that an MSR moderated with graphite could hardly achieve a compromise on five major constraints: safety (feedback coefficients), breeding, ^{233}U initial inventory, moderator (graphite) longevity and feasibility of reprocessing. The work by the CNRS on optimization documented

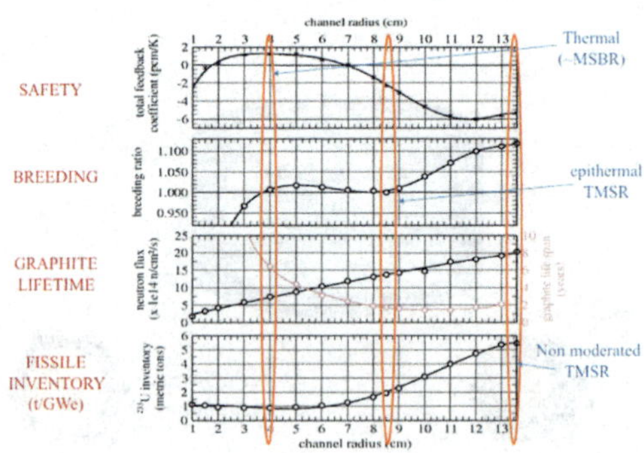

FIG. 98. Main performance parameters of the TMSR (reproduced from Ref. [62] with permission).

in Ref. [447] showed that no configuration with graphite could satisfy the requirements of breeding, limited graphite swelling and safety (see Fig. 98).

III.2. WORK ON NON-MODERATED MSRs, 2004–2008

The CNRS proposed the non-moderated thorium molten salt reactor (TMSR) [62–64] in 2004, which uses a fluoride salt and is an iso-breeder (see Fig. 99).

In the same time frame, EDF defined a non-moderated iso-breeder MSR, the REBUS-3700 [65, 66], with a classical depleted U–Pu cycle. This reactor uses a chloride salt ($38UCl_3$–$7TRUCl_3$–55NaCl mol%) as a preliminary study showed that $^{238}U/^{239}Pu$ breeding is difficult to reach with a fluoride salt, since the neutron moderation by fluoride is too high. This design, considered preliminary at that time, has very good safety coefficients (0.006%Δk/k/°C) and an initial fissile (plutonium) mass comparable to other fast breeder reactors (11.9 t TRU/GW(e)). The initial fissile loading is 17.8 t of TRU coming from the used fuel from a pressurized water reactor, representing 15% of the heavy nuclei. The fuel reprocessing is then adjusted to be as slow as possible while keeping the reactor operating as a breeder (36 L/day representing 0.065% of the primary salt or a reprocessing period of 1538 days). The fuel burnup is approximately 4.5% at equilibrium. The REBUS-3700 reactor achieves the main design goals: a breeding gain equal to zero and a strong negative temperature coefficient of reactivity. Accordingly, the REBUS-3700 is an MSR with a fast neutron spectrum and the fuel is dissolved in a chloride salt. Table 17 gives its major technical parameters.

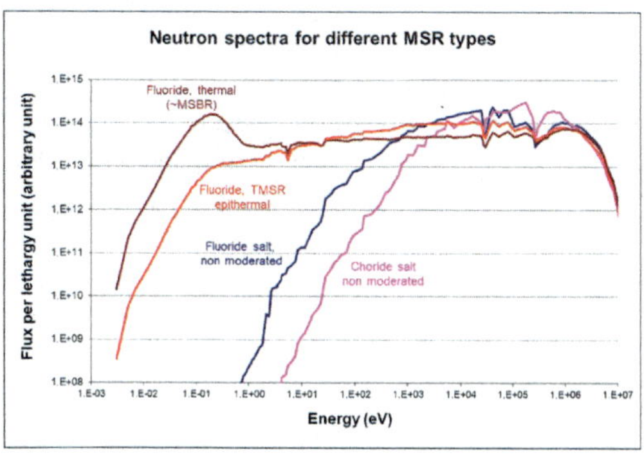

FIG. 99. Neutron spectrum in different MSRs (courtesy of D. Lecarpentier, Électricité de France).

TABLE 17. MAJOR TECHNICAL PARAMETERS OF THE REBUS-3700

Parameter	Value
Technology developer, country of origin	EDF (France)
Thermal/electrical capacity, MW(th)/MW(e)	3700/1500
Moderator	None
Conversion ratio	1.03
Fuel processing	Full
Kilowatt per total fissile load (kW(th)/kg of fissile material)	483
Primary circulation	Forced (13.6 m^3/s)
Salt speed	1.2 m/s
Core inlet/exit temperatures (°C)	650/730
Fuel type	Chloride (38%UCl_3–7%$TRUCl_3$–55%NaCl)
Fuel enrichment (%)	None. It is started with 15.6% TRU from a pressurized water reactor and then fuelled with depleted uranium. Equivalent fissile enrichment ~10%
Fuel loading/top-up scheme	Continuous fuel make-up and fuel salt cleanup in batches with removal time of 1538 equivalent full power days for soluble fission products
Main reactivity control mechanism	Large negative feedback temperature and void coefficients, and removal of fuel from the core by thermal dilation
Core diameter/height (m)	3.8/3.25
Core/primary salt volume (m^3)	36.8/56.3
Salt density at 690°C (g/cm^3)	3.6
Design status	Concept

At that time, the chloride salt seemed to be less well known to researchers at EDF than the fluoride salt for its use as reactor fuel. However, around 2015, TerraPower, Elysium Industries and Moltex Energy proposed non-moderated MSRs with chloride salts based on the experience of the integral fast reactor of reprocessing metal fuel, that is, compatibility of chloride with a standard alloy. Hence, the titanium vessel of the REBUS could be replaced with a less expensive material like steel.

Finally, two very similar non-moderated TRU burners on inert support (without ^{238}U or ^{232}Th) were defined in Europe. The MOSART 2400 MW(th) concept defined by the Kurchatov Institute in the Russian Federation [448] (15LiF–27BeF$_2$–58NaF mol% dissolving trifluoride of plutonium and minor actinides produced by a pressurized water reactor and lithium enriched to 99.99% ^{7}Li) and the SPHINX 1340 MW(th) project [449] defined by the Nuclear Research Institute Řež (Czech Republic) (fluoride salt 35^{7}LiF–38BeF$_2$–27NaF mol% dissolving plutonium and minor actinides produced by a pressurized water reactor).

The ALISIA project concluded in 2008 that non-moderated MSRs had significant potential because of their stability owing to negative feedback coefficients of reactivity and the absence of graphite waste [450]. The two most interesting concepts are the non-moderated concepts, one breeder (non-moderated TMSR) and one burner (MOSART). Similar behaviour in accidental transients is expected because both have a 0.005%Δk/k/°C temperature coefficient of reactivity of the salt and the same recirculation of delayed neutron precursors.

Appendix IV

HISTORY OF MSR TECHNOLOGY IN THE RUSSIAN FEDERATION

The Russian Federation carried out R&D activities on container materials, heat and mass transport properties of molten salts and the irradiation of fuel salt. Related test facilities and loops were established from 1976 to 1986. Reference [451] offers information about these activities, and in general about molten salt fuels for nuclear waste transmutation in accelerator driven systems. Some of the activities are described in more detail in the following paragraphs.

General requirements of materials containing fuel and coolant salts were formulated as follows:

— The materials need to be easily produced, that is, easily subjected to treatment by pressure, and suitable for cutting, rolling and welding. The fabrication of complex components of the reactor structure using these materials also has to be possible.

— Pressure on walls in a typical MSR fuel circuit is less than 2 MPa. Thus, guided by this limiting value and taking a tenfold safety factor, the candidate material needs to have a limiting value for lasting strength of more than 20 MPa for 30 years of operation at a temperature of 750°C.

— The peak neutron flux and neutron fluence in the reactor vessel of an MSR are estimated as 10^{20} and 5×10^{21} n/cm^2 for fast (>0.5 MeV) and thermal neutrons, respectively.

— To keep the effects of corrosion at an acceptable level (i.e. maintaining the wall thickness of the heat exchangers, molten salt content and mass transfer), the depth of corrosion should not exceed 10 mm/a and the metal should not be subjected to local corrosion (pitting or intergranular cracking).

A list of in-reactor tests regarding the radiation stability of different fluoride molten salt fuels carried out at the Kurchatov Institute is given in Table 18. This table presents measured values of radiation chemical yield G(F_2), which is the number of F_2 molecules evolving per 100 eV of absorbed energy.

Several molten salt test loops operating at high temperature with forced and natural circulation were created and successfully tested in the Russian MSR programme. In the laboratory and in reactor molten salt test loops operated for 500–3500 h at temperatures of 500–800°C, the working capacity of loop components and systems was shown. The modes of startup and shutdown of the loops were completed, and also approaches for removing impurities and measuring redox potential were improved. In addition, the interaction with

TABLE 18. IN-REACTOR TESTS OF THE RADIATION STABILITY OF FLUORIDE MOLTEN SALT FUELS

Fuel salt (mol%)	Liquid phase		Solid phase	
	T (°C)	$G(F_2)$[a] (10^{-5} mol/100 eV)	T (°C)	$G(F_2)$ (10^{-2} mol/100 eV)
$66LiF–33BeF_2–1UF_4$	615	7	50	1
$69LiF–31BeF_2$	680	2	50	0.2
$71.7LiF–16BeF_2–12ThF_4–0.3UF_4$	740	3	25	0.6
$65.6^7LiF–34.1BeF_2–0.3UF_4$	740	0.2	25	n.a.[b]
$73.6iF–25.9ThF_4–0.5UF_4$	1200	2	n.a.	2.5
$74NaF–25.5ThF_4–0.5UF_4$	1150	0.15	50	2

[a] *G*: evolution.

[b] n.a.: not applicable.

construction materials, radiation resistance and heat and mass transfer in fluoride molten salts were studied. Table 19 presents the molten salt test loops operated by the Kurchatov Institute in the Russian Federation.

The loop SOLARIS for forced convection is operated at the Kurchatov Institute to evaluate the compatibility of stainless steel with Li, Na, K/F eutectics and the operability of main components and systems, including centrifugal pumps, salt–air heat exchangers and salt valves. All piping joints in contact with salt are made by argon arc welding.

The reactor loop KURS-2 for natural convection is operated by the Kurchatov Institute to evaluate the behaviour of stainless steel construction material with fuel Li,Be,U/F salt at temperatures, neutron fluxes and flow rates similar to those in a typical MSR primary circuit. Investigations were carried out at the experimental reactor VVR-SM with a neutron flux up to 0.76×10^{14} n/(cm^2 s) and an operation time of 3500 h. The test section was cooled by reactor water. Heat generated by nuclear reactions directly in the fuel salt and the thermosyphon walls is transferred from the core into the cooling zone by natural convection over the central insertion through flow distributors. In the downflow section, the heat is removed by the reactor water. In the KURS-2 loop, the possibility is provided for a preliminary fluoride passivation of gas supply lines, a mass spectrometric analysis of gas samples and an inline control of pressure in gas volumes.

TABLE 19. LIST OF THE MOLTEN SALT TEST LOOPS OPERATED BY THE KURCHATOV INSTITUTE

Loop type	Melt (mol%)	Volume (l)	T_{max} (°C)	ΔT (°C)	Operation (h)	Alloy
Forced convection	46.5LiF–11.5NaF–42KF	90	620	20	3500	12H18N10T
Natural convection	$92NaBF_4$–8NaF	6	630	100	1000	HN80MT
	72LiF–$16BeF_2$–$12ThF_4$ + UF_4	6	750	70	1000	HN80MTY
	LiF–NaF–BeF_2 + PuF_3	8	700	100	1600	Based on Ni
	LiF–NaF–BeF_2 + Cr_3Te_4	12	650	10	500	Based on Ni
Natural convection in reactor	66LiF–$34BeF_2$ + UF_4	19	630	100	500	12H18N10T
	66LiF–$34BeF_2$ + UF_4	19	750	250	750	12H18N10T

Appendix V

DESCRIPTION OF MSR CONCEPTS

V.1. INTRODUCTION

This appendix presents brief descriptions of MSR concepts that are currently being developed. Concepts that were proposed in the past are presented in Section 2 and Appendices I–IV.

Almost all current MSR developers were contacted and each was requested to provide a summary of its MSR concept. Unfortunately, the summaries that were not received in time for publication could not be included.

A detailed description of MSR concepts can be found in the IAEA's on-line database, Advanced Reactors Information System (ARIS) (if the developer of the concept has provided a description for the database) [405]. This database is updated periodically. The recent IAEA publication Advances in Small Modular Reactor Technology Developments [452] is a supplement to the database, and includes a section describing 13 MSRs that are SMRs.

V.2. DESCRIPTION OF MSR CONCEPTS

This subsection presents the summaries of MSR concepts. The concepts are organized according to the six major MSR families[24] and then sorted alphabetically within each family. The six major MSR families are listed below:

- — I.1. Fluoride salt cooled reactors;
- — I.2. Graphite moderated MSRs;
- — II.3. Homogeneous fluoride fast MSRs;
- — II.4. Homogeneous chloride fast MSRs;
- — III.5. Non-graphite moderated MSRs;
- — III.6. Heterogeneous chloride fast MSRs.

[24] See Section 4 for the taxonomy used in this publication.

V.2.1. Mark 1 Pebble Bed Fluoride Salt Cooled High Temperature Reactor (Mk1 PB-FHR)

Class: Graphite based MSRs (I)
Family: Fluoride salt cooled reactors (I.1)
Type: Salt cooled reactor with pebble bed fuel

TABLE 20. MAIN CHARACTERISTICS OF Mk1 PB-FHR

Characteristic	Value
Fuel state	Solid
Reactor spectrum	Thermal
Salt type	Fluorides
Coolant	Salt
Actinide	U–Pu

TABLE 21. MAJOR TECHNICAL PARAMETERS OF Mk1 PB-FHR

Parameter	Value
Technology developer, country of origin	University of California, Berkeley, USA
Thermal/electrical capacity, MW(th)/MW(e)	236/100
Moderator	Graphite and salt
Conversion ratio	Less than 1
Fuel processing	Once through
Kilowatt per total fissile load (kW(th)/kg of fissile material)	1869
Primary circulation	Forced circulation
Primary pressure (MPa)	0.3

TABLE 21. MAJOR TECHNICAL PARAMETERS OF Mk1 PB-FHR (cont.)

Parameter	Value
Core inlet/exit temperatures (°C)	600/700
Fuel type	TRISO with enriched uranium kernel
Fuel enrichment (%)	19.9
Fuel loading/top-up scheme	On-line refuelling
Main reactivity control mechanism	Control rods
Approach to engineered safety systems	Passive
Design life (years)	60
Plant footprint (m^2)	60
Reactor vessel's height/diameter (m)	12/3.5
Seismic design	Base isolation
Distinguishing features	Pebbles are inserted from the bottom of the core and extracted at the top
Design status	Pre-conceptual design

The Mk1 PB-FHR is a small, modular graphite moderated reactor. Fluoride salt cooled high temperature reactors are distinguished from other reactor technologies because they use high temperature, coated particle fuels, cooled by the fluoride salt FLiBe. Fluoride salt coolants have uniquely high volumetric heat capacity, low chemical reactivity with air and water, very low volatility at high temperature, effective natural circulation heat transfer and high retention of most fission products. The Mk1 PB-FHR is a non-water-cooled SMR that does not use an intermediate coolant loop. It instead directly heats the power conversion fluid. It eliminates the conventional reactor guard vessel used in sodium fast reactors and instead uses a refractory reactor cavity liner system. The design of all components for the Mk1 PB-FHR are modular to facilitate construction and are transportable by rail. The reference configuration for the Mk1 PB-FHR site is based on the 12 units that are capable of producing 1200 MW(e) of baseload electricity and ramping up to a peak power output of 2900 MW(e) through the addition of burning natural gas or other combustible fluid as a topping cycle in the gas turbine. The core incorporates 3 cm diameter spherical pebble fuel elements with coated uranium particles in an annular fuel zone and a low density graphite core. One pebble of the reactor contains 1.5 g of uranium enriched in ^{235}U to 19.9%, which is encapsulated inside 4730 coated particles. The very low circulating power for the coolant in salt cooled reactors, compared with helium cooled reactors, makes it practical to use 3 cm pebbles. This small pebble design doubles the pebble surface area per unit volume and halves the thermal diffusion length, enabling a substantial increase in power density while maintaining relatively low peak fuel particle temperature. Low fuel temperature reduces the thermal transient caused by hypothetical anticipated transient without scram events.

The design uses a buoyant control rod system for normal reactivity control. It also provides a passive shutdown function because the buoyant rods will only insert if the reactor coolant temperature in the control rod channel exceeds the buoyant stability limit of 615°C. If electrical power to the motor of the drive mechanism for the control rod and shutdown blade cable drums is interrupted, the control rods will insert and shut the reactor down. The insertion of the shutdown blades therefore provides a backup shutdown function. The safety function of emergency decay heat removal also works in passive mechanisms. The design employs a passive check valve to activate heat transport driven by natural circulation from the primary coolant to a set of three direct reactor auxiliary cooling system loops and ultimately to thermosyphon cooled heat exchangers upon loss of flow condition (see Fig. 100). In addition to the passive emergency decay heat removal provided by the direct reactor auxiliary cooling system, the power conversion system and the normal shutdown cooling system provide heat removal capability and defence in depth in assuring the adequate removal of core heat.

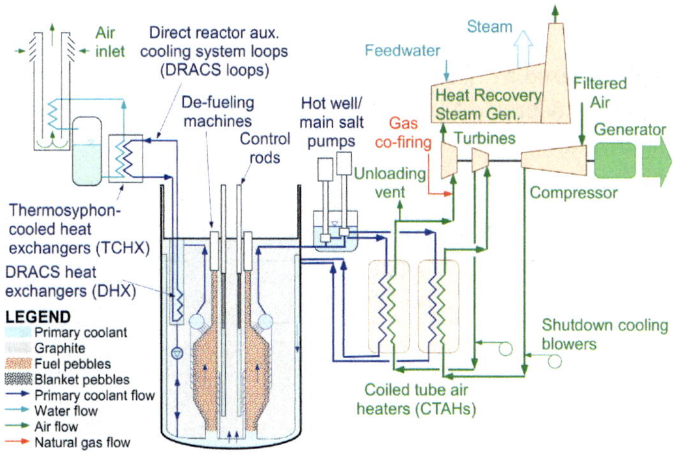

FIG. 100. Schematic of the flow of the Mk1 PB-FHR (reproduced from Ref. [452]).

V.2.2. FUJI

Class: Graphite based MSRs (I)
Family: Graphite moderated MSRs (I.2)
Type: Single-fluid Th–U breeder

TABLE 22. MAIN CHARACTERISTICS OF FUJI

Characteristic	Value
Fuel state	Liquid
Reactor spectrum	Thermal
Salt type	Fluorides
Coolant	Fuel salt
Actinide	Th–U or Pu–Th

TABLE 23. MAJOR TECHNICAL PARAMETERS OF FUJI

Parameter	Value
Technology developer, country of origin	International Thorium Molten-Salt Forum, Japan
Thermal/electrical capacity, MW(th)/MW(e)	450/200
Moderator	Graphite
Conversion ratio	Equal to 1
Fuel processing	Off-site reprocessing
Kilowatt per total fissile load (kW(th)/kg of fissile material)	400
Primary circulation	Forced circulation
Primary pressure (MPa)	0.5 (by pump head)

TABLE 23. MAJOR TECHNICAL PARAMETERS OF FUJI (cont.)

Parameter	Value
Core inlet/exit temperatures (°C)	565/704
Fuel type	Molten salt with Th and U
Fuel enrichment (%)	2 (0.24% U-233 + 12.0% Th) Pu or low enriched uranium can be used
Fuel loading/top-up scheme	Slight adjustment every month
Main reactivity control mechanism	Control rod, pump speed or fuel concentration
Approach to engineered safety systems	Passive
Design life (years)	30
Plant footprint (m^2)	<5000
Reactor vessel's height/diameter (m)	5.4/5.34 (inner)
Seismic design	0.6–2.0g, depending on reactor location
Distinguishing features	High safety, high economic performance, contribution to non-proliferation, fuel cycle flexibility
Design status	Three experimental MSRs were built [339]. Detailed design will be started within 2–3 years

The typical electric output of the FUJI MSR is 200 MW(e) with 44% thermal efficiency. Furthermore, FUJI's power is flexible, from 100 MW(e) to 1000 MW(e) [339]. In addition to the above high thermal efficiency, the low manufacturing cost owing to the simple core structure and high fuel efficiency owing to the high conversion ratio contribute to high economic performance. FUJI can consume plutonium as the fissile material and can contribute to reducing the proliferation risk caused by plutonium from light water reactor spent fuel. It can also be used to transmute long lived minor actinides to shorter ones. Reactor vessel, pumps and heat exchangers are enclosed in the high temperature containment. FUJI has very favourable safety characteristics that essentially exclude the possibility of severe accidents. It is expected that FUJI can be deployed in less than ten years [338]. FUJI is based on the results obtained by ORNL in the 1960s and has been optimized as a small sized plant and further simplified by removing the on-line reprocessing facility. Based on the experience of operating the three experimental MSRs at ORNL, it has been verified that FUJI is feasible. The steam generator is a major unverified component, but it could be developed based on experience from the fast breeder reactor and the recent supercritical power station technology.

FUJI adopts a passive safety system to improve the plant's safety and also its economics. Molten fuel salt can be drained to a subcritical drain tank through a freeze valve. Since gaseous fission products are always removed from molten fuel salt, the risk of releasing such products during postulated accidents is reduced. FUJI is operated at very low pressure and a thick reactor vessel and pipes are not required. There are no fuel assemblies, and the plant does not have a complex core internal structure. The only component within the reactor vessel is the graphite moderator. With these design principles, in-factory fabrication would be simple. Figure 101 presents a view of FUJI.

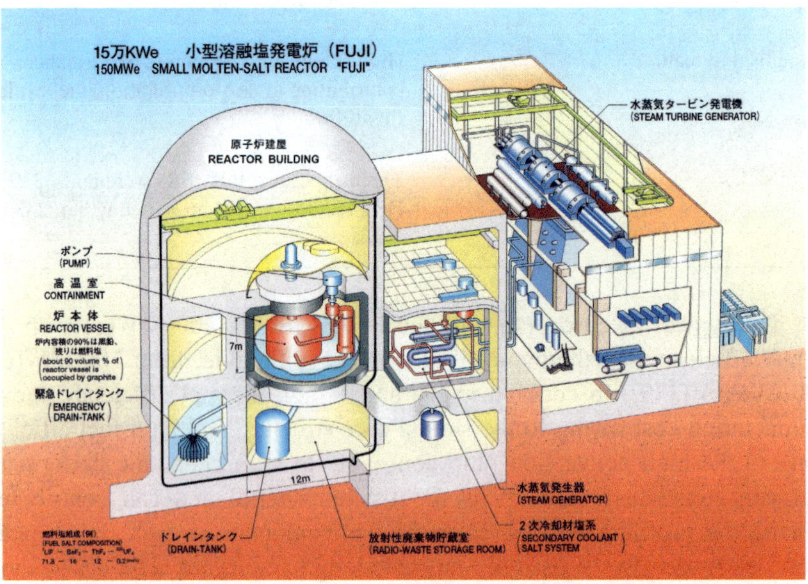

FIG. 101. View of the FUJI molten salt reactor (reproduced from Ref. [452]).

V.2.3. Integral Molten Salt Reactor (IMSR)

Class: Graphite based MSRs (I)
Family: Graphite moderated MSRs (I.2)
Type: Uranium converters and other concepts

TABLE 24. MAIN CHARACTERISTICS OF THE IMSR

Characteristic	Value
Fuel state	Liquid
Reactor spectrum	Thermal
Salt type	Fluorides
Coolant	Fuel salt
Actinide	U–Pu

TABLE 25. MAJOR TECHNICAL PARAMETERS OF THE IMSR

Parameter	Value
Technology developer, country of origin	Terrestrial Energy, Canada
Thermal/electrical capacity, MW(th)/MW(e)	~440/~195
Moderator	Graphite
Conversion ratio	Less than 1
Fuel processing	Once through
Kilowatt per total fissile load (kW/kg)	—[a]
Primary circulation	Forced circulation
Primary pressure (MPa)	Hydrostatic, near atmospheric
Core inlet/exit temperatures (°C)	~620/~700

TABLE 25. MAJOR TECHNICAL PARAMETERS OF THE IMSR (cont.)

Parameter	Value
Fuel type	UF_4 standard assay low enriched uranium
Fuel enrichment (%)	~2% startup, 4.95% make-up fuel
Fuel loading/top-up scheme	Initial ~2% beginning-of-life fuel load with periodic make-up fuel additions over 7 year core unit lifetime. No fuel removal during lifetime of 7 years
Main reactivity control mechanism	Negative reactivity coefficient of temperature and shutdown rods
Approach to engineered safety systems	Passive
Design life (years)	7 years per core unit, 56 years for plant
Plant footprint (m^2)	45 000
Reactor vessel's height/diameter (m)	~18/4.1
Seismic design	0.3g for North-East US siting
Distinguishing features	Passive safety approach, integral primary systems within a 7 year replaceable core unit
Design status	Basic engineering

[a] —: data not available.

The IMSR is an advanced, 440 MW(th) liquid fuelled small modular MSR. It features a sealed integral core unit that houses the liquid fuel, moderator, pumps, primary heat exchangers and shutdown rods. The entire core unit is replaced at the end of its seven year life. This allows factory production levels of quality control and economy, while avoiding the need to open and service the reactor vessel at the power plant site. A non-fuelled liquid fluoride salt loop transfers heat from the core unit to a third salt loop that, using solar nitrate salts, transfers the heat to a separate building where it generates superheated steam for electric power generation or is used for industrial process heat applications. Figure 102 depicts the IMSR core unit and its major components shown in an operating silo.

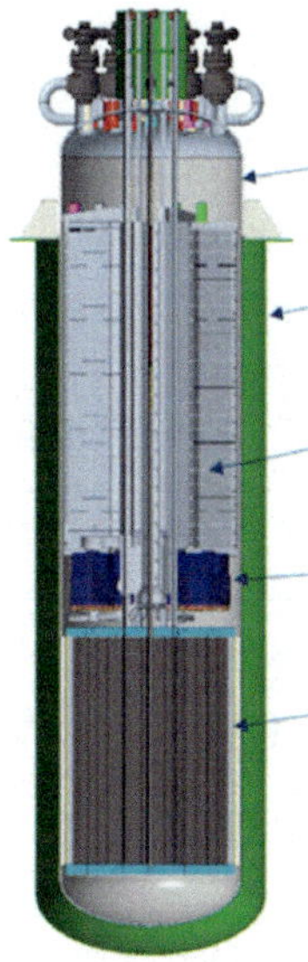

FIG. 102. The core unit of the IMSR and its major components shown in an operating silo (reproduced from Ref. [452]).

Using liquid fuel, the IMSR realizes a naturally strong negative reactivity coefficient of temperature. This desirable safety characteristic provides a self-governing stable temperature regime and establishes an inherently safe operating profile. When power is demanded in the short term, reactor power is inherently controlled without the need to manipulate any reactivity control device. In the long term, reactivity is controlled by small routine manual additions of fuel salt. There are no restrictive neutron flux limits in contrast to limits imposed on traditional water cooled reactors relating to fuel cladding integrity. Hence, neutron flux can transiently increase by a significant amount without any negative effects on core integrity. For these reasons, the IMSR does not require any control rods, nor does it require automatic flux control algorithms for reactor power control. The rapid response to temperature changes and the low fission product poison density also enable load following capability, which facilitates the IMSR to be a backup for variable wind and solar power generation.

In summary, the liquid fuel IMSR allows a significant simplification of the power plant's design, which eases construction and commissioning capital cost burdens and improves overall operating cost performance.

V.2.4. smTMSR-400

Class: Graphite based MSRs (I)
Family: Graphite moderated MSRs (I.2)
Type: Uranium converters and other concepts

TABLE 26. MAIN CHARACTERISTICS OF smTMSR-400

Characteristic	Value
Fuel state	Liquid
Reactor spectrum	Thermal
Salt type	Fluorides
Coolant	Fuel salt
Actinide	Th–U

TABLE 27. MAJOR TECHNICAL PARAMETERS OF smTMSR-400

Parameter	Value
Technology developer, country of origin	SINAP, Chinese Academy of Sciences, China
Thermal/electrical capacity, MW(th)/MW(e)	400/168
Moderator	Graphite
Conversion ratio	~0.6
Fuel processing	Limited
Kilowatt per total fissile load (kW(th)/kg of fissile material)	~1000
Primary circulation	Forced circulation
Primary pressure (MPa)	0.5

TABLE 27. MAJOR TECHNICAL PARAMETERS OF smTMSR-400 (cont.)

Parameter	Value
Core inlet/exit temperatures (°C)	650/700
Fuel type	LiF–BeF_2–UF_4–ThF_4
Fuel enrichment (%)	19.75
Fuel loading/top-up scheme	Th–U loading initial. Low enriched uranium addition on-line. Fission product gas removal on-line. Batch reprocessing off-line
Main reactivity control mechanism	Control rods. Negative reactivity feedback. On-line fuel addition. Draining off fuel salt
Approach to engineered safety systems	Passive
Design life (years)	60
Plant footprint (m^2)	—[a]
Reactor vessel's height/diameter (m)	~10/3.8
Seismic design	0.3g
Distinguishing features	Replaceable reactor pressure vessel (8–10 years). Passive safety. Greater than 40% power contributed by thorium
Design status	Pre-conceptual design

[a] —: data not available.

The smTMSR-400 is a 400 MW(th)/168 MW(e) small modular thorium molten salt demonstration reactor. It is designed as a thorium convertor and in situ burner driven by low enriched uranium and will be applied as a heat source at high temperature, which not only can be used for electricity generation, but also can satisfy diversified energy demands. The smTMSR-400 consists of a reactor module, heat transfer system, heat storage system, heat utilization system and other auxiliary systems, as shown in Fig. 103.

The reactor module is designed as a compact loop structure with the reactor core, three plate salt–salt heat exchangers, one centrifugal pump and connecting

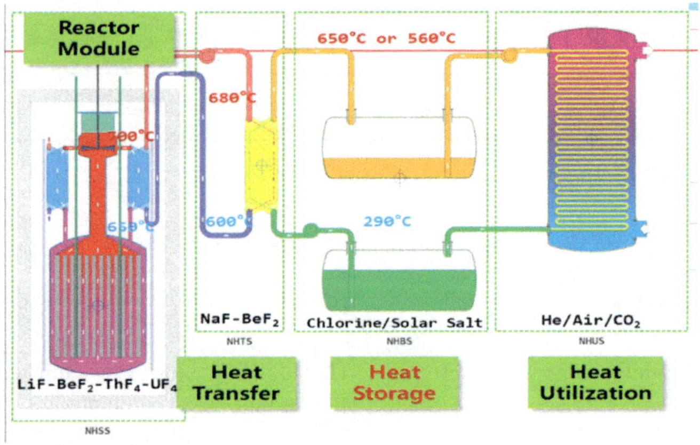

FIG. 103. Diagram of the smTMSR-400 (reproduced from Ref. [452]). NHSS — nuclear heat supply system. NHTS — nuclear heat transfer system. NHBS — nuclear heat bulk storage. NHUS — nuclear heat utilization system.

pipelines. This module is inside a safety vessel that provides an additional radioactive confinement barrier. The diameter of the reactor module is about 3.8 m, so it is suitable for railway transportation. The heat transfer system is designed to transfer the heat to the heat storage system and to isolate a radioactive release. The secondary coolant is NaF–BeF_2 because of its good thermal properties, good chemical stability and compatibility with the fuel salt. The operating temperature of this coolant is 600–680°C. The heat storage system is in standby mode and its design employs a double tank structure for meeting demands for electricity during peak times and heat recovery. A chlorine salt or solar salt will be used as coolant and storage media. The temperatures of the cold and hot tanks are 290°C and 650°C, respectively. The heat utilization system can be designed for electricity generation by using a helium or air or CO_2 Brayton power cycle and for other non-electric applications, such as seawater desalination, heat supply and hydrogen production.

Other auxiliary systems include the fuel management system, on-line fission gas and tritium removal system and two passive decay heat removal systems. The fuel management system is used for initial fuel loading, on-line fuel addition and fuel salt discharge into the drain tank under normal and accident conditions. Fission gases and some noble metals are removed from the reactor module by entrained cover gas. They then flow into a fission gas removal system for decay and separation. If the heat removal used during normal operation fails, two kinds of passive decay heat removal systems will provide cooling of fuel salt for the long term. One is located around the safety vessel in the silo and the other is in the fuel salt drain tank below the reactor vessel.

V.2.5. ThorCon

Class: Graphite based MSRs (I)
Family: Graphite moderated MSRs (I.2)
Type: Uranium converters and other concepts

TABLE 28. MAIN CHARACTERISTICS OF ThorCon

Characteristic	Value
Fuel state	Liquid
Reactor spectrum	Thermal
Salt type	Fluorides
Coolant	Fuel salt
Actinide	Th–U

TABLE 29. MAJOR TECHNICAL PARAMETERS OF ThorCon

Parameter	Value
Technology developer, country of origin	ThorCon US, USA
Thermal/electrical capacity, MW(th)/MW(e)	557/250
Moderator	Graphite
Conversion ratio	Less than 1
Fuel processing	Limited, future uranium re-enrichment
Kilowatt per total fissile load (kW(th)/kg of fissile material)	1000
Primary circulation	Forced
Primary pressure (MPa)	1.2
Core inlet/exit temperatures (°C)	560/704

TABLE 29. MAJOR TECHNICAL PARAMETERS OF ThorCon (cont.)

Parameter	Value
Fuel type	$NaF–BeF_2–ThF_4–UF_4$ (76–12–10.2–1.8 mol%)
Fuel enrichment (%)	19.85
Fuel loading/top-up scheme	High assay low enriched uranium make-up fuel from the tank within the Can
Main reactivity control mechanism	Negative temperature coefficient, salt flow rate; fissile/fertile additions
Approach to engineered safety systems	Intrinsic, passive, using natural circulation, water evaporation
Design life (years)	Can life: 4, plant life: 80
Plant footprint (m^2)	20 000 for 500 MW(e)
Reactor vessel's height/diameter (m)	5.590/4.916
Seismic design	Hull on shear limited sand; reactor containing the Can on elastomeric bearings, stabilized by dash-pots
Distinguishing features	Low cost, full passive safety, short construction time
Design status	Preliminary design

ThorCon is a nuclear power plant in a hull. The plant uses an MSR and the fuel is in molten salt circulated by a pump and passively drained in the event of an accident. The reactor operates at garden hose pressures using normal pipe thicknesses and it is easily automated. The plant is in the style of a ship, with steel plate construction. Cooling ponds passively remove decay heat. From each power module, water naturally circulates from a cold wall around the 'Can' containing the reactor (Pot). Intermediate molten salt loops transfer heat to the steam generator. Basement water below the steam generator provides a third, passive, long term decay heat sink. The balance of the plant is similar to that of a supercritical coal fired power plant. The yellow rectangles in the figure are hatches for access by the cranes. Figure 104 presents the plant, and Fig. 105 shows the Can and the Pot.

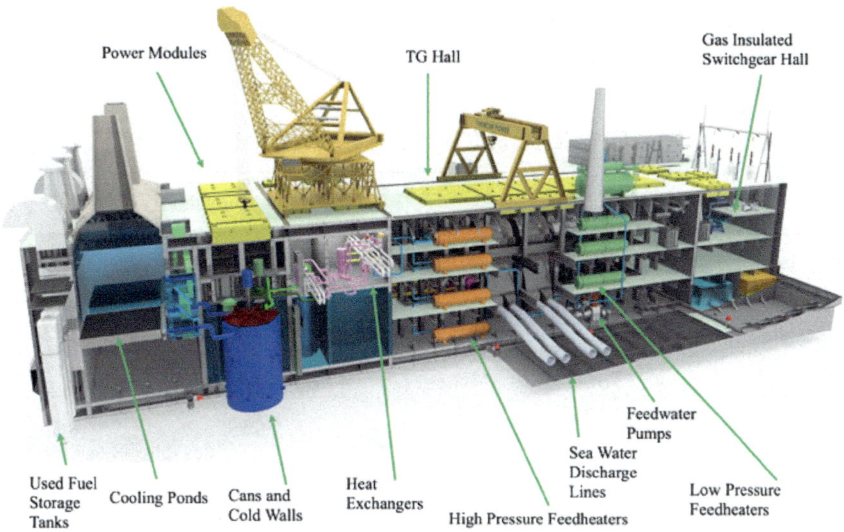

FIG. 104. ThorCon plant (courtesy of R. Hargraves, ThorCon International). TG — turbine generators.

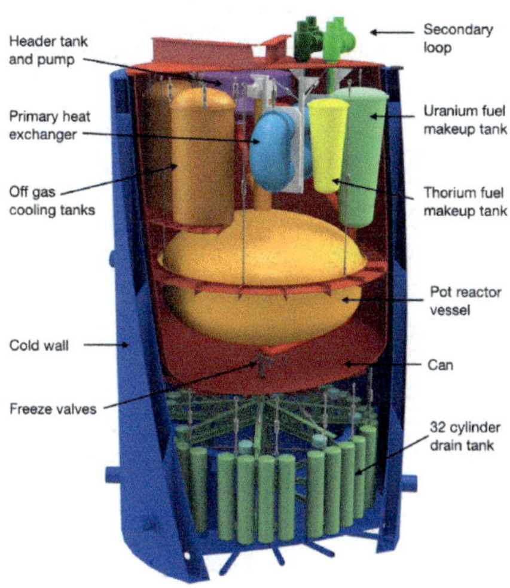

FIG. 105. ThorCon's Pot in the Can in the cold wall (courtesy of R. Hargraves, ThorCon International).

V.2.6. Molten Salt Actinide Recycler and Transformer system with and without Th–U support (MOSART)

Class: Homogeneous MSRs (II)
Family: Homogeneous fluoride fast MSRs (II.3)
Type: Plutonium containing fluoride fast reactor

TABLE 30. MAIN CHARACTERISTICS OF MOSART

Characteristic	Value
Fuel state	Liquid
Reactor spectrum	Fast/epithermal
Salt type	Li,Be,An/F (An denotes actinide)
Coolant	Fuel salt
Actinide	TRU–Th–U

TABLE 31. MAJOR TECHNICAL PARAMETERS OF MOSART

Parameter	Value
Technology developer, country of origin	Rosatom/National Research Centre, Kurchatov Institute, Russian Federation
Thermal/electrical capacity, MW(th)/MW(e)	2400/1000
Moderator	None
Conversion ratio	Up to 1
Fuel processing	Full
Kilowatt per total fissile load (kW(th)/kg of fissile material)	68.5 (per kg of TRU)
Primary circulation	Forced
Primary pressure (MPa)	1.5

TABLE 31. MAJOR TECHNICAL PARAMETERS OF MOSART (cont.)

Parameter	Value
Core inlet/exit temperatures (°C)	630/750
Fuel type	TRU–Th–U fluorides
Fuel enrichment (%)	Minor actinides to TRU ratio (MA/TRU) ≈0.1 (Pu-238: 3.18%; Pu-239: 43.93%; Pu-240: 21.27%; Pu-241: 13.52%; Pu-242: 7.88%; Np-237: 6.42%; Am-241: 0.55%; Am-243: 2.33%)
Fuel loading/top-up scheme	Continuous fuel make-up and fuel salt cleanup in batches with removal time of 300 equivalent full power days for soluble fission products
Main reactivity control mechanism	Fuel salt make-up and control rods
Approach to engineered safety systems	Inherent and passive safety
Design life (years)	50
Plant footprint (m^2)	—[a]
Reactor vessel's height/diameter (m)	11.2/4.5
Seismic design	Seismic capacity for SL2 level is 7 (per Medvedev–Sponheuer–Karnik scale)
Distinguishing features	Flexibility of the fuel cycle, significant fuel quantities outside the core
Design status	Conceptual design

[a] —: data not available.

The MOSART system represents a concept of nuclear power reactors using a molten $^{73}LiF–27BeF_2–AnF_n$ salt mixture (in mol%) in the primary circuit. Molten salt serves as both fuel and primary coolant. The secondary coolant salt does not contain fissile or fertile materials. The core would generate about 2.4 GW(th) at conditions affording efficient transmutation and recycling of TRUs from uranium and mixed oxide fuels from VVER-1000 spent nuclear fuel. The reference core uses a fast neutron spectrum; it does not have a moderator. Radial, bottom and top reflectors are attached to the reactor vessel. This leaves a gap filled with fuel salt surrounding the core to cool the bottom, radial and top reflectors and reactor vessel. Cooling of the upper and lower reflectors is achieved by circulating fuel salt through openings in the lower and upper collectors, respectively. In nominal conditions, the fuel salt passes over the bottom reflector and enters the core at 630°C. The core outlet temperature increases up to 750°C, and the primary salt transfers the heat from the core to the secondary salt in the primary heat exchanger. The fuel salt mixture is circulated through the core by four pumps operating in parallel. This design is being studied in different configurations that consider core configurations with single and two fluids, with and without Th–U support for startup loadings and make-up scenarios with different minor actinides to TRU ratios, as well as the addition of uranium after spent nuclear fuel reprocessing. This system can serve as an efficient burner of TRU elements from spent nuclear fuel and has a breeding capability when using ^{232}Th as a fertile addition. Figure 106 shows the fuel circuit.

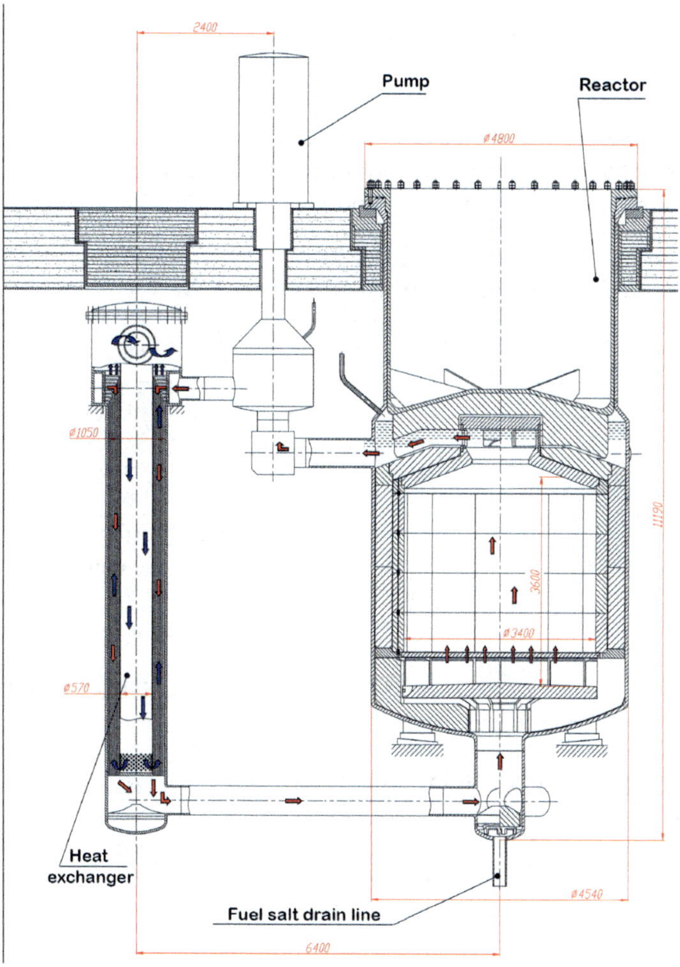

FIG. 106. Fuel circuit of the MOSART system (courtesy of M. Gurov).

V.2.7. Molten Salt Fast Reactor (MSFR)

Class: Homogeneous MSRs (II)
Family: Homogeneous fluoride fast MSRs (II.3)
(optionally: homogeneous chloride fast MSRs (II.4))
Type: Fluoride fast Th–U breeder (optionally: chloride fast U–Pu breeder)

TABLE 32. MAIN CHARACTERISTICS OF THE MSFR

Characteristic	Value
Fuel state	Liquid
Reactor spectrum	Fast
Salt type	Fluorides (LiF–(HN)F_4) or chlorides (NaCl–(HN)Cl_3)
Coolant	Fuel salt
Actinide	U–Pu, Th–U, TRU–Th or TRU

TABLE 33. MAJOR TECHNICAL PARAMETERS OF THE MSFR

Parameter	Value
Technology developer, country of origin	CNRS, France
Thermal/electrical capacity, MW(th)/MW(e)	3000 /—[a]
Moderator	None
Conversion ratio	Greater than 1
Fuel processing	Full
Kilowatt per total fissile load (kW(th)/kg of fissile material)	400
Primary circulation	Forced (5 m^3/s)
Primary pressure (MPa)	<0.5

TABLE 33. MAJOR TECHNICAL PARAMETERS OF THE MSFR (cont.)

Parameter	Value
Core inlet/exit temperatures (°C)	650/750
Fuel type	Fluoride (LiF–(HN)F_4)
Fuel enrichment (%)	None (U-233), or 13% if started with TRU
Fuel loading/top-up scheme	Fuel salt volume is 18 m^3 — continuous refuelling and fuel processing during reactor operation
Main reactivity control mechanism	Large negative feedback temperature and void coefficients, and removal of fuel from the core by thermal dilation
Approach to engineered safety systems	Combination of passive and active systems
Design life (years)	>60
Plant footprint (m^2)	—
Reactor vessel's height/diameter (m)	6/6
Seismic design	The compactness of the fuel circuit inside the core vessel leads to seismic robustness
Distinguishing features	Fuel composition can be changed during operation. Passive cooling after shutdown
Design status	Concept (preliminary design for safety analysis leading to safety by design)

[a] —: data not available.

The MSFR is a homogeneous reactor concept using a circulating fuel salt also used as a coolant, with a core that contains no moderator and exhibits a fast neutron spectrum. The circulation of the fuel salt in pipes is avoided because of an integrated geometry of the fuel circuit, where heat exchangers and pumps are integrated in the core vessel. Heat is extracted from the core by an intermediate molten salt. In the case of failure of the cooling systems (fuel or coolant circuits), the fuel is passively drained into a dedicated emergency draining system where it is passively cooled by natural gas or air convection. Volatile species are continuously extracted by gas bubbling and stored and processed in specific gas casings inside the fuel casing. No volatile fluid is present in the reactor, and thus, depressurization of gas-containing casings would only release a very limited amount of gases in the reactor building. Figure 107 shows a schematic layout of the MSFR fuel circuit.

The MSFR has low core reactivity because of the fuelling and fuel processing during reactor operation. Control rods or neutron poisons are thus not necessary, also because of the large negative feedback temperature and void coefficients which allow a reactivity control based on the balance between the power generated in the fuel salt and the power extracted in the heat exchangers. The absence of rods for absorbing neutrons simplifies reactor operation and eliminates some accident initiators (e.g. ejection of a control rod).

The 'reference MSFR' configuration produces 3 GW(th) and uses a fluoride salt, has been studied for more than 15 years at the CNRS and in European projects, and is mainly adapted to the Th–U cycle in breeder mode. In this configuration, the fuel salt is based on the $LiF–ThF_4$ eutectic and contains UF_4 and PuF_3 as fissile salts. Lithium is enriched in ^7Li to limit tritium production. Uranium-233 is bred from ^{232}Th in the fuel salt and in a separate $LiF–ThF_4$ blanket salt (breeder version). The fissile content in the fuel salt can be adjusted without stopping the reactor by liquid salt transfer. The fuel and the blanket salts are processed on-site at a rate of about 10 L per day and per GW(th) to limit the fission product concentration in the fuel and extract the fissile matter produced in the blanket.

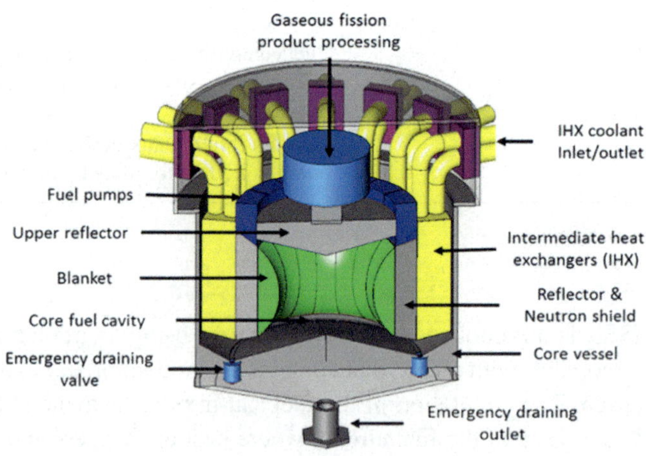

FIG. 107. Schematic layout of the MSFR fuel circuit (courtesy of E. Merle, Laboratoire de Physique Subatomique et de Cosmologie).

V.2.8. Compact Molten Salt Reactor (CMSR)

Class: Heterogeneous MSRs (III)
Family: Non-graphite moderated MSRs (III.5)
Type: Liquid moderator heterogeneous MSR

TABLE 34. MAIN CHARACTERISTICS OF THE CMSR

Characteristic	Value
Fuel state	Liquid
Reactor spectrum	Thermal
Salt type	Fluorides
Coolant	Fuel salt and other
Actinide	U (or other cycle)

TABLE 35. MAJOR TECHNICAL PARAMETERS OF THE CMSR

Parameter	Value
Technology developer, country of origin	Seaborg Technologies, Denmark
Thermal/electrical capacity, MW(th)/MW(e)	250/100
Moderator	NaOH
Conversion ratio	—[a]
Fuel processing	Once through
Kilowatt per total fissile load (kW(th)/kg of fissile material)	200
Primary circulation	Forced
Primary pressure (MPa)	<1
Core inlet/exit temperatures (°C)	600/670

TABLE 35. MAJOR TECHNICAL PARAMETERS OF THE CMSR (cont.)

Parameter	Value
Fuel type	Fluoride salt
Fuel enrichment (%)	High assay low enriched uranium
Fuel loading/top-up scheme	Batchwise, 12 year fuel cycle
Main reactivity control mechanism	Control rods
Approach to engineered safety systems	Passive and active
Design life (years)	12 (reactor module), 24 (plant), 36 (plant life extension)
Plant footprint (m^2)	98.4 m × 32 m (L × B) for a 200 MW(e) plant
Reactor vessel's height/diameter (m)	5.5/2.5
Seismic design	—
Distinguishing features	No refuelling or reprocessing, liquid moderator
Design status	Conceptual design

[a] —: data not available.

The CMSR is an advanced small and modular MSR with a thermal neutron spectrum. It is characterized by the use of a proprietary alkali–hydroxide moderator salt. The moderator salt, containing hydrogen, allows for a compact form factor that simplifies central manufacturing and decommissioning of the complete power plant, and enables simple transport to and from the site where power is produced. The liquid moderator also does not accumulate irradiation damage, enabling a long reactor service lifetime of 12 years, which is also the plant refuelling interval. One unit produces up to 250 MW(th) in the form of superheated steam, which can be used as industrial heat or to produce 100 MW(e), or for both purposes. The reactor does not require, nor does it allow, any access to the fuel salt during the service lifetime of 12 years. A conceptual illustration of the CMSR is provided in Fig. 108. A guiding principle in the development of the

CMSR design is pragmatism. Consequently, the use of established solutions and components is preferred to the extent possible to facilitate rapid deployment and scale-up to importantly displace fossil fuels in time frames relevant for global warming. Inherent safety is provided by passive means, supplemented with active reactor control and protection systems, in accordance with the principles of defence in depth.

Multiple CMSRs will be installed together in floating reactor barges that are assembled and tested in a shipyard. The reactor barge is of modular build and can hold 2–10 CMSRs for a total thermal power capacity of 500–2500 MW(th), corresponding to 200–1000 MW(e). Upon completion and testing, a reactor barge is towed to the production site and moored in sheltered waters. A reactor barge has a nominal service lifetime of 24 years, during which it needs to be refuelled once.

FIG. 108. Concept illustration of the CMSR primary circuit and fuel salt drain tank (reproduced from Ref. [452]).

V.2.9. Copenhagen Atomics Waste Burner

Class: Heterogeneous MSRs (III)
Family: Non-graphite moderated MSRs (III.5)
Type: Solid moderator heterogeneous MSR

TABLE 36. MAIN CHARACTERISTICS OF THE COPENHAGEN ATOMICS WASTE BURNER

Characteristic	Value
Fuel state	Liquid
Reactor spectrum	Thermal
Salt type	Fluorides
Coolant	Fuel salt
Actinide	Th–U, TRU–Th

TABLE 37. MAJOR TECHNICAL PARAMETERS OF THE COPENHAGEN ATOMICS WASTE BURNER

Parameter	Value
Technology developer, country of origin	Copenhagen Atomics, Denmark
Thermal/electrical capacity, MW(th)/MW(e)	100/n.a.[a]
Moderator	Heavy water/^{7}LiOD
Conversion ratio	Greater than 1 (when fuel composition has converged)
Fuel processing	Mechanical processing
Kilowatt per total fissile load (kW/kg)	—[b]
Primary circulation	Forced circulation
Primary pressure (MPa)	0.05–0.25

TABLE 37. MAJOR TECHNICAL PARAMETERS OF THE COPENHAGEN ATOMICS WASTE BURNER (cont.)

Parameter	Value
Core inlet/exit temperatures (°C)	600/650–700
Fuel type	LiF–ThF_4–PuF_4
Fuel enrichment (%)	None
Fuel loading/top-up scheme	None
Main reactivity control mechanism	Level of liquid moderator
Approach to engineered safety systems	Passive
Design life (years)	5–10
Plant footprint (m^2)	—
Reactor vessel's height/diameter (m)	12/2.4
Seismic design	n.a.
Distinguishing features	Liquid moderator, low fissile inventory, active electromagnetic bearing canned pumps
Design status	Conceptual

[a] n.a.: not applicable.
[b] —: data not available.

The Copenhagen Atomics Waste Burner is a heavy water moderated, single-fluid, fluoride salt based, thermal spectrum and autonomously controlled MSR. The core, fission product extraction and separation systems, dump tank, primary heat exchanger, pumps, valves and compressors are all contained in a leaktight stainless steel containment, the size of a shipping container. Figure 109 presents this reactor concept.

The heavy water is unpressurized, thermally insulated from the salt and continuously drained and cooled to 50°C. A molten ^{7}LiOD moderator variant is also being researched as a liquid moderator. Both moderator options allow for

FIG. 109. Copenhagen Atomics Waste Burner (courtesy of A. Stubsgaard, Copenhagen Atomics).

an excellent neutron economy and low fissile inventory. The liquid moderator level is used as the main reactivity control mechanism and to compensate for fluctuations in reactivity due to fuel composition changes.

Active development is focused on the testing and validation of materials, flanges, valves, pumps, heat exchangers, salt purification, chemistry control systems, measurement systems and autonomous control systems. All components are designed to last for the lifetime of the reactor (5–10 years) without any service; for example, this is achieved for the pumps by using active electromagnetic bearings and a canned pump design.

The mechanical separation of fission products is achieved through vacuum spraying, where the salt leaving the core is sprayed into a chamber that is at a partial vacuum pressure, and the volatility of intermediate decay products is used to extract a large fraction of the fission products. The extracted fission products are separated and stored in tanks inside a leaktight stainless steel containment.

During the early use of the reactor, starting with the initial $LiF–ThF_4–PuF_4$ fuel composition, ^{233}U production benefits from the high number of excess neutrons from the fission of plutonium. Later, the breeding process benefits from the superior neutron economy of ^{233}U as the fuel converges towards its equilibrium composition of $LiF–ThF_4–^{233}UF_4$. Thus, when the fuel has reached this composition, the Copenhagen Atomics Waste Burner is expected to be a breeder reactor.

V.2.10. Moltex Energy SSR-W300

Class: Heterogeneous MSRs (III)
Family: Heterogeneous chloride fast MSRs (III.6)
Type: Heterogeneous salt cooled fast MSR

TABLE 38. MAIN CHARACTERISTICS OF SSR-W300

Characteristic	Value
Fuel state	Liquid
Reactor spectrum	Fast
Salt type	Chlorides
Coolant	Other
Actinide	TRU

TABLE 39. MAJOR TECHNICAL PARAMETERS OF SSR-W300

Parameter	Value
Technology developer, country of origin	Moltex Energy, UK and Canada
Thermal/electrical capacity, MW(th)/MW(e)	750/300 baseload, 750/900 peak
Moderator	None
Conversion ratio	Less than 1
Fuel processing	Full
Kilowatt per total fissile load (kW(th)/kg of fissile material)	~100
Primary circulation	Forced
Primary pressure (MPa)	~0.1
Core inlet/exit temperatures (°C)	575/625

TABLE 39. MAJOR TECHNICAL PARAMETERS OF SSR-W300 (cont.)

Parameter	Value
Fuel type	Molten chloride salt fuel
Fuel enrichment (%)	Reactor grade plutonium
Fuel loading/top-up scheme	Refuelling at power
Main reactivity control mechanism	Regulation through fuel temperature coefficients; refuelling mechanism; shutdown with boron carbide assemblies
Approach to engineered safety systems	Inherent safety features, dedicated passive safety systems
Design life (years)	60
Plant footprint (m^2)	22 500
Reactor vessel's height/diameter (m)	14/7
Seismic design	0.3g peak ground acceleration
Distinguishing features	Molten salt fuel in conventional fuel assemblies; burning of nuclear waste; thermal energy storage to allow operation as a peaking plant; very low cost
Design status	Conceptual design; Canadian vendor design review in progress

A specific feature of the Stable Salt Reactor Wasteburner (SSR-W300) is the adoption of molten salt fuel instead of metallic pellets in fuel assemblies used in traditional water cooled reactors. This brings safety advantages, including the ability to operate at a low operating pressure, eliminating the need for high pressure reactor vessels; no concern of fission products leakage from cladding; and because the fuel is already in molten phase, no risk of fuel melts.

The reactor core of SSR-W300 is composed of very low purity, reactor grade plutonium recycled from stocks of spent uranium oxide fuel and produced by the waste to stable salts (WATSS) process that reduces the cost. The reactor generates heat as a stream of molten nitrate salts that can be stored in large

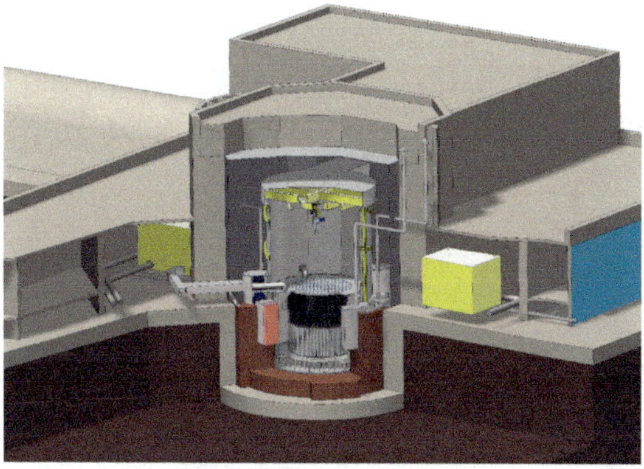

FIG. 110. A sectional view of the SSR-W300 (courtesy of K. Chen, Moltex Energy).

volume at low cost, enabling the SSR-W300 to serve as a low cost peak load power plant, not only for baseload operation.

The turbine island of SSR-W300 adopts a typical Rankine steam cycle identical to that in a traditional fossil fuelled power plant. It can also be operated independently from the nuclear power plant. Figure 110 presents a sectional view of this reactor concept.

The molten chloride salt fuel is contained within vented fuel tubes. The tubes are arranged in a hexagonal array. Molten ZrF_4–KF salt is used as the primary reactor coolant. As for the secondary system coolant, molten nitrate salt is used. The SSR-W300 is designed for countries with significant stocks of spent nuclear fuel. The reactor burns the higher actinide component leaving a waste stream that contains only relatively short lived fission products. The fuel cost is expected to be negative, net of the reduced liability cost for disposal of the original spent fuel. The SSR-W300 is designed to generate peaking electrical power economically and efficiently while operating at constant power. It therefore complements intermittent renewable energy sources and enables flexible lower carbon generation of national power systems.

REFERENCES

[1] Strengthening the Agency's Activities Related to Nuclear Science, Technology and Applications, Resolution GC(60)/RES/12, IAEA, Vienna (2016).

[2] Strengthening the Agency's Activities Related to Nuclear Science, Technology and Applications, Resolution GC(61)/RES/11, IAEA, Vienna (2017).

[3] Strengthening the Agency's Activities Related to Nuclear Science, Technology and Applications, Resolution GC(62)/RES/9, IAEA, Vienna (2018).

[4] LANE, J.A., MACPHERSON, H.G., MASLAN, F., Fluid Fuel Reactors, 1st edn, Addison-Wesley, Reading, MA (1958).

[5] NORDHEIM, L.W., The Case for an Enriched Pile, Rep. ORNL-CF-44-11-236, Oak Ridge Natl Lab., TN (1944).

[6] ROSENTHAL, M.W., KASTEN, P.R., BRIGGS, R.B., Molten-salt reactors — history, status, and potential, Nucl. Appl. Technol. **8** 2 (1970) 107–117.

[7] CANTOR, S., Molten salt chemistry at ORNL — A short history, ORNL Review **10** 4 (1977) 33–39.

[8] BRIANT, R.C., WEINBERG, A.M., Molten fluorides as power reactor fuels, Nucl. Sci. Eng. **2** 6 (1957) 797–803.

[9] SCATCHARD, G., CLARK, H.M., GOLDEN, S., BOLTAX, A., SCHUHMANN, R., Jr., Chemical Problems of Non-Aqueous Fluid-Fuel Reactors, Rep. MIT-5001, Massachusetts Institute of Technology, MA (1952).

[10] BULMER, J.J., et al., Fused Salt Fast Breeder, Rep. ORNL-CF-56-8-204, Oak Ridge Natl Lab., TN (1956).

[11] MACPHERSON, H.G., et al., A Preliminary Study of Molten Salt Power Reactors. Rep. ORNL-CF-57-4-27, Oak Ridge Natl Lab., TN (1957).

[12] WEHMEYER, D.B., et al., Study of a Fused Salt Breeder Reactor for Power Production, Rep. CF-53-10-25, Oak Ridge School of Reactor Technology, TN (1953).

[13] JOSEPH, E.F., DALEY, F.L., HANNAFORD, B.A., YOUNGBLOOD, E.L., Aircraft Nuclear Propulsion Fluoride Fuel Preparation Facility, Rep. ORNL-CF-54-6-126, Oak Ridge Natl Lab., TN (1954).

[14] MANLY, W.D., et al., Aircraft Reactor Experiment — Metallurgical Aspect, Rep. ORNL-2349, Oak Ridge Natl Lab., TN (1957).

[15] SAVAGE, H.W., WHITMAN, G.D., COBB, W.G., MCDONALD, W.B., Components of the Fused-Salt and Sodium Circuits of the Aircraft Reactor Experiment, Rep. ORNL-2348, Oak Ridge Natl Lab., TN (1958).

[16] GRIMES, W.R., CUNEO, D.R., BLANKENSHIP, F.F., "Fused salt systems", Reactor Handbook, AECD-3646, Vol. 2 (HOGERTON, J.F., GRASS, R.C., Eds) US Atomic Energy Commission, Washington, DC (1955) Sec. 6, https://www.osti.gov/biblio/4377036-reactor-handbook-vol-engineering

[17] THOMA, R.E., GRIMES, W.R., Phase Equilibrium Diagrams for Fused Salt Systems, Rep. ORNL-2295, Oak Ridge Natl Lab., TN (1957).

[18] ADAMSON, G.M., CROUSE, R.S., MANLY, W.D., Interim Report on Corrosion by Zirconium-Base Fluorides, Rep. ORNL-2338, Oak Ridge Natl Lab., TN (1961).

[19] RICHARDSON, L.S., VREELAND, D.C., MANLY, W.D., Corrosion by Molten Fluorides, Rep. ORNL-1491, Oak Ridge Natl Lab., TN (1953).

[20] MCCOY, H.E., The INOR-8 story, ORNL Rev. **3** 2 (1969) 35–49.

[21] UNITED STATES ATOMIC ENERGY COMMISSION, Report of the Fluid Fuel Reactors Task Force to the Division of Reactor Development, Rep. TID-8507, US Atomic Energy Commission Technical Information Service, DC (1959).

[22] MACPHERSON, H.G., Molten-Salt Reactor Program, Quarterly Progress Report for Period Ending July 31, 1960, Rep. ORNL-3014, Oak Ridge Natl Lab., TN (1960).

[23] HAUBENREICH, P.N., ENGEL, J.R., Experience with the molten-salt reactor experiment, Nucl. Appl. Technol. **8** (1970) 118–136.

[24] STEFFY, R.C., Jr., Inherent Neutron Source in MSRE with Clean ^{233}U Fuel, Rep. ORNL-TM-2685, Oak Ridge Natl Lab., TN (1969).

[25] UNITED STATES ATOMIC ENERGY COMMISSION, An Evaluation of the Molten Salt Breeder Reactor, Rep. WASH-1222, US Atomic Energy Commission, DC (1972).

[26] MCCOY, H.E., Jr., Status of Materials Development for Molten Salt Reactors, Rep. ORNL-TM-5920, Oak Ridge Natl Lab., TN (1978).

[27] GRIMES, W.R., Molten salt reactor chemistry, Nucl. Appl. Technol. **8** 2 (1970) 137–155.

[28] ROSS, R.G., GRIMES, W.R., BARTON, C.J., BAMBERGER, C.E., BAES, C.F., Jr., The reductive extraction of protactinium and uranium from molten LiF-BeF_2-ThF_4 mixtures into bismuth, Nucl. Met. **15** (1969) 363–374.

[29] HIGHTOWER, J.R., Jr., Engineering Development Studies for Molten-Salt Breeder Reactor Processing No. 21, Rep. ORNL-TM-4894, Oak Ridge Natl Lab., TN (1976).

[30] MAYS, G.T., SMITH, A.N., ENGEL, J.R., Distribution and Behavior of Tritium in the Coolant-Salt Technology Facility, Rep. ORNL-TM-5759, Oak Ridge Natl Lab., TN (1977).

[31] FOSTER WHEELER CORPORATION, Design Studies of Steam Generators for Molten Salt Reactors, Rep. ND/74/66, Foster Wheeler Corp., NJ (1974), http://technicalreports.ornl.gov/1974/3445605702442.pdf

[32] ENGEL, J.R., GRIMES, W.R., RHOADES, W.A., DEARING, J.F., Molten-Salt Reactors for Efficient Fuel Utilization Without Plutonium Separation, Rep. ORNL-TM-6413, Oak Ridge Natl Lab., TN (1978).

[33] THE AMERICAN PRESIDENCY PROJECT, Nuclear Power Policy Remarks and a Question-and-Answer Session With Reporters on Decisions Following a Review of U.S. Policy (1977), https://www.presidency.ucsb.edu/node/243132

[34] ENGEL, J.R., et al., Conceptual Design Characteristics of a Denatured Molten-Salt Reactor with Once-Through Fueling, Rep. ORNL-TM 7207, Oak Ridge Natl Lab., TN (1980).

[35] TAUBE, M., Plutonium fused salts fuels for fast breeder reactor; nuclear and chemical criterion, Nukleonika **6** (1961) 565–585.

[36] TAUBE, M., MIELCARSKI, M., POTURAJ-GUTNIAK, S., KOWALEW, A., New boiling salt fast breeder reactor concepts, Nucl. Eng. Des. **5** 2 (1967) 109–112.

[37] TAUBE, M., SCHUMACHER, H., Carbide-Fuelled Fast Breeder Reactor with Molten Chloride as Fuel Bonding Material and Chloride Vapour as Coolant, Rep. EIR-167, Eidg. Institut für Reaktorforschung, Würenlingen, Switzerland (1969).

[38] TAUBE, M., LIGOU J., Molten Chlorides Fast Breeder Reactor Problems and Possibilities, Rep. EIR-215, Eidg. Institut für Reaktorforschung, Würenlingen, Switzerland (1972).

[39] TAUBE, M., Thorium-Uranium Fast/Thermal Breeding System with Molten Salt Fuel, Rep. EIR-253, Eidg. Institut für Reaktorforschung, Würenlingen, Switzerland (1974).

[40] TAUBE, M., HEER, W., Reactor with Very Low Fission Product Inventory, Rep. EIR-411, Eidg. Institut für Reaktorforschung, Würenlingen, Switzerland (1980).

[41] OTTEWITTE, E., Configuration of a Molten Chloride Fast Reactor on a Thorium Fuel Cycle to Current Nuclear Fuel Cycle Concerns, PhD Thesis, Univ. of California (1982).

[42] KASTEN, P.R., "The MOSEL reactor concept", Proc. Third Int. Conf. on the Peaceful Uses of Atomic Energy, Geneva, 1964, United Nations, New York (1964) 363–369.

[43] KASTEN, P.R., GAT, U., HORN, S.S., VORNHUSEN, H.W., Design concepts for the core structure of a MOSEL (molten salt experimental) reactor, Nucl. Struct. Eng. **2** (1965) 224–232.

[44] HARDER, B.R., LONG, G., STANAWAY, W.P., Compatibility and processing problems in the use of molten uranium chloride-alkali chloride mixtures as reactor fuels, Nucl. Met. **15** (1969) 405–432.

[45] SMITH, J., et al., An Assessment of a 2500 MWe Molten Chloride Salt Fast Reactor, Rep. AEEW-R956, UK Atomic Energy Authority, Winfrith (1974).

[46] VRIESEMA, B., Aspects of Molten Fluorides as Heat Transfer Agents for Power Generation, PhD Thesis, Delft Univ. of Technology (1979).

[47] DELFT UNIVERSITY OF TECHNOLOGY, Molten Salt Setup in the Laboratory for Energy Supply and Nuclear Reactors of the Delft University of Technology (1969),
https://repository.tudelft.nl/view/MMP/uuid:b268aa77-dd5c-4f5c-8650-599dc94c9907

[48] LATZKO, D.G.H., KROECKEL, H., DE MELKER, P., SPIEWAKOWSKI, W., SWART, W., Delft Molten Salt Project Design Summary Report, Rep. DMSP-G-8, Delft Univ. of Technology, Netherlands (1966).

[49] SHI, L., Zero power experiment device in Shanghai Institute of Nuclear Research, Chin. J. Nucl. Sci. Eng. **4** 2 (1984) (in Chinese).

[50] MSR GROUP OF PROJECT-728, Criticality Experiments Summary Report of Molten Salt–Graphite Zero Power Reactor, Rep. D151-692-01, Shanghai Institute of Nuclear Research (1972) (in Chinese).

[51] MSR GROUP OF PROJECT-728, Control Rod Calibration Experiment Summary Report of Molten Salt–Graphite Zero Power Reactor, Rep. D151-692-02, Shanghai Institute of Nuclear Research (1972) (in Chinese).

[52] MSR GROUP OF PROJECT-728, Neutron Flux Measurement Experiments Summary Report of Molten Salt–Graphite Zero Power Reactor, Rep. D151-693-02, Shanghai Institute of Nuclear Research (1972) (in Chinese).

[53] MSR GROUP OF PROJECT-728, Thorium-Uranium Conversion Ratio Measurement Experiments Summary Report of Molten Salt–Graphite Zero Power Reactor, Rep. D151-693-01, Shanghai Institute of Nuclear Research (1972) (in Chinese).

[54] FURUKAWA, K., TSUKADA, K., NAKAHARA, Y., Single-fluid-type accelerator molten-salt breeder concept, J. Nucl. Sci. Technol. **18** 1 (1981) 79–81.

[55] FURUKAWA, K., MINAMI, K., MITACHI, K., KATOH, Y., Compact molten-salt fission power stations (FUJI-series) and their developmental program, ECS Proc. **1987** 1 (1987) 896–905.

[56] HERY, M., LECOCQ, A., Réacteurs à Sels Fondus — Synthese des Etudes Réalisées Entre 1973 et 1983, Rep. EDF HT/12/74/83 and CEA/DGR/SEP – 83/365, Paris (1984).

[57] GOVERNMENT OF FRANCE, Loi n° 91-1381 du 30 décembre 1991 relative aux recherches sur la gestion des déchets radioactifs, Official Journal (1991) (in French),
https://www.legifrance.gouv.fr/loda/id/JORFTEXT000000356548

[58] BOWMAN, C., et al., Nuclear Energy Generation and Waste Transmutation Using Accelerator Driven Intense Thermal Neutron Source, Rep. LA-UR-91-2601, Los Alamos Natl Lab., NM (1991).

[59] SLESSAREV, I., BERTHOU, V., SALVATORES, M., TCHISTIAKOV, A., "Concept of thorium fuelled accelerator driven subcritical system for both energy production and TRU incineration — 'TASSE'", Proc. 3rd Int. Conf. on Accelerator Driven Transmutation Technologies and Applications (ADTTA'99), Prague, Czech Republic, 1999.

[60] BERTHOU, V., Optimisation d'Options Nouvelles d'Incinération par Interactions d'Actinides et de Produits de Fission, PhD Thesis, Evry Univ. (2000) (in French).

[61] VALADE, M., Optimisation du potentiel d'incinération en système hybride, PhD Thesis, Univ. Louis Pasteur Strasbourg (2000) (in French).

[62] MATHIEU, L., et al., The thorium molten salt reactor: moving on from the MSBR, Prog. Nucl. Energy **48** 7 (2006) 664–679.

[63] MERLE-LUCOTTE, E., et al., "Molten salt reactors and possible scenarios for future nuclear power deployment", Proc. Int. Conf. Physics of Reactors (PHYSOR 2004) — The Physics of Fuel Cycles and Advanced Nuclear Systems: Global Developments, Chicago, 2004, American Nuclear Society, La Grange Park, IL (2004) 1–12.

[64] MERLE-LUCOTTE, E., et al., "Fast thorium molten salt reactors starting with plutonium", Proc. Int. Conf. Advances in Nuclear Power Plants (ICAPP 2006), Reno, NV, 2006, American Nuclear Society, La Grange Park, IL (2006) 6132–6139.

[65] MOUROGOV, A., BOKOV, P., "Fast spectrum molten salt reactor concept: REBUS-3700", paper presented at CAPRA CADRA Int. Sem., Aix-en-Provence, 2004.

[66] MOUROGOV, A., BOKOV, P., Potentialities of the fast spectrum molten salt reactor concept: REBUS-3700, Energy Convers. Manage. **47** 17 (2006) 2761–2771.

[67] NOVIKOV, V., IGNATIEV, V., FEDULOV, V., CHEREDNIKOV, V., Molten Salt Nuclear Power Systems — Perspectives and Problems, Energoatomizdat, Moscow (1990).

[68] IGNATIEV, V., "Historical MSR programme in Russia", presented at MSR Summer School, Lecco, Italy, 2017.

[69] BELOUSOV, I.G., et al., Features layout of VTRS for technological purpose, VANTS **16** 3 (1983) 13–14.

[70] BRAIKO, V.D., et al., Tests on a setup with natural circulation of LiF-BeF_2-UF_4 liquid salt fuel, Sov. At. Energy **69** 4 (1990) 817–821.

[71] MIGNACCA, B., LOCATELLI, G., Economics and finance of molten salt reactors, Prog. Nucl. Energy **129** (2020).

[72] LAUREAU, A., et al., Transient coupled calculations of the molten salt fast reactor using the transient fission matrix approach, Nucl. Eng. Des. **316** (2017) 112–124.

[73] HEUER, D., et al., Towards the thorium fuel cycle with molten salt fast reactors, Ann. Nucl. Energy **64** (2014) 421–429.

[74] AGUIAR, J.A., et al., Bringing nuclear materials discovery and qualification into the 21st century, Nat. Commun. **11** (2020).

[75] KEDL, R., The Migration of a Class of Fission Products (Noble Metals) in the Molten Salt Reactor Experiment, Rep. ORNL-TM-3884, Oak Ridge Natl Lab., TN (1972).

[76] PEEBLES, F.N., Removal of Xenon-135 from Circulating Fuel Salt by Mass Transfer to Helium Bubble, Rep. ORNL-TM-2245, Oak Ridge Natl Lab., TN (1968).

[77] MCFARLANE, J., et al., Fission Product Volatility and Off-Gas Systems for Molten Salt Reactors, Rep. ORNL/TM-2019/1266, Oak Ridge Natl Lab., TN (2019).

[78] LIU, J., THALLAPALLY, P.K., STRACHAN, D., Metal–organic frameworks for removal of Xe and Kr from nuclear fuel reprocessing plants, Langmuir **28** 31 (2012) 11584–11589.

[79] PRICE, T., TAYLOR, Z., CHVALA, O., A review of circulating voids and gaseous fission product behavior in molten salt reactors, J. Nucl. Eng. Radiat. Sci. **6** 4 (2020).

[80] MCFARLANE, J., WEBER, C.F., GREENWOOD, M.S., QUALLS, A.L., Thermochemical and transport properties important to molten salt reactor operation: Off-gas performance and the fission product mechanistic source term, Rep. ORNL/TM-2018/958, Oak Ridge Natl Lab., TN (2018).

[81] AMERICAN SOCIETY OF MECHANICAL ENGINEERS, ASME Boiler and Pressure Vessel Code, Section III, Rules for Construction of Nuclear Facility Components, Division 5, High Temperature Reactors (BPVC-III-5 – 2021), New York (2021).

[82] AFCEN, RCC-MRx 2022 Design and Construction Rules for Mechanical Components of Nuclear Installations: High Temperature, Research and Fusion Reactors, AFCEN, Courbevoie (2022).

[83] TERRAPOWER, Molten nuclear fuel salts and related systems and methods, US Patent US 2016/0189813 A1 (2016).

[84] EUROPEAN UNION, Horizon 2020 – Work Programme 2014-2015, General Annex G – Technology Readiness Levels (TRL) (2014), https://ec.europa.eu/research/participants/data/ref/h2020/wp/2014_2015/annexes/h2020-wp1415-annex-g-trl_en.pdf

[85] TAUBE, M., Fast Reactors Using Molten Chloride Salts as Fuel — Final Report (1972–1977), Rep. EIR-332, Eidg. Institut für Reaktorforschung, Würenlingen, Switzerland (1978).

[86] FRAAS, A.P., SAVOLAINEN, A.W., ORNL Aircraft Nuclear Power Plant Design, Rep. ORNL-1721, Oak Ridge Natl Lab., TN (1954).

[87] HOMBOURGER, B.A., Conceptual Design of a Sustainable Waste Burning Molten Salt Reactor, PhD Thesis, École Polytechnique Fédérale de Lausanne (Swiss Federal Institute of Technology Lausanne) (2018).

[88] HUDDLE, R.A., GOUGH, J.R., BEUTLER, H., Coated Particle Fuel for the Dragon Reactor Experiment — Project Dragon, Rep. DP-Report-116, Atomic Energy Establishment, Winfrith, UK (1962).

[89] FRATONI, M., GREENSPAN, E., Neutronic feasibility assessment of liquid salt-cooled pebble bed reactors, Nucl. Sci. Eng. **168** 1 (2011) 1–22.

[90] ANDREADES, C., et.al., Design summary of the Mark-I pebble-bed, fluoride salt-cooled, high-temperature reactor commercial power plant, Nucl. Technol. **195** 3 (2016) 223–238.

[91] GREENE, S. R., et.al., Pre-Conceptual Design of a Small Modular Fluoride Salt-Cooled High Temperature Reactor (SmAHTR), Rep. ORNL/TM-2010/199, Oak Ridge Natl Lab., TN (2010).

[92] FORSBERG, C., et al., Fluoride-salt-cooled high-temperature reactor (FHR) using British advanced gas-cooled reactor (AGR) refueling technology and decay heat removal systems that prevent salt freezing, Nucl. Technol. **205** 9 (2019) 1127–1142.

[93] FORSBERG, C.W., PETERSON, P.F., PICKARD, P.S., Molten-salt-cooled advanced high-temperature reactor for production of hydrogen and electricity, Nucl. Technol. **144** 3 (2003) 289–302.

[94] CASINO, W.A., Jr., "Investigation of alternative fuel form for the liquid salt very high temperature reactor (LS-VHTR)", Proc. Int. Congr. Advances in Nuclear Power Plants (ICAPP 2006), Reno, NV, 2006, American Nuclear Society, La Grange Park, IL (2006) 618–626.

[95] ELLIS, C.B., THOMPSON, W.E., The Aircraft Nuclear Propulsion Project, Quarterly Progress Report for Period Ending August 31, 1950, Rep. ORNL-0858, Oak Ridge Natl Lab., TN (1950).

[96] BRIANT, R.C., et al., The Aircraft Nuclear Propulsion Project, Quarterly Progress Report for Period Ending December 10, 1952, Rep. ORNL-1439, Oak Ridge Natl Lab., TN (1955).

[97] JORDAN, W.H., et al., The Aircraft Nuclear Propulsion Project, Quarterly Progress Report for Period Ending September 10, 1955, Rep. ORNL-1771, Oak Ridge Natl Lab., TN (1954).

[98] BRIANT, R.C., et al., The Aircraft Nuclear Propulsion Project, Quarterly Progress Report for Period Ending March 10, 1953, Rep. ORNL-1515, Oak Ridge Natl Lab., TN (1953).

[99] COTTRELL, W.B., et al., The Aircraft Nuclear Propulsion Project, Quarterly Progress Report for Period Ending June 10, 1953, Rep. ORNL-1536, Oak Ridge Natl Lab., TN (1953).

[100] COTTRELL, W.B., et al., The Aircraft Nuclear Propulsion Project, Quarterly Progress Report for Period Ending September 10, 1953, Rep. ORNL-1609, Oak Ridge Natl Lab., TN (1953).

[101] JORDAN, W.H., et al., The Aircraft Nuclear Propulsion Project, Quarterly Progress Report for Period Ending December 10, 1955, Rep. ORNL-1816, Oak Ridge Natl Lab., TN (1954).

[102] BRIANT, R.C., MILLER, A.J., COTTRELL, W.B., The Aircraft Nuclear Propulsion Project, Quarterly Progress Report for Period Ending December 10, 1953, Rep. ORNL-1649, Oak Ridge Natl Lab., TN (1954).

[103] SAVOLAINEN, A.W., et al., The Aircraft Nuclear Propulsion Project, Quarterly Progress Report for Period Ending March 10, 1954, Rep. ORNL-1692, Oak Ridge Natl Lab., TN (1954).

[104] MACPHERSON, H.G., Molten-Salt Reactor Program, Quarterly Progress Report for Period Ending October 31, 1959, Rep. ORNL-2626, Oak Ridge Natl Lab., TN (1959).

[105] ROSENTHAL, M.W., et al., The Development Status of Molten-Salt Breeder Reactors, Rep. ORNL-4812, Oak Ridge Natl Lab., TN (1972).

[106] MACPHERSON, H.G., Molten-Salt Reactor Program, Status Report, Rep. ORNL-2634, Oak Ridge Natl Lab., TN (1958).

[107] ALEXANDER, L.G., et al., Nuclear Characteristics of Spherical, Homogeneous, Two-region, Molten-Fluoride-Salt Reactors, Rep. ORNL-2751, Oak Ridge Natl Lab., TN (1959).

[108] ROBERTSON, R.C., et al., Conceptual Design of a Single-Fluid Molten-Salt Breeder Reactor, Rep. ORNL-4541, Oak Ridge Natl Lab., TN (1971).

[109] INTERNATIONAL ENERGY AGENCY, NUCLEAR ENERGY AGENCY, INTERNATIONAL ATOMIC ENERGY AGENCY, Innovative Nuclear Reactor Development, OECD, Paris (2002).

[110] VERGNES, J., LECARPENTIER, D., GARZENNE, C., "The AMSTER Concept", Proc. 6th Information Exchange Mtg on Actinide and Fission Product Partitioning and Transmutation, Madrid, 2000, OECD, Paris (2001) Session II.

[111] XU, H., "Status and perspective of TMSR in China", presented at the Molten Salt Reactor Workshop, Paul Scherrer Institute, Switzerland, 2017.

[112] BRIGGS, R.B., Molten-Salt Reactor Program, Semiannual Progress Report for Period Ending July 31, 1964, Rep. ORNL-3708, Oak Ridge Natl Lab., TN (1964).

[113] SOWDER, A., et al., Program on Technology Innovation: Technology Assessment of a Molten Salt Reactor Design, The Liquid-Fluoride Thorium Reactor (LFTR), Rep. EPRI-3002005460 (2015).

[114] SCOTT, I., "Stable salt fast reactor", Molten Salt Reactors and Thorium Energy (DOLAN, T.J., Ed.), Woodhead Publishing, Duxford, UK (2017) Ch. 21.

[115] ROSENTHAL, M.W., et al., Molten-Salt Reactor Program, Semiannual Progress Report for Period Ending August 31, 1967, Rep. ORNL-4191, Oak Ridge Natl Lab., TN (1967).

[116] ROSENTHAL, M.W., et al., Molten-Salt Reactor Program, Semiannual Progress Report for Period Ending August 31, 1970, Rep. ORNL-4622, Oak Ridge Natl Lab., TN (1971).

[117] CHOE, J., IVANOVA, M., LEBLANC, D., MOHAPTRA, S., ROBINSON, R., "Fuel cycle flexibility of Terrestrial Energy's Integral Molten Salt Reactor (IMSR®)", Proc. 38th Annual Conf. of the Canadian Nuclear Society and 42nd Annual CNS/CNA Student Conf., Saskatoon, 2018, Canadian Nuclear Society, Toronto (2018).

[118] JORGENSEN, L., "ThorCon reactor", Molten Salt Reactors and Thorium Energy (DOLAN, T.J., Ed.), Woodhead Publishing, Duxford, UK (2017) Ch. 19.

[119] ROSENTHAL, M.W., et al., Molten-Salt Reactor Program, Semiannual Progress Report for Period Ending August 31, 1971, Rep. ORNL-4728, Oak Ridge Natl Lab., TN (1971).

[120] FURUKAWA, K., MITACHI, K., KATO, Y., Small molten-salt reactors with a rational thorium fuel-cycle, Nucl. Eng. Des. **136** 1–2 (1992) 157–165.

[121] ROSENTHAL, M.W., BRIGGS, R.B., KASTEN, P.R., Molten-Salt Reactor Program, Semiannual Progress Report for Period Ending August 31, 1968, Rep. ORNL-4344, Oak Ridge Natl Lab., TN (1968).

[122] BRIGGS, R.B., Molten-Salt Reactor Program, Semiannual Progress Report for Period Ending February 28, 1966, Rep. ORNL-3739, Oak Ridge Natl Lab., TN (1966).

[123] KASTEN, P.R., et al., Design Studies of 1000-Mw(e) Molten-Salt Breeder Reactors, Rep. ORNL-3996, Oak Ridge Natl Lab., TN (1966).

[124] MACPHERSON, H.G., Molten-Salt Reactor Program, Quarterly Progress Report for Period Ending January 31, 1958, Rep. ORNL-2474, Oak Ridge Natl Lab., TN (1958).

[125] IGNATIEV, V., et al., Molten salt actinide recycler and transforming system without and with Th–U support: fuel cycle flexibility and key material properties, Ann. Nucl. Energy **64** (2014) 408–420.

[126] VIJAYAN, P.K., et al., Conceptual design of Indian molten salt breeder reactor, PRAMANA, J. Phys. **85** 3 (2015) 539–554.

[127] IGNATIEV, V., et al., "Progress in development of Li, Be, Na/F Molten Salt Actinide Recycler and Transmuter Concept", Proc. Int. Congr. Advances in Nuclear Power Plants (ICAPP 2007), Nice, France, Vol. 5, Société française d'énergie nucléaire (SFEN), Paris (2007).

[128] DEGTYAREV, A, MYASNIKOV, A., PONOMAREV, L., Molten salt fast reactor with U–Pu fuel cycle, Prog. Nucl. Energy **82** (2015) 33–36.

[129] HIROSE, Y., MITACHI, K., SHIMAZU, Y., "Operation control of molten salt U–Pu fast breeder reactor", Proc. Int. Congr. Advances in Nuclear Power Plants (ICAPP 2016), San Francisco, CA, American Nuclear Society, IL (2016).

[130] GOODMAN, C., et al., Nuclear Problems of Non-aqueous Fluid Fuel Reactors, Rep. MIT-5000, Massachusetts Institute of Technology, Cambridge, MA (1952).

[131] NAUMANN, D., Laborstudie zur Chlorierenden Aufarbeitung. Neutronenbestrahlter Urankernbrennstoffe, 1. Mitteilung, Kernenergie **5** 2 (1962) 118.

[132] MOORE, R.V., FAWCETT, S., "Present and future types of fast breeder reactors", Proc. London Conf. on Fast Breeder Reactors organized by the British Nuclear Energy Society, 1966, Vol. 103 (EVANS, P.V., Ed.), Elsevier (2013) 99.

[133] NELSON, P.A., BUTLER, D.K., CHASANOV, M.G., MENEGHETTI, D., Fuel properties and nuclear performance of fast reactors fueled with molten chlorides, Nucl. Technol. **3** 9 (1967) 540–547.

[134] TAUBE, M., OTTEWITTE, E.H., LIGOU, J., A High-Flux Fast Molten Salt Reactor for the Transmutation of Caesium-137 and Strontium-90, Rep. EIR-259, Eidg. Institut für Reaktorforschung, Würenlingen, Switzerland (1975).

[135] HOMBOURGER, B., KREPEL, J., MIKITYUK, K., PAUTZ, A., "Fuel cycle analysis of a molten salt reactor for breed-and-burn mode", Proc. Int. Congr. Advances in Nuclear Power Plants (ICAPP 2015: Nuclear Innovations for a Low-Carbon Future) Nice, France, 2015, Société française d'énergie nucléaire (SFEN), Paris (2015) 910–916.

[136] LATKOWSKI, J., TerraPower and the Molten Chloride Fast Reactor, MSR — 2015 Workshop on Molten Salt Reactor Technologies, Oak Ridge Natl Lab., TN (2015).

[137] PHEIL, E., "Elysium Molten Chloride Salt Fast Reactor (MCSFR)", presented at 8th Thorium Energy Alliance Conf., St. Louis, MO, 2017.

[138] RAFFUZZI, V., KREPEL, J., "Simulation of breed and burn fuel cycle operation of Molten Salt Reactor in batch-wise refueling mode", Proc. Int. Conf. Physics of Reactors (PHYSOR 2020), Cambridge, UK, Nuclear Energy Group, Cambridge (2020) 1185.

[139] DE OLIVEIRA, R.G., HOMBOURGER, B.A., "Fuel tap: a simplified breed-and-burn MSR", Proc. Int. Conf. Physics of Reactors (PHYSOR 2020), Cambridge, UK, Nuclear Energy Group, Cambridge (2020) 1547.

[140] BRIANT, R.C., Aircraft Nuclear Propulsion Program: Quarterly Progress Report for Period Ending December 10, 1950, Rep. ORNL-0919, Oak Ridge Natl Lab., TN (1951).

[141] MASSIE, M., DEWAN, L.C., Nuclear Reactors and Related Methods and Apparatus, US Patent Office, US 20130083878 A1, April 4, 2013.

[142] PEDERSEN, T.J., "A walkthrough of the Copenhagen Atomics Waste Burner design", presented at Int. Thorium Energy Conference (ThEC15), Mumbai, India, 2015.

[143] SCHÖNFELDT, T., et al., Molten Salt Reactor, AWA Denmark patent WO2018229265, PCT/EP2018/065989, Copenhagen (2018).

[144] WU, J., et al., A novel concept for a molten salt reactor moderated by heavy water, Ann. Nucl. Energy **132** (2019) 391–403.

[145] TEETER, CH.E., LECKY, J.A., MARTENS, J.H., Catalog of Nuclear Reactor Concepts: Part I, Section III, Rep. ANL-7092, Argonne Natl Lab., Argonne, IL (1965).

[146] WEINBERG, A.M., Aircraft Nuclear Propulsion Program: Quarterly Progress Report for Period Ending November 30, 1949, Rep. ORNL-0528, Oak Ridge Natl Lab., TN (1950).

[147] JORDAN, W.H., CROMER, S.J., MILLER, A.J., Aircraft Nuclear Propulsion Program: Quarterly Progress Report for Period Ending March 10, 1956, Rep. ORNL-2061, Oak Ridge Natl Lab., TN (1956).

[148] BRIANT, R.C., Aircraft Nuclear Propulsion Project: Quarterly Progress Report for Period Ending December 10, 1951, Rep. ORNL-1170, Oak Ridge Natl Lab., TN (1952).

[149] TRÉMILLON, B., Chemistry in Non-Aqueous Solvents, Springer, Dordrecht (1974).

[150] TRÉMILLON, B. Potential-acidity (pO^{2-}) equilibrium diagrams in molten hydroxides alkali salts (Diagrammes d'équilibre potentiel-acidité (pO^{2-}) dans les sels et hydroxydes alcalins fondus), Pure Appl. Chem. **25** 2 (1971) 395–428.

[151] ALEXANDER, L.G., "Molten-salt fast reactors", Proc. Conf. Breeding, Economics and Safety in Large Fast Power Reactors, Rep. ANL-6792, Argonne Natl Lab., IL (1963).

[152] TAUBE, M., A Molten Salt Fast Thermal Reactor System with no Waste, Rep. EIR-249, Eidg. Institut für Reaktorforschung, Würenlingen, Switzerland (1974).

[153] KASAM, A., SHWAGERAUS, E., Neutronic Feasibility of a Breed & Burn Molten Salt Reactor, Serpent User Group Mtg 2016, Milan (2016).

[154] KASAM, A., LEE, J.I., SHWAGERAUS, E., Thermal-hydraulic design methodology and trade-off studies for a dual-salt breed-and-burn molten salt reactor, Nucl. Eng. Des. **360** (2020).

[155] SCOTT, I., et al., "Stable Salt Reactor design concept", Proc. Thorium Energy Conf. 2015 (ThEC15), Mumbai, India, 2015, Springer, Singapore (2019).

[156] SCOTT, I., et al., A Practical Molten Salt Fission Reactor, UK patent GB 2508537 (2014).

[157] HUKE, A., et al., The Dual Fluid Reactor — a novel concept for a fast nuclear reactor of high efficiency, Ann. Nucl. Energy **80** (2015) 225–235.

[158] ANDREI, A., "HSR — Hard Spectrum Reactor", presented at Thorium Energy Conf. 2018, Brussels, Belgium 2018.

[159] BURCH, W.D., et al., Immiscible-liquid-cooled Fluid-fuel Reactor, Rep. CF-55-8-188, Oak Ridge School of Reactor Technology, TN (1955).

[160] GAT, U., Cooling concepts for a compact MOSEL (molten salt) reactor, Nucl. Eng. Des. **5** 2 (1967) 113–122.

[161] BOWMAN, C.D., "Basis and objectives of the Los Alamos accelerator driven transmutation technology project", Accelerator Driven Systems: Energy Generation and Transmutation of Nuclear Waste — Status Report, IAEA-TECDOC-985, IAEA, Vienna (1997) 135–153.

[162] VENNERI, F., WILLIAMSON, M.A., NING, L., "The Los Alamos accelerator driven transmutation of nuclear waste (ATW) concept development of the ATW target/blanket system", Accelerator Driven Systems: Energy Generation and Transmutation of Nuclear Waste — Status Report, IAEA-TECDOC-985, IAEA, Vienna (1997) 154–178.

[163] KATSUTA, H., "Liquid target/fuel and associated fuel cycle technologies", Accelerator Driven Systems: Energy Generation and Transmutation of Nuclear Waste — Status Report, IAEA-TECDOC-985, IAEA, Vienna (1997) 17–22.

[164] HRON, M., Project SPHINX spent hot fuel incinerator by neutron flux (the development of a new reactor concept with liquid fuel based on molten fluorides), Prog. Nucl. Energy **47** 1–4 (2005) 347–353.

[165] TODREAS, N.E., et al., Flexible conversion ratio fast reactors: overview, Nucl. Eng. Des. **239** 12 (2009) 2582–2595.

[166] LIN, M., CHENG, M., DAI, Z., Feasibility of an innovative long-life molten chloride-cooled reactor, Nucl. Sci. Tech. **31** (2020) 33.

[167] ZOU, Y., "Research progress of TMSR design, Shanghai Institute of Applied Physics, CAS", presented at SAMOFAR Final Mtg, Delft, Netherlands, 2019.

[168] KŘEPEL, J., et al., "Hybrid spectrum molten salt reactor", Proc. Int. Conf. Physics of Reactors (PHYSOR 2014), Kyoto, 2014, Japan Atomic Energy Agency, Tokai (2015) Session 10.16 SS1.

[169] BETZLER, B.R., et al., Fuel cycle and neutronic performance of a spectral shift molten salt reactor design, Ann. Nucl. Energy **119** (2018) 396–410.

[170] ELLIS, C.B., The Atomic Powered Aircraft, Rep. ORNL-0684, Oak Ridge Natl Lab., TN (1950).

[171] MACPHERSON, H.G., The molten salt adventure, Nucl. Sci. Eng. **90** 4 (1985) 374–380.

[172] GOVERNMENT OF CANADA, Canada's Mid-Century Long-Term Low-Greenhouse Gas Development Strategy (2020), https://unfccc.int/files/focus/long-term_strategies/application/pdf/canadas_mid-century_long-term_strategy.pdf

[173] GOVERNMENT OF CANADA, Canadian SMR Roadmap (2020), https://smrroadmap.ca/

[174] GOVERNMENT OF CANADA, Canada's Small Modular Reactor Action Plan (2020), https://www.nrcan.gc.ca/our-natural-resources/energy-sources-distribution/nuclear-energy-uranium/canadas-small-modular-reactor-action-plan/21183

[175] CANADIAN NUCLEAR SAFETY COMMISSION, Pre-Licensing Vendor Design Review (2020), https://www.nuclearsafety.gc.ca/eng/reactors/power-plants/pre-licensing-vendor-design-review/index.cfm

[176] GOVERNMENT OF CANADA, Government of Canada Invests in Innovative Small Modular Reactor Technology (2020), https://www.canada.ca/en/innovation-science-economic-development/news/2020/10/government-of-canada-invests-in-innovative-small-modular-reactor-technology.html

[177] CANADIAN NUCLEAR LABORATORIES, CNL's Canadian Nuclear Research Initiative (2020), https://www.cnl.ca/en/home/commercial/cnl-s-canadian-nuclear-research-initiative-.aspx

[178] CANADIAN NUCLEAR LABORATORIES, CNL partners with Kairos Power on SMR research (2020), https://www.cnl.ca/cnl-partners-with-kairos-power-on-smr-research/

[179] CANADIAN NUCLEAR LABORATORIES, CNL & Moltex Energy partner on SMR fuel research (2020),
https://www.cnl.ca/cnl-moltex-energy-partner-on-smr-fuel-research/

[180] CANADIAN NUCLEAR LABORATORIES, CNL & Terrestrial Energy partner on SMR fuel research (2020),
https://www.cnl.ca/cnl-terrestrial-energy-partner-on-smr-fuel-research/

[181] BALE, C.W., et al., Reprint of: FactSage thermochemical software and databases, 2010–2016, CALPHAD **55** (2016) 1–19.

[182] PIRO, M.H.A., SIMUNOVIC, S., BESMANN, T.M., LEWIS, B.J., THOMPSON, W.T., The thermochemistry library Thermochimica, Comp. Mater. Sci. **67** (2013) 266–272.

[183] GOVERNMENT OF NEW BRUNSWICK, $10 Million committed for nuclear research cluster (2018),
https://www2.gnb.ca/content/gnb/en/news/news_release.2018.06.0832.html

[184] SERP, J., et al., The molten salt reactor (MSR) in generation IV: overview and perspectives, Prog. Nucl. Energy **77** (2014) 308–319.

[185] HARGRAVES, R.F., MOIR, R., Liquid fluoride thorium reactors — an old idea in nuclear power gets reexamined, Am. Sci. **98** (2010) 304–313.

[186] ZHANG, D., et al., Review of conceptual design and fundamental research of molten salt reactors in China, Int. J. Energy Res. **42** 5 (2018) 1834–1848.

[187] WU, Y., et al., Experimental and numerical studies on natural circulation behavior of heat exchanger for molten salt loop, J. Nucl. Sci. Technol. **57** 3 (2020) 236–242.

[188] WANG, Z., et al., The evaporation behaviors of rare-earth-doped FLiNaK melts during low-pressure distillation, J. Radioanal. Nucl. Chem. **311** (2017) 637–642.

[189] ZENG, Y., et al., Tritium transport analysis in a 2-mw liquid-fueled molten salt experimental reactor with the code TMSR-TTAC, Nucl. Technol. **205** 4 (2019) 582–591.

[190] LOSA, E., et al., Neutronic tests of fluoride salt based MSR/FHR coolants, Trans. Am. Nucl. Soc. **116** 1 (2017) 1167–1169.

[191] KOŠŤÁL, M., et al., Study of graphite reactivity worth on well-defined cores assembled on LR-0 reactor, Ann. Nucl. Energy **87** (2016) 601–611.

[192] UHLÍŘ, J., STRAKA, M., MAREČEK, M., SZATMÁRY, L., “Progress in experimental development of MSR fuel cycle technology”, Proc. GLOBAL 2019, Seattle, WA, American Nuclear Society, La Grange Park, IL (2019) 275–278.

[193] STRAKA, M., SZATMÁRY, L., MAREČEK, M., KORENKO, M., Uranium recovery from $LiF–CaF_2–UF_4–GdF_3$ system on Ni electrode, J. Radioanal. Nucl. Chem. **298** 1 (2013) 393–397.

[194] KOUKOLÍKOVÁ, M., et al., Research of materials and components for molten salt reactors, Trans. Am. Nucl. Soc. **116** 1 (2017) 523–526.

[195] SCHØNFELDT, T., KLINKBY, E., “Molten salt thermal wasteburner”, Molten Salt Reactors and Thorium Energy (DOLAN, T.J., Ed.), Woodhead Publishing, Duxford, UK (2017) Ch. 24.

[196] CORDIS, Final Report Summary — EVOL (Evaluation and Viability of Liquid Fuel Fast Reactor System) (2019), https://cordis.europa.eu/project/id/249696/reporting

[197] KLOOSTERMAN, J.L., Activities in Euratom projects SAMOFAR and SAMOSAFER, presented at 2020 Molten Salt Reactor Workshop (virtual), Oak Ridge Natl Lab., 2020.

[198] MASCHEK, W., et al., "The SIMMER-III and SIMMER-IV Code Family: 2-D and 3-D Mechanistic Simulation Tools for Reactor Transients and Accidents", Proc. Tech. Mtg on use and development of coupled computer codes for the analysis of accidents at nuclear power plants, Vienna, 2003, IAEA, Vienna (2007).

[199] AUFIERO, M., et al., Development of an OpenFOAM model for the molten salt fast reactor transient analysis, Chem. Eng. Sci. **111** (2014) 390–401.

[200] WELLER, H.G., TABOR, G., JASAK, H., FUREBY, C., A tensorial approach to computational continuum mechanics using object-oriented techniques, Comput. Phys. **12** 6 (1998) 620.

[201] FIORINA, C., et al., GeN-Foam: a novel OpenFOAM® based multi-physics solver for 2D/3D transient analysis of nuclear reactors, Nucl. Eng. Des. **294** (2015) 24–37.

[202] FIORINA, C., KERKAR, N., MIKITYUK, K., RUBIOLO, P., PAUTZ, A., Development and verification of the neutron diffusion solver for the GeN-Foam multi-physics platform, Ann. Nucl. Energy **96** (2016) 212–222.

[203] AUFIERO, M., RUBIOLO, P., FRATONI, M., Monte Carlo/CFD coupling for accurate modeling of the delayed neutron precursors and compressibility effects in molten salt reactors, Trans. Am. Nucl. Soc. **116** (2017).

[204] TIBERGA, M., LATHOUWERS, D., KLOOSTERMAN, J.L., "A discontinuous Galerkin FEM multi-physics solver for the molten salt fast reactor", Proc. Int. Conf. on Mathematics and Computational Methods Applied to Nuclear Science and Engineering (M&C 2019), Portland, OR, 2019, American Nuclear Society (2019) 1818–1827.

[205] VOZÁROVÁ, N., et al., Thermodynamic determination and assessment of the CsF-ThF_4 system, J. Chem. Thermodyn. **114** (2017) 71–82.

[206] LAUREAU, A., Development of Neutronic Models for the Thermal Hydraulics Coupling of the Molten Salt Fast Reactor and for the Calculation of Effective Kinetic Parameters (Développement de Modèles Neutroniques pour le Couplage Thermohydraulique du MSFR et le Calcul de Paramètres Cinétiques Effectifs), PhD Thesis, Grenoble Alpes Univ. (2015).

[207] LEPPÄNEN, J., Serpent, The Serpent Monte Carlo code: Status, development and applications in 2013, Ann. Nucl. Energy **82** (2015) 142–150.

[208] SOUČEK, P., et al., Synthesis of UF_4 and ThF_4 by HF gas fluorination and re-determination of the UF_4 melting point, J. Fluorine Chem. **200** (2017) 33–40.

[209] SOUČEK, P., et al., Chemistry of molten salt reactor fuel salt candidates, Trans. Am. Nucl. Soc. **118** (2018) 114–117.

[210] TOSOLIN, A., et al., Synthesis of plutonium trifluoride by hydro-fluorination and novel thermodynamic data for the PuF_3-LiF system, J. Nucl. Mater. **503** (2018) 171–177.

[211] SOUČEK, P., et al., "Synthesis of pure actinide chlorides for thermodynamic studies on molten salt reactor fuel candidates", to be presented at 16th Information Exchange Mtg. on Actinide and Fission Product Partitioning and Transmutation (16IEMPT), Paris, France, 2023.

[212] CASSAYRE, L., et al., Electrochemistry of thorium in LiCl-KCl eutectic melts, Electrochim. Acta **52** 26 (2007) 7432–7437.

[213] MASSET, P., et al., Electrochemical behaviour of neptunium in the molten LiCl-KCl eutectic, J. Electroanal. Chem. **603** 2 (2007) 166–174.

[214] MASSET, P., et al., Electrochemistry of uranium in molten LiCl-KCl eutectic, J. Electrochem. Soc. **152** 6 (2005) A1109–A1115.

[215] MASSET, P., et al., Thermochemical properties of lanthanides (Ln=La, Nd) and actinides (An=U, Np, Pu, Am) in the molten LiCl-KCl eutectic, J. Nucl. Mater. **344** 1–3 (2005) 173–179.

[216] SERP, J., et al., Electrochemical behaviour of americium ions in LiCl-KCl eutectic melts, Electrochim. Acta **51** 19 (2006) 4024–4032.

[217] SERP, J., et al., Electrochemical behaviour of plutonium ion in LiCl-KCl eutectic melts, J. Electroanal. Chem. **561** (2004) 143–148.

[218] GLATZ, J.-P., et al., "Development of pyrochemical separation processes for recovery of actinides from spent nuclear fuel in molten LiCl-KCl", Molten Salts Chemistry — From Lab to Applications (LANTELME, F., GROULT, H., Eds) Elsevier, Oxford (2013) Ch. 26.

[219] SOUČEK, P., MALMBECK, R., "Pyrochemical processes for recovery of actinides from spent nuclear fuels", Reprocessing and Recycling of Spent Nuclear Fuel (TAYLOR, R., Ed.), Woodhead Publishing, Oxford (2015) Ch. 17.

[220] CONOCAR, O., et al., Promising pyrochemical actinide/lanthanide separation processes using aluminum, Nucl. Sci. Eng. **153** 3 (2006) 253–261.

[221] SERP, J., et al., Electroseparation of actinides from lanthanides on solid aluminum electrode in LiCl-KCl eutectic melts, J. Electrochem. Soc. **152** 3 (2005) C167–C172.

[222] SOUČEK, P., et al., Separation of actinides from irradiated An–Zr based fuel by electrorefining on solid aluminium cathodes in molten LiCl–KCl, J. Nucl. Mater. **459** (2015) 114–121.

[223] SOUČEK, P., et al., Electrorefining of irradiated experimental metallic fuel at high lanthanides concentration in the LiCl–KCl electrolyte, J. Nucl. Mater. **526** (2019).

[224] SOUČEK, P., et al., Electrochemical measurements of LiF-CaF_2-ThF_4 and LiF-ThF_4 melt and activity coefficient of ThF_4 in LiF-CaF_2 eutectic melt, Electrochim. Acta **380** (2021).

[225] SOUČEK, P., et al. "Development of electrochemical characterisation of actinides in molten fluoride salts in JRC Karlsruhe", paper presented at 11th Int. Symp. Molten Salts Chemistry and Technology, Orléans, France, 2019.

[226] BENEŠ, O., et al., The high-temperature heat capacity of the (Li,Na)F liquid solution, J. Chem. Thermodyn. **41** 8 (2009) 899–903.

[227] BENEŠ, O., SIERIG, M., KONINGS, R., DOCKENDORF, A., WURZER S., A DSC study of the NaNO3–KNO3 system using an innovative encapsulation technique, Thermochim. Acta **509** 1–2 (2010) 62–66.

[228] CAPELLI, E., et al., Thermodynamic investigation of the LiF–ThF4 system, J. Chem. Thermodyn. **58** (2013) 110–116.

[229] CAPELLI, E., BENEŠ, O., KONINGS, R.J.M., Thermodynamic assessment of the LiF–ThF_4–PuF_3–UF_4 system, J. Nucl. Mater. **462** (2015) 43–53.

[230] TOSOLIN, A., et al., Vaporization behaviour of the molten salt fast reactor fuel: The LiF-ThF_4-UF_4 system, J. Nucl. Mater. **508** (2018) 319–328.

[231] BEILMANN, M., et al., Excess heat capacity in liquid binary alkali-fluoride mixtures, Inorg. Chem. **52** 5 (2013) 2404–2411.

[232] CAPELLI, E., BENEŠ, O., KONINGS, R.J.M., Excess heat capacity of the $(Li_{1-x}Ca_x)F_{1+x}$ liquid solution determined by differential scanning calorimetry and drop calorimetry, J. Chem. Thermodyn. **81** (2015) 160–166.

[233] TOSOLIN, A., et al., Isobaric heat capacity of solid and liquid thorium tetrafluoride, J. Chem. Eng. Data **64** 9 (2019) 3945–3950.

[234] TOSOLIN, A., et al., Vaporization behaviour of a PuF_3-containing fuel mixture for the molten salt fast reactor, J. Nucl. Mater. **527** (2019).

[235] BENEŠ, O., et al., Cesium and iodine release from fluoride-based molten salt reactor fuel, Phys. Chem. Chem. Phys. **23** (2021) 9512–9523.

[236] SHEINDLIN, M., et al., Advances in the use of laser-flash techniques for thermal diffusivity measurement, Rev. Sci. Instrum. **69** (1998) 1426.

[237] VLAHOVIC, L., STAICU, D., KÜST, A., KONINGS, R.J.M., Thermal diffusivity of UO2 up to the melting point. J. Nucl. Mater. **499** (2018) 504–511.

[238] BENEŠ, O., KONINGS, R.J.M., Thermodynamic evaluation of the (LiF-NaF-BeF_2-PuF_3) system: An actinide burner fuel, J. Chem. Thermodyn. **41** 10 (2009) 1086–1095.

[239] BENEŠ, O., KONINGS, R.J.M., Thermodynamic evaluation of the MF-LaF_3 (M=Li, Na, K, Rb, Cs) systems, CALPHAD **32** 1 (2008) 121–128.

[240] BENEŠ, O., BEILMANN, M., KONINGS, R.J.M., Thermodynamic assessment of the LiF-NaF-ThF_4-UF_4 system, J. Nucl. Mater. **405** 2 (2010) 186–198.

[241] BEILMANN, M., et al., Thermodynamic assessment of the (LiF-UF_3) and (NaF-UF_3) systems, J. Chem. Thermodyn. **57** (2013) 22–31.

[242] CAPELLI, E., BENEŠ, O., KONINGS, R.J.M., Thermodynamics of soluble fission products cesium and iodine in the molten salt reactor, J. Nucl. Mater. **501** (2018) 238–252.

[243] BENEŠ, O., KONINGS, R.J.M., Thermodynamic evaluation of the NaCl-$MgCl_2$-UCl_3-$PuCl_3$ system, J. Nucl. Mater. **375** 2 (2008) 202–208.

[244] HANIA, P.R., et al., "SALIENT-01: Preparation and start of irradiation of thorium-bearing molten fluoride salt in graphite crucibles", Proc. GLOBAL 2019, Seattle, WA, American Nuclear Society, La Grange Park, IL (2019) 285–292.

[245] SRIDHARAN, K., ALLEN, T.R., "Corrosion in molten salts", Molten Salts Chemistry — From Lab to Applications (LANTELME, F., GROULT, H., Eds) Elsevier, Amsterdam, (2013) Ch. 12.

[246] TOTH, L.M., FELKER, L.K., Fluorine generation by gamma radiolysis of a fluoride salt mixture, Radiat. Eff. Defects Solids **112** 4 (1990) 201–210.

[247] HAUBENREICH, P.N., Fluorine Production and Recombination in Frozen MSR Salts after Reactor Operation, Rep. ORNL/TM-3144, Oak Ridge Natl Lab., TN (1970).

[248] GENERATION IV INTERNATIONAL FORUM, Annual Report (2008), https://www.gen-4.org/gif/upload/docs/application/pdf/2013-10/gif_2008_annual_report.pdf

[249] GERARDIN, D., et al., "Design evolutions of the Molten Salt Fast Reactor", Proc. Int. Conf. on Fast Reactors and Related Fuel Cycles: Next Generation Nuclear Systems for Sustainable Development (FR17), Yekaterinburg, 2017, IAEA, Vienna (2018).

[250] ALLIBERT, M., et al., "Molten Salt Fast Reactors", Handbook of Generation IV Nuclear Reactors (PIORO, I., Ed.), Woodhead Publishing Series in Energy, Cambridge, UK (2015) Ch. 7.

[251] MERLE-LUCOTTE, E., HEUER, D., ALLIBERT, M., DOLIGEZ, X., GHETTA, V., "Simulation tools and new developments of the Molten Salt Fast Reactor", Proc. European Nuclear Conf. (ENC2010), Barcelona, 2010, European Nuclear Society, Brussels (2010).

[252] DOLIGEZ, X., HEUER, D., MERLE LUCOTTE, E., ALLIBERT, M., GHETTA, V., Coupled study of the molten salt fast reactor core physics and its associated reprocessing unit, Ann. Nucl. Energy **64** (2014) 430–440.

[253] HEUER, D., et al., Towards the thorium fuel cycle with molten salt fast reactors, Ann. Nucl. Energy **64** (2014) 421–429.

[254] BROVCHENKO, M., et al., Neutronic benchmark of the molten salt fast reactor in the frame of the EVOL and MARS collaborative projects, EPJ Nucl. Sci. Technol. **5** 2 (2019) 1.

[255] ALLIBERT, M., et al., Preliminary proliferation study of the molten salt fast reactor, EPJ Nucl. Sci. Technol. **6** (2020) 5.

[256] IGNATIEV, V.V., et al., "Molten Salt Fast Reactor in Generation IV: proliferation challenges", Proc. Symp. Int. Safeguards, Vienna 2018, IAEA, Vienna (2018).

[257] GIOT, L., PITOIS, H., SAVALOINEN, P., GUO, Z., "Decay heat calculations with JEFF libraries", presented at JEFF Meetings, NEA, Paris, France, 2019.

[258] LAUREAU, A., AUFIERO, M., RUBIOLO, P., MERLE-LUCOTTE, E., HEUER, D., Transient fission matrix: kinetic calculation and kinetic parameters beta-eff and lambda-eff calculation, Ann. Nucl. Energy **85** (2015) 1035–1044.

[259] LAUREAU, A., BUIRON, L., FONTAINE, B., Local correlated sampling Monte Carlo calculations in the TFM neutronics approach for spatial and point kinetics applications, EPJ Nucl. Sci. Technol. **3** 16 (2017) 1.

[260] LAUREAU, A., BUIRON, L., FONTAINE, B., Towards spatial kinetics in a low void effect sodium fast reactor: core analysis and validation of the TFM neutronic approach, EPJ Nucl. Sci. Technol. **3** 17 (2017) 1.

[261] LAUREAU, A., BUIRON, L., FONTAINE, B., "Fission matrix interpolation for the TFM approach based on a local correlated sampling technique for fast spectrum heterogeneous reactors", Proc. Int. Conf. Mathematics and Computational Methods (M&C 2017), Jeju, Korea, 2017, Korean Nuclear Society, Daejeon (2017).

[262] LAUREAU, A., LEDERER, Y., KRAKOVICH, A., BUIRON, L., FONTAINE, B., "Transient coupled neutronics-thermalhydraulics study of ULOF accidents in sodium fast reactors using spatial kinetics: Comparison of the TFM Monte Carlo and S_N approaches", Proc. Int. Congr. Advances in Nuclear Power Plants (ICAPP 2017) Kyoto, 2017 Atomic Energy Society of Japan (2017).

[263] BLAISE, P., LAUREAU, A., ROS, P., LECONTE, P., ROUTSONIS, K., Transient fission matrix approach for assessing complex kinetics behavior in the ZEPHYR ZPR coupled core configurations, Ann. Nucl. Energy **128** (2019) 390–397.

[264] LAUREAU, A., et al., "The LiCore Power Plant Simulator of the Molten Salt Fast Reactor", Proc. Int. Conf. Physics of Reactors (PHYSOR 2020), Cambridge, UK, 2020, EDP Sciences, Paris (2021).

[265] HEUER, D., LAUREAU, A., MERLE-LUCOTTE, E., ALLIBERT, M., GERARDIN, D., "A starting procedure for the MSFR: Approach to criticality and incident analysis", Proc. Int. Congr. Advances in Nuclear Power Plants (ICAPP 2017) Kyoto, 2017, Atomic Energy Society of Japan (2017).

[266] BROVCHENKO, M., Preliminary Safety Studies of the MSFR Molten Salt Reactor (Etudes préliminaires de sûreté du réacteur à sels fondus MSFR), PhD Thesis, Grenoble Institute of Technology (2013).

[267] GENERATION 4 INTERNATIONAL FORUM (GIF), An Integrated Safety Assessment Methodology (ISAM) for Generation IV Nuclear Systems (2011), https://www.gen-4.org/gif/upload/docs/application/pdf/2013-09/gif_rsgw_2010_2_isamrev1_finalforeg17june2011.pdf

[268] GERARDIN, D., et al., Identification of the postulated initiating events of the molten salt fast reactor, Nucl. Eng. Technol. **51** 4 (2019) 1024–1031.

[269] UGGENTI, A.C., et al., "Preliminary functional safety assessment for Molten Salt Fast Reactors in the framework of the SAMOFAR project", Proc. Int. Topical Mtg on Probabilistic Safety Assessment and Analysis (PSA 2017), Pittsburgh, 2017, American Nuclear Society, IL (2017).

[270] GERARDIN, D., Development of Numerical Tools and Studies for the Operation and Safety of the MSFR Molten Salt Reactor (Développement d'outils numériques et réalisation d'études pour le pilotage et la sûreté du réacteur à sels fondus MSFR), PhD Thesis, Grenoble Alpes Univ., France (2018).

[271] RODRIGUES, D., Solvation of Thorium by Fluorides in a Molten Salt Medium at High Temperature. Application to the Extractive Reduction Process for the MSFR Concept (Solvatation du thorium par les fluorures en milieu sel fondu à haute température. Application au procédé d'extraction réductrice pour le concept MSFR), PhD Thesis, Paris-Sud Univ. (2015).

[272] DURAN-KLIE, G., Study of the Uranium and Iodine Behavior in the Molten Fluoride Mix LiF-ThF4 at 650°C (Etude du comportement de l'uranium et de l'iode dans le mélange de fluorures fondus LiF-ThF_4 à 650°C), PhD Thesis, Paris-Saclay Univ. (2017).

[273] DELPECH, S., RODRIGUES, D., DURAN-KLIE, G., "Extraction of zirconium in the Molten Salt Fast Reactor", Proc. Int. Congr. Advances in Nuclear Power Plants (ICAPP 2019) Juan-Les-Pins, 2019, Société française d'énergie nucléaire (SFEN), Paris (2019).

[274] MARCHIX, A., DIEUAIDE, M., Molten salt fast reactor SAMOFAR: Safety issues of the chemical plant, EPJ Web Conf. **239** (2020).

[275] BETTIS, E.S., ROBERTSON, R.C., The design and performance features of a single-fluid molten salt breeder reactor, Nucl. Appl. Technol. **8** 2 (1970) 190–207.

[276] WHATLEY, M.E., MCNEESE, L.E., CARTER, W.L., FERRIS, L.M., NICHOLSON, E.L., Engineering development of the MSBF fuel recycle, Nucl. Appl. Technol. **8** 2 (1970) 170–178.

[277] DELPECH, S., CABET, C., SLIM, C., PICARD, G., Molten fluorides for nuclear applications, Mater. Today **13** (2010) 34–41.

[278] DELPECH, S., Possible routes for pyrochemical separations: Focus on the reductive extraction in fluoride media, Pure Appl. Chem. **85** 1 (2012) 71–87.

[279] JASKIEROWICZ, S., et al., "Pyrochemical reprocessing of thorium-based fuel", Proc. Int. Congr. Advances in Nuclear Power Plants (ICAPP 2011) Nice, France, 2011, Société française d'énergie nucléaire (SFEN), Paris (2011).

[280] RODRIGUES, D., DURAN, G., DELPECH, S., Pyrochemical reprocessing of molten salt fast reactor fuel: focus on the reductive extraction step, Nukleonika **60** (2015) 907.

[281] LAPLACE, A., et al., Elaboration of solid solutions of actinides and lanthanides oxides in a molten salt medium: application to a new fuel remanufacturing process by pyrochemical method (Elaboration de solutions solides d'oxydes d'actinides et de lanthanides en milieu sels fondus: application à un nouveau procédé de refabrication du combustible par voie pyrochimique), Patent filed on 26/09/2011, no. FR 11/58572 (2011).

[282] DELPECH, S., et al., Description of the chemical plant required for MSFR fuel reprocessing, Deliverable 5.2, Work Package 5, SAMOFAR European Project, Grant Agreement No. 661891, Horizon 2020 Program (2018).

[283] CHMAKOFF, A., Influence of Redox Potential on Material Corrosion in Molten Salt Media (Influence du potentiel redox sur la corrosion des matériaux en milieux sels fondus), MSc Thesis, IPN-Orsay, France (2019).

[284] RUBIOLO, P., et al., "Design of close and open channel experiments to study molten salt flows", Proc. Int. Congr. Advances in Nuclear Power Plants (ICAPP 2018) Charlotte, NC, 2018, American Nuclear Society, IL (2018).

[285] TANO-RETAMALES, M., Development of Multi-physical Multiscale Models for Molten Salts at High Temperature and their Experimental Validation, PhD Thesis, Grenoble Alpes Univ. (2018).

[286] TANO-RETAMALES, M., RUBIOLO, P., GIRAUD, J., GHETTA, V., "Multiphysics study of the draining transients in the Molten Salt Fast Reactor", Proc. Int. Congr. Advances in Nuclear Power Plants (ICAPP 2018) Charlotte, NC, 2018, American Nuclear Society, IL (2018).

[287] TANO-RETAMALES, M., RUBIOLO, P., DOCHE, O., Progress in modeling solidification in molten salt coolants, Model. Simul. Mater. Sci. Eng. **25** (2017).

[288] GIRAUD, J., GHETTA, V., RUBIOLO, P., TANO-RETAMALES, M., Development of a cold plug valve with fluoride salt, EPJ Nucl. Sci. Technol. **5** (2019) 9.

[289] INTERNATIONAL ATOMIC ENERGY AGENCY, Near Term and Promising Long Term Options for Deployment of Thorium Based Nuclear Energy (2016), https://www.iaea.org/projects/crp/t12026.

[290] LUZZI, L., DI MARCELLO, V., CAMMI, A., Multi-Physics Approach to the Modelling and Analysis of Molten Salt Reactors, Nova Science Publishers, New York, NY (2012).

[291] CAMMI, A., DI MARCELLO, V., GUERRIERI, C., LUZZI, L., Transfer function modeling of zero-power dynamics of circulating fuel reactors, J. Eng. Gas Turbines Power **133** 5 (2011) 1–8.

[292] ZANETTI, M., CAMMI, A., FIORINA, C., LUZZI, L., A geometric multiscale modelling approach to the analysis of MSR plant dynamics, Prog. Nucl. Energy **83** (2015) 82.

[293] CAMMI, A., FIORINA, C., GUERRIERI, C., LUZZI, L., Dimensional effects in the modelling of MSR dynamics: Moving on from simplified schemes of analysis to a multi-physics modelling approach, Nucl. Eng. Des. **246** (2012) 12.

[294] GUERRIERI, C., CAMMI, A., LUZZI, L., An approach to the MSR dynamics and stability analysis, Prog. Nucl. Energy **67** (2013) 56.

[295] TRIPODO, C., DI RONCO, A., LORENZI, S., CAMMI, A., Development of a control-oriented power plant simulator for the molten salt fast reactor, EPJ Nucl. Sci. Technol. **5** 13 (2019).

[296] GUERRIERI, C., CAMMI, A., LUZZI, L., A preliminary approach to the MSFR control issues, Ann. Nucl. Energy **64** (2014) 472.

[297] TRIPODO, C., DI RONCO, A., LORENZI, S., CAMMI, A., Definition of model-based control strategies for the molten salt fast reactor nuclear power plant, Nucl. Eng. Des. **373** (2021).

[298] AUFIERO, M., et al., Calculating the effective delayed neutron fraction in the molten salt fast reactor: Analytical, deterministic and Monte Carlo approaches, Ann. Nucl. Energy **65** (2014) 78.

[299] AUFIERO, M., et al., An extended version of the SERPENT-2 code to investigate fuel burn-up and core material evolution of the molten salt fast reactor, J. Nucl. Mater. **441** 1–3 (2013) 473.

[300] FIORINA, C., et al., Investigation of the MSFR core physics and fuel cycle characteristics, Prog. Nucl. Energy **68** (2013) 153–168.

[301] CAMMI, A., DI MARCELLO, V., LUZZI, L., MEMOLI, V., RICOTTI, M.E., A multi-physics modelling approach to the dynamics of molten salt reactors, Ann. Nucl. Energy **38** 6 (2011) 1356.

[302] FIORINA, C., et al., Modelling and analysis of the MSFR transient behaviour, Ann. Nucl. Energy **64** (2014) 485.

[303] CERVI, E., LORENZI, S., CAMMI, A., LUZZI, L., Development of a multiphysics model for the study of fuel compressibility effects in the molten salt fast reactor, Chem. Eng. Sci. **193** (2019) 379.

[304] CERVI, E., LORENZI, S., CAMMI, A., LUZZI, L., Development of an SP 3 neutron transport solver for the analysis of the molten salt fast reactor, Nucl. Eng. Des. **346** (2019) 209.

[305] CERVI, E., LORENZI, S., LUZZI, L., CAMMI, A., Multiphysics analysis of the MSFR helium bubbling system: A comparison between neutron diffusion, SP3 neutron transport and Monte Carlo approaches, Ann. Nucl. Energy **132** (2019) 227.

[306] TIBERGA, M., et al., Results from a multi-physics numerical benchmark for codes dedicated to molten salt fast reactors, Ann. Nucl. Energy **142** (2020).

[307] LUZZI, L., AUFIERO, M., CAMMI, A., FIORINA, C., "Thermo-hydrodynamics of internally heated molten salts for innovative nuclear reactors", Hydrodynamics — Theory and Model (ZHENG, J., Ed.), InTechOpen, Rijeka, Croatia (2012) Ch. 6.

[308] FIORINA, C., et al., Thermal-hydraulics of internally heated molten salts and application to the molten salt fast reactor, J. Phys. Conf. Ser. **501** 1 (2014).

[309] LECCE, F.D., et al., "CFD-based correlation for forced convection heat transfer in circular ducts of internally heated molten salts", Proc. Int. Conf. on Nuclear Engineering (ICONE 26), London, 2018, American Society of Mechanical Engineers, New York (2018).

[310] RUIZ, D.E., CAMMI, A., LUZZI, L., Dynamic stability of natural circulation loops for single phase fluids with internal heat generation, Chem. Eng. Sci. **126** (2015) 573.

[311] CAMMI, A., LUZZI, L., PINI, A., The influence of the wall thermal inertia over a single-phase natural convection loop with internally heated fluids, Chem. Eng. Sci. **153** (2016) 411.

[312] PINI, A., CAMMI, A., LORENZI, S., CAUZZI, M.T., LUZZI, L., A CFD-based simulation tool for the stability analysis of natural circulation systems, Prog. Nucl. Energy **117** (2019).

[313] PINI, A., CAMMI, A., LUZZI, L., Analytical and numerical investigation of the heat exchange effect on the dynamic behaviour of natural circulation with internally heated fluids, Chem. Eng. Sci. **145** (2016) 108.

[314] PINI, A., CAMMI, A., CAUZZI, M., FANALE, F., LUZZI, L., "An experimental facility to investigate the natural circulation dynamics in presence of distributed heat sources", Proc. 71st Conf. Italian Thermal Machines Engineering Association (ATI 2016), Turin, Italy, 2016, Elsevier, Amsterdam (2016) 10.

[315] COPPA, G., et al., Enhanced quasi-static method for reactor dynamics, Trans. Am. Nucl. Soc. **74** (1996) 150–151.

[316] RAVETTO, P., Perturbation theory for fluid-core recirculated systems, Trans. Am. Nucl. Soc. **78** (1998) 259.

[317] LAPENTA, G., RAVETTO, P., Basic reactor physics problems in fluid-fuel recirculated reactors, Kerntechnik **65** 5 (2000) 250–253.

[318] LAPENTA, G., MATTIODA, F., RAVETTO, P., Point kinetic model for fluid-fuel systems, Ann. Nucl. Energy **28** 17 (2001) 1759–1772.

[319] MATTIODA, F., RAVETTO, P., RITTER, G., Effective delayed neutron fraction for fluid-fuel systems, Ann. Nucl. Energy **27** 16 (2000) 1523–1532.

[320] DULLA, S., ROSTAGNO, M.M., RAVETTO, P., Quasi-static method for the time-dependent neutronics of fluid-fuel systems, Trans. Am. Nucl. Soc. **89** (2003) 446–448.

[321] DULLA, S., RAVETTO, P., ROSTAGNO, M.M., Neutron kinetics of fluid-fuel systems by the quasi-static method, Ann. Nucl. Energy **31** 15 (2004) 1709–1733.

[322] DELPECH, M., et al., "Benchmark of dynamic simulation tools for molten salt reactors", Proc. GLOBAL 2003, New Orleans, LA, 2003, American Nuclear Society, IL (2003) 2182.

[323] DULLA, S., RAVETTO, P., Interaction between fluid-dynamics and neutronic phenomena in the physics of molten-salt systems, Nucl. Sci. Eng. **155** 3 (2007) 475–488.

[324] DULLA, S., NICOLINO, C., RAVETTO, P., "Reactivity oscillations in molten salt reactors", Joint Int. Topical Mtg on Mathematics & Computation and Supercomputing in Nuclear Applications (M&C + SNA 2007), Monterey, CA, 2007, American Nuclear Society, IL (2007).

[325] DULLA, S., NICOLINO, C., RAVETTO, P., "Feedback effects in the dynamics of molten salt reactors", Proc. 12th Int. Conf. Emerging Nuclear Energy Systems (ICENES'05), Brussels, Belgium, 2005, Belgian Nuclear Research Centre (SCK CEN), Mol, Belgium (2005).

[326] DULLA, S., NICOLINO, C., LAPENTA, G., RAVETTO, P., Thermal and fluid-dynamic effects in the physics of molten salt reactors, Trans. Am. Nucl. Soc. **95** 1 (2006) 765–767.

[327] NICOLINO, C., DULLA, S., RAVETTO, P., LAPENTA, G., "Multi-physics approach to the analysis of molten salt reactors", Proc. Int. Conf. Physics of Reactors (PHYSOR 2008), Interlaken, Switzerland, 2008, Paul Scherrer Institut (PSI), Villigen (2008).

[328] NICOLINO, C., LAPENTA, G., DULLA, S., RAVETTO, P., Coupled dynamics in the physics of molten salt reactors, Ann. Nucl. Energy **35** 2 (2008) 314–322.

[329] ALCARO, F., DULLA, S., RAVETTO, P., Neutronic evaluations for the EVOL molten salt reactor, Trans. Am. Nucl. Soc. **108** 1 (2013) 927–930.

[330] MASCHEK, W., et al., Report on intermediate results of the IAEA CRP on studies of advanced reactor technology options for effective incineration of radioactive wastes, Energy Convers. Manage. **49** 7 (2008) 1810–1819.

[331] SANCHEZ, R., GANAPOL, B.D., DULLA, S., RAVETTO, P., "Analysis of a simplified molten-fuel reactor", Proc. 18th Int. Conf. Transport Theory, Rio de Janeiro, Brazil, Universidade do Estado, Brazil (2003) 196–204.

[332] DULLA, S., SANCHEZ, R., RAVETTO, P., "Neutron transport and diffusion models for molten-salt reactor dynamics", Int. Conf. Physics of Reactors (PHYSOR 2006), Vancouver, 2006, American Nuclear Society, IL (2006).

[333] DULLA, S., PRINJA, A.K., RAVETTO, P., Random effects on reactivity in molten salt reactors, Ann. Nucl. Energy **64** (2014) 353–364.

[334] ABRATE, N., AUFIERO, M., DULLA, S., FIORITO, L., "Nuclear data uncertainty quantification in molten salt reactors with xGPT", Proc. Int. Conf. Mathematics Computational Methods and Reactor Physics (M&C 2019), Portland, OR, 2019, American Nuclear Society, IL (2019) 2348.

[335] INTERNATIONAL ELECTROTECHNICAL COMMISSION, IEC 61508-1:2010 Functional safety of electrical/electronic/programmable electronic safety-related systems — Part 1: General requirements (2010).

[336] GÉRARDIN, D., et al., A methodology for the identification of the postulated initiating events of the molten salt fast reactor, Nucl. Eng. Technol. **51** 4 (2019) 1024–1031.

[337] MERLE, E., et al., "Design and safety studies of the Molten Salt Fast Reactor concept in the frame of the SAMOFAR H2020 Project", Proc. Generation IV Int. Forum Symp., Paris, France (2018) 299–304.

[338] FURUKAWA, K., et al., A road map for the realization of global-scale thorium breeding fuel cycle by single molten fluoride flow, Energy Convers. Manage. **49** (2008) 1832–1848.

[339] YOSHIOKA, R., KINOSHITA, M., "Liquid Fuel, Thermal Neutron Spectrum Reactors", Molten Salt Reactor and Thorium Energy (DOLAN, T.J., Ed.), Woodhead Publishing, Duxford, UK (2017) Ch. 11.

[340] FURUKAWA, K., et al., Single-fluid-type accelerator molten-salt breeder concept, J. Nucl. Sci. Technol. **18** 1 (1981) 79–81.

[341] HARMS, A.A., HEINDLER, M., Nuclear Energy Synergetics: An Introduction to Conceptual Models of Integrated Nuclear Energy Systems, Springer (1982).

[342] FURUKAWA, K., et al., Thorium molten-salt nuclear energy synergetics, J. Nucl. Sci. Technol. **27** 12 (1990) 1157–1178.

[343] INTERNATIONAL ATOMIC ENERGY AGENCY, Unconventional Options for Plutonium Disposition, IAEA, Vienna (1995).

[344] STEMPNIEWICZ, M., SPECTRA Sophisticated Plant Evaluation Code for Thermal-Hydraulic Response Assessment, Version 3.61, Volume 1 — Program Description, Rep. K6202/18.149165 MSt-181130, NRG, Arnhem (2018), available on request from:
https://www.ensuringnuclearperformance.com/producten/safe-operations/veiligheidsanalyses/spectra

[345] STEMPNIEWICZ, M., ibid., Volume 2 — User's Guide.

[346] STEMPNIEWICZ, M., ibid., Volume 3 — Verification and Validation.

[347] KEDL, R., A Model for Computing the Migration of Very Short Lived Noble Gases into MSRE Graphite, Rep. ORNL-TM-1810, Oak Ridge Natl Lab., TN (1967).

[348] KEDL, R., HOUTZEEL, A., Development of a Model for Computing Xe-135 Migration in MSRE, Rep. ORNL-4069, Oak Ridge Natl Lab., TN (1967).

[349] CAPELLI, E., et al., "Noble metal behaviour and extraction in molten salt reactor", presented at Thorium Energy Conf. 2018 (ThEC18), Brussels, Belgium, 2018.

[350] ZHENG, G., et al., High-temperature corrosion of UNS N10003 in molten Li2BeF4 (FLiBe) salt, Corrosion **71** 10 (2015) 1257–1266.

[351] ROELOFS, F., STEMPNIEWICZ, M., "Molten salt modelling capabilities in SPECTRA and application to MSRE", Proc. Int. Topical Mtg. Advances in Thermal Hydraulics (ATH 2020), Paris, France, 2020, American Nuclear Society, IL (2020).

[352] AHMED, K., SCARLAT, R., HU, R., "Benchmark simulation of natural circulation cooling systems with salt working fluid using SAM", Proc. 17th Int. Topical Mtg. Nuclear Reactor Thermal Hydraulics (NURETH17), Vol. 10, Xi'an, China, 2017, American Nuclear Society, IL (2017) 5421.

[353] STEMPNIEWICZ, M., DE GEUS, E., ROELOFS, F., "Analysis of molten salt natural circulation in Mk1 PB-FHR with SPECTRA", presented at Thorium Energy Conf. 2018 (ThEC18), Brussels, Belgium, 2018.

[354] ALCARO, F., STH analysis of the MSRE Thermal-Convection Heat Removal Test with SPECTRA, NRG note 24597/20.173962, Arnhem, Netherlands (2020).

[355] ALSAYYARI, F., Adaptive Data-Driven Reduced Order Modelling Techniques for Nuclear Reactor Analysis, PhD Thesis, Delft Univ. of Technology (2020).

[356] CAPELLI, E., Thermodynamic Characterization of Salt Components for Molten Salt Reactor Fuel, PhD Thesis, Delft Univ. of Technology (2016).

[357] CHRISTIE, S., Quantifying Nuclide Contributions to Reactor Behaviour over Time, PhD Thesis, Delft Univ. of Technology (2014).

[358] GILLI, L., Uncertainty Quantification in Reactor Physics Using Adjoint/Perturbation Techniques and Adaptive Spectral Methods, PhD Thesis, Delft Univ. of Technology (2013).

[359] MASTROMARINO, S., Determination of Thermodynamic Properties of Molten Salts, PhD Thesis, Delft Univ. of Technology (2020).

[360] NAGY, K., Dynamics and Fuel Cycle Analysis of a Moderated Molten Salt Reactor, PhD Thesis, Delft Univ. of Technology (2012).

[361] PERKÓ, Z., Sensitivity and Uncertainty Analysis of Coupled Reactor Physics Problems — Method Development for Multi-Physics in Reactors, PhD Thesis, Delft Univ. of Technology (2015).

[362] TIBERGA, M., Development of a High-Fidelity Multi-Physics Simulation Tool for Liquid-Fuel Fast Nuclear Reactors, PhD Thesis, Delft Univ. of Technology (2020).

[363] FIORINA, C., et al., Modeling and analysis of the MSFR transient behaviour, Ann. Nucl. Energy **64** (2014) 485–498.

[364] TIBERGA, M., LATHOUWERS, D., KLOOSTERMAN, J.L., A multi-physics solver for liquid-fueled fast systems based on the discontinuous Galerkin FEM discretization, Prog. Nucl. Energy **127** (2020).

[365] TIBERGA, M., et al., Results of the CNRS benchmark (dataset), Zenodo (2019).

[366] ALSAYYARI, F., PERKO, Z., LATHOUWERS, D., KLOOSTERMAN, J.L., A nonintrusive reduced order modelling approach using proper orthogonal decomposition and locally adaptive sparse grids, J. Comput. Phys. **399** (2019).

[367] ALSAYYARI, F., PERKO, Z., LATHOUWERS, D., A nonintrusive adaptive reduced order modelling approach for a molten salt reactor system, Ann. Nucl. Energy **141** (2020).

[368] PERKÓ, Z., GILLI, L., LATHOUWERS, D., KLOOSTERMAN, J.L., Grid and basis adaptive polynomial chaos techniques for sensitivity and uncertainty analysis, J. Comput. Phys. **260** (2014) 54–84.

[369] PERKÓ, Z., LATHOUWERS, D., KLOOSTERMAN, J.L., VAN DER HAGEN, T., Large scale applicability of a fully adaptive non-intrusive spectral projection technique: sensitivity and uncertainty analysis of a transient, Ann. Nucl. Energy **71** (2014) 272–292.

[370] IGNATIEV, V., et al., Key experimental results of the PYROSMANI project, Procedia Chem. **21** (2016) 417–422.

[371] IGNATIEV, V., et al., Molten salt reactor for nuclear fuel cycle closure on all actinides, At. Energy **125** 5 (2019) 279–283.

[372] HOMBOURGER, B., KREPEL, J., PAUTZ, A., Breed-and-burn fuel cycle in molten salt reactors, EPJ Nucl. Sci. Technol. **5** (2019).

[373] KREPEL, J., KRAMER, K.J., "Molten Chloride Fast Reactors (MCFRs)", Encyclopedia of Nuclear Energy, Elsevier (2021), 625–642.

[374] KŘEPEL, J., et al., Fuel cycle advantages and dynamics features of liquid fueled MSR, Ann. Nucl. Energy **64** (2013) 380–397.

[375] KŘEPEL, J., et al., EQL3D: ERANOS based equilibrium fuel cycle procedure for fast reactors, Ann. Nucl. Energy **36** 5 (2009) 550–561.

[376] RIMPAULT, G., et al., "The ERANOS code and data system for fast reactor neutronic analyses", Proc. Int. Conf. Physics of Reactors (PHYSOR 2002), Seoul, Korea, American Nuclear Society, IL (2002).

[377] HOMBOURGER, B., Parametric Lattice Study for Conception of a Molten Salt Reactor in Closed Thorium Fuel Cycle, MSc Thesis, EPFL–ETH Zürich, Switzerland (2013).

[378] HOMBOURGER, B., et al., Parametric lattice study of a graphite-moderated molten salt reactor, J. Nucl. Eng. Radiat. Sci. **1** (2015).

[379] HOMBOURGER, B., et al., "The EQL0D procedure for fuel cycle studies in molten salt reactors", Proc. Int. Congr. Advances in Nuclear Power Plants (ICAPP 2016), San Francisco, 2016, American Nuclear Society, IL (2016).

[380] PUSA, M., LEPPÄNEN, J., Solving linear systems with sparse Gaussian elimination in the Chebyshev rational approximation method, Nucl. Sci. Eng. **175** 3 (2013) 250–258.

[381] KŘEPEL, J., LOSA, E., Closed U–Pu and Th–U cycle in sixteen selected reactors: Comparison of major equilibrium features, Ann. Nucl. Energy **128** (2019) 341–357.

[382] KŘEPEL, J., et al., "Molten salt reactor with simplified fuel recycling and delayed carrier salt cleaning", Proc. 22nd Int. Conf. on Nuclear Engineering (ICONE 22), Prague, Czech Republic, 2014, American Society of Mechanical Engineers, New York (2014).

[383] KŘEPEL, J., et al., "Comparison of several recycling strategies and relevant fuel cycles for molten salt reactor", Int. Congr. Advances in Nuclear Power Plants (ICAPP 2015), Nice, France, 2015, Société française d'énergie nucléaire (SFEN), Paris (2015).

[384] SHCHERBINA, N., KIVEL, N., GÜNTHER-LEOPOLD, I., KULIK, D.A., BERTSCH, J., HERACLES Project and Thermodynamic Database (2012), http://www.psi.ch/heracles/heracles

[385] KULIK, D.A., et al., GEM-Selektor geochemical modeling package: revised algorithm and GEMS3K numerical kernel for coupled simulation codes, Comput. Geosci. **17** (2013) 1–24.

[386] WAGNER, T., KULIK, D.A., HINGERL, F.F., DMYTRIEVA, S.V., GEM-Selektor geochemical modeling package: TSolMod library and data interface for multicomponent phase models, Can. Mineral. **50** (2012) 1173–1195.

[387] NICHENKO, S., STREIT, M., "Thermodynamic modelling of molybdenum behaviour in chloride molten salt", Proc. Topical Mtg on the reactor fuel performance (TopFuel-2015), Zürich, Switzerland, 2015, European Nuclear Society, Brussels (2015).

[388] MELCOR Computer Code Manuals, Vol. 2: Reference Manual, Version 2.2.9541, Rep. SAND 2017-0876 O, Sandia Natl Lab., NM (2017).

[389] KALILAINEN, J., NICHENKO, S., KREPEL, J., Evaporation of materials from the molten salt reactor fuel under elevated temperatures, J. Nucl. Mater. **533** (2020).

[390] DAI, J., LI, X., ZHAO, L., SUN, H., Enthalpies of mixing predicted using molecular dynamics simulations and OPLS force field, Fluid Phase Equilib. **289** (2010) 156–165.

[391] DAI, J., WU, C., BAO, X., SUN, H., Prediction of the heat of mixing for binary fluids using molecular dynamics simulation, Fluid Phase Equilib. **236** (2005) 78–85.

[392] NICHENKO, S., STAICU, D., Molecular dynamics study of the mixed oxide fuel thermal conductivity, J. Nucl. Mater. **439** 1–3 (2013) 93–98.

[393] NICHENKO, S., STAICU, D., Thermal conductivity of porous UO_2: Molecular dynamics study, J. Nucl. Mater. **454** 1–3 (2014) 315–322.

[394] BENEŠ, O., KONINGS, R.J.M., Thermodynamic properties and phase diagrams of fluoride salts for nuclear applications, J. Fluorine Chem. **130** 1 (2009) 22–29.

[395] GUO, X., et al., Theoretical evaluation of microscopic structural and macroscopic thermo-physical properties of molten AF-ThF4 systems (A+ = Li+, Na+ and K+), J. Mol. Liq. **277** 1 (2019) 409–417.

[396] FIORINA, C., MIKITYUK, K., "Application of the new GeN-Foam multi-physics solver to the European Sodium Fast Reactor and verification against available codes", Proc. Int. Congr. Advances in Nuclear Power Plants (ICAPP 2015), Nice, France, 2015, Société française d'énergie nucléaire (SFEN), Paris (2015).

[397] ARIU, V., Heat Exchanger Analysis for Innovative Molten Salt Fast Reactor, MSc Thesis, EPFL–ETH Zürich, Switzerland (2014).

[398] ZANETTI, M., et al., "Extension of the FAST code system for the modelling and simulation of MSR dynamics", Proc. Int. Congr. Advances in Nuclear Power Plants (ICAPP 2015), Nice, France, 2015, Société française d'énergie nucléaire (SFEN), Paris (2015).

[399] KIM, H., Static and Transient Analysis of Molten Salt Reactor Experiment Using Serpent/TRACE/PARCS Codes, MSc Thesis, EPFL–ETH Zürich, Switzerland (2015).

[400] CHOE, J., Empirical Decay Heat Correlation and Fission Product Behavior in MSR, MSc Thesis, EPFL–ETH Zürich, Switzerland (2015).

[401] BAO, J., Development of the Model for the Multi-Physics Analysis of Molten Salt Reactor Experiment Using GeN-Foam Code, MSc Thesis, EPFL–ETH Zürich, Switzerland (2016).

[402] PETTERSEN, E., Coupled Multi-Physics Simulations of the MSFR Using TRACE-PARCS, MSc Thesis, Univ. Paris-Saclay (2016).

[403] UNITED STATES DEPARTMENT OF ENERGY, Nuclear Energy Research and Development Roadmap, Report to Congress (2010),
https://www.energy.gov/sites/prod/files/NuclearEnergy_Roadmap_Final.pdf

[404] TOLSON, G.M., TABOADA, A., MSRE Control Elements: Manufacture, Inspection, Drawings, and Specifications, Rep. ORNL-4123, Oak Ridge Natl Lab., TN (1967).

[405] INTERNATIONAL ATOMIC ENERGY AGENCY, Advanced Reactors Information System (ARIS),
https://aris.iaea.org

[406] NUCLEAR ENERGY INSTITUTE, Addressing the Challenges with Establishing the Infrastructure for the Front-end of the Fuel Cycle for Advanced Reactors, NEI White Paper, Washington, DC (2018).

[407] NUCLEAR REGULATORY COMMISSION, Hypothetical Accident Conditions, 10 CFR 71, US Govt Printing Office, Washington, DC (1995).

[408] FLANAGAN, G.F., QUALLS, L.A., Establish Fuel Qualification Expectations, Rep. ORNL/LTR-2018/495, Oak Ridge Natl Lab., TN (2018).

[409] RILEY, B.J., et al., Identification of Potential Waste Processing and Waste Form Options for Molten Salt Reactors, Rep. PNNL-27723, Pacific Northwest Natl Lab., WA (2018).

[410] ARM, S.T., HOLCOMB, D.E., HOWARD, R., RILEY, B.J., Status of Fast Spectrum Molten Salt Reactor Waste Management Practice, Rep. PNNL-30739, Pacific Northwest Natl Lab., WA (2020).

[411] SIEMER, D.D., Molten salt breeder reactor waste management, Nucl. Technol. **185** 1 (2014) 100–108.

[412] SIEMER, D.D., Improving the integral fast reactor's proposed salt waste management system, Nucl. Technol. **178** 3 (2012) 341–352.

[413] ROCKWOOD, L., Legal Framework for IAEA Safeguards, IAEA, Vienna (2013).

[414] The Structure and Content of Agreements Between the Agency and States Required in Connection with the Treaty on the Non-Proliferation of Nuclear Weapons, INFCIRC/153 (Corrected), IAEA, Vienna (1972).

[415] KOVACIC, D.N., et al., "Safeguards challenges for molten salt reactors", Proc. INMM Annual Mtg, Baltimore, MD, 2018, Institute of Nuclear Materials Management, Oakbrook Terrace, IL (2018) 1573.

[416] INTERNATIONAL ATOMIC ENERGY AGENCY, Safeguards Techniques and Equipment: 2011 Edition, International Nuclear Verification Series No. 1 (Rev. 2), IAEA, Vienna (2011).

[417] BEDDINGFIELD, D.H., HORI, M., "Nuclear safeguards challenges at reactors types that defy traditional item counting", presented at JAEA-IAEA Workshop on Advanced Safeguards Technology for the Future Nuclear Fuel Cycle, Tokai-mura, Japan, 2007.

[418] DURST, P.C., et al., Nuclear Safeguards Considerations for the Pebble Bed Modular Reactor (PBMR), Rep. INL/EXT-09-16782, Idaho Natl Lab., ID (2009).

[419] NUCLEAR ENERGY INSTITUTE, Proposed Physical Security Requirements for Advanced Reactor Technologies, NEI White Paper, Washington, DC (2016), https://www.nrc.gov/docs/ML1702/ML17026A474.pdf

[420] BLUMBERG, R., HISE, E.C., MSRE Design and Operations Report, Part X: Maintenance Equipment and Procedures, Rep. ORNL-TM-910, Oak Ridge Natl Lab., TN (1968).

[421] THOMA, R.E., Chemical Aspects of MSRE Operations, Rep. ORNL-4658, Oak Ridge Natl Lab., TN (1971).

[422] MCCOY, H.E., Jr., WEIR, J.R., Jr., Materials Development for Molten-Salt Breeder Reactors, Rep. ORNL-TM-1854, Oak Ridge Natl Lab., TN (1967).

[423] TAUBE, M., LIGOU, J., Molten plutonium chloride fast breeder reactor cooled by molten uranium chloride, Ann. Nucl. Sci. Eng. **1** (1974) 277–281.

[424] TAUBE, M., FURRER, M., Information über die Anwendungsmöglichkeiten von Salzschmelzen in der Energiewirtschaft, Rep. AN-HL-234, Eidg. Institut für Reaktorforschung, Würenlingen, Switzerland (1975).

[425] TAUBE, M., LIGOU, J., BUCHER, K.H., The Transmutation of Fission Products (Cs-137, Sr-90) in a Liquid Fuelled Fast Fission Reactor with Thermal Column, Rep. EIR-270, Eidg. Institut für Reaktorforschung, Würenlingen, Switzerland (1975).

[426] OTTEWITTE, E., Fast molten chloride reactor on the thorium cycle, Trans. Am. Nucl. Soc. **28** (1978) 725–726.

[427] IANOVICI, E., TAUBE, M., Chemical State of Sulphur Obtained by the ^{35}CI (n, p) ^{35}S Reaction During In Pile Irradiation, Rep. EIR-267, Eidg. Institut für Reaktorforschung, Würenlingen, Switzerland (1974).

[428] LIU, Y., et al., Criticality properties and control rod worth of the critical experiment device for MSR research, Nucl. Technol. **204** 2 (2018) 203–212.

[429] LIU, Y., et al., Neutron flux distribution and conversion ratio of critical experiment device for molten salt reactor research, Ann. Nucl. Energy **133** (2019) 707–717.

[430] VERGNES, J., et al., "Limiting plutonium and minor actinides inventory: comparison between accelerator driven systems (ADS) and critical reactors", Proc. Int. Conf. on Future Nuclear Systems (GLOBAL 1999), Jackson Hall, 1999, American Nuclear Society, La Grange Park, IL (1999) Vol. 29.

[431] LECARPENTIER, D., VERGNES, J., BARBRAULT, P., TETART, P., MOUNEY, H., "The AMSTER concept (Actinide Molten Salt TransmutER)", Proc. Int. Conf. Physics of Reactors (PHYSOR 2000), Pittsburgh, 2000, American Nuclear Society, La Grange Park, IL (2000).

[432] VERGNES, J., GARZENNE, C., LECARPENTIER, D., MOUNEY, H., DELPECH, M., "The AMSTER Concept: a Configuration Generating its own Uranium with a Mixed Thorium and Uranium Support", Proc. Int. Conf. on Back-End of the Fuel Cycle: from Research to Solutions (GLOBAL 2001), Paris (2001) CD-ROM.

[433] LECARPENTIER, D., Contribution aux Travaux sur la Transmutation des Déchets Nucléaires, Voie des Réacteurs a Sel Fondu: le Concept AMSTER, Aspects Physiques et Sûreté, PhD Thesis, Conservatoire National des Arts et Métiers (2001).

[434] VERGNES, J., LECARPENTIER, D., The AMSTER concept (actinides molten salt transmutER), Nucl. Eng. Des. **216** 1–3 (2002) 43–67.

[435] FINNE, J., Chimie des Mélanges de Sels Fondus — Application à l'Extraction Réductrice d'Actinides et de Lanthanides par un Métal Liquid, PhD Thesis, Paris 6 Univ. (2005).

[436] WALLE, E., et al., A computational tool for selective pyrochemical process based on molten salts in nuclear industry, J. Nucl. Mater. **344** 1–3 (2005) 158–164.

[437] WALLE, E., SIMON, N., HUGUET, A., GASTALDI, O., PENIT, T., "Development of simulation tools for the on-line reprocessing of the Molten Salt Reactor concept", Proc. Int. Conf. on Nuclear Energy Systems for Future Generation and Global Sustainability (GLOBAL 2005), Tsukuba, Japan, 2005, Atomic Energy Society of Japan, Tokyo (2005).

[438] SIMON, N., et al., "Simulation tool of the on-line reprocessing unit of a molten salt reactor", Proc. European Nuclear Conf — Nuclear Power for the 21st Century: from Basic Research to High-Tech Industry (ENC 2005), Versailles, France, 2005, Société française d'énergie nucléaire (SFEN), Paris (2006).

[439] FINNE, J., et al., Molten salt/liquid metal extraction: electrochemical determination of activity coefficients in liquid metals, J. Nucl. Mater. **344** 1–3 (2005) 165–168.

[440] FINNE, J., et al., "Electrochemical technique for accurate determination of solvation properties for liquid metals at high temperature", Proc. 7th Int. Symp. on Molten Salts Chemistry and Technology (MS 7), Toulouse, France, 2005, Elsevier, Amsterdam (2007).

[441] NUTTIN, A., Potentialités du Concept de Réacteur a Sels Fondus pour une Production Durable d'Énergie Nucléaire Basée sur le Cycle du Thorium en Spectre Epithermique, PhD Thesis, Univ. Joseph Fourier de Grenoble (2002).

[442] RENAULT, C., et al., "The MOST Project: Key-points and Challenges for the Feasibility of Molten Salt Reactors", Proc. Int. Congr. on Advances in Nuclear Power Plants (ICAPP 2005), Seoul, Korea, 2005, American Nuclear Society, La Grange Park, IL (2005) 15–19.

[443] LECARPENTIER, D., CARPENTIER, V., A neutronic program for critical and nonequilibrium study of mobile reactors: the Cinsf1D code, Nucl. Sci. Eng. **143** 1 (2003) 33–46.

[444] PRINCE, B.E., BALL, S.J., ENGEL, J.R., HAUBENREICH, P.N., KERLIN, T.W., Zero-power experiments on the molten salt-reactor experiment, Rep. ORNL-4233, Oak Ridge Natl Lab., TN (1968).

[445] LECARPENTIER, D., GARZENNE, C., HEUER, D., NUTTIN, A., DELPECH, M., "Temperature Feedbacks of a Thermal Molten Salt Reactor: Compromise between Stability and Breeding Performances", Proc. Int. Congr. on Advances in Nuclear Power Plants (ICAPP 2003), Cordoba, Spain, 2003, Vol. 3, American Nuclear Society, La Grange Park, IL (2003) 4–7.

[446] SANCHEZ, R., STANKOVSKY, Z., COSSIC, A., MONDOT, J., ZMIJAREVIC, I., "APOLLO II: a user-oriented, portable, modular code for multigroup transport assembly calculations", Proc. Int. Mtg on Advances in reactor physics, mathematics and computation, Vol. 1, Paris, 1987, American Nuclear Society, La Grange Park, IL (1987) 1563–1577.

[447] MATHIEU, L., et al., "Proposal for a simplified thorium molten salt reactor", Proc. Int. Conf. on Nuclear Energy Systems for Future Generation and Global Sustainability (GLOBAL 2005), Tsukuba, Japan, 2005, Atomic Energy Society of Japan, Tokyo (2005) 9–13.

[448] IGNATIEV, V., FEYNBERG, O., MERZLYAKOV, A., "Integrated study of molten Na,Li,Be/F salts for LWR waste burning in accelerator driven and critical systems", Proc. Int. Conf. on Nuclear Energy Systems for Future Generation and Global Sustainability (GLOBAL 2005), Tsukuba, Japan, 2005, Atomic Energy Society of Japan, Tokyo (2005).

[449] VOCKA, R., "Closing the PWR fuel cycle with a molten-salt incinerator", Proc. Int. Conf. Physics of Reactors (PHYSOR 2004) — The Physics of Fuel Cycles and Advanced Nuclear Systems: Global Developments, Chicago, 2004, American Nuclear Society, La Grange Park, IL (2004).

[450] BENES, O., et al., European Specific Support Actions: Assessment of Liquid Salts for Innovative Applications (ALISIA), Review Report on Liquid Salts for Various Applications (2009),
https://cordis.europa.eu/docs/projects/files/44/44809/107605351-6_en.pdf

[451] IGNATIEV, V.V., "Molten salt fuels for nuclear waste transmutation in accelerator driven systems", Review of National Accelerator Driven System Programmes for Partitioning and Transmutation, IAEA-TECDOC-1365, IAEA, Vienna (2003) 119–129.

[452] INTERNATIONAL ATOMIC ENERGY AGENCY, Advances in Small Modular Reactor Technology Developments — A Supplement to: IAEA Advanced Reactors Information System (ARIS), 2022 Edition, IAEA, Vienna (2022).

ABBREVIATIONS

AMSTER	Actinide Molten Salt Transmuter
ANP	aircraft nuclear propulsion
An(s)	actinide(s)
CALPHAD	calculation of phase diagrams (code)
CANDU	Canada Deuterium Uranium
CEA	French Alternative Energies and Atomic Energy Commission (Commissariat à l'énergie atomique et aux énergies alternatives) (France)
CFD	computational fluid dynamics
CMSR	Compact Molten Salt Reactor
CNL	Canadian Nuclear Laboratories
CNRS	National Center for Scientific Research (Centre national de la recherche scientifique) (France)
CORYS	Compagnie de Réalisation Industrielle de Simulateurs
DTU	Technical University of Denmark
DYNASTY	dynamics of natural circulation for molten salt internally heated
EDF	Électricité de France
EIR	Federal Institute for Reactor Research (Eidgenössisches Institut für Reaktorforschung) (Switzerland)
Euratom	European Atomic Energy Community
EVOL	evaluation and viability of liquid (fuel reactors) (project)
FFFER	forced fluoride flow for experimental research
FHR	fluoride salt cooled high temperature reactor
FLiBe	LiF–BeF_2 mixture (2LiF–BeF_2)
FLiNaK	LiF–NaF–KF mixture
GEMS	Gibbs energy minimization software
GPC	generalized polynomial chaos
IMSR	Integral Molten Salt Reactor
IRSN	Institute for Radiological Protection and Nuclear Safety (Institut de Radioprotection et de Sûreté Nucléaire) (France)
ISTC	International Science and Technology Center
JRC	Joint Research Centre (European Commission)
JRCMSD	Joint Research Centre Molten Salt Database
LPSC	Laboratory of Subatomic Physics and Cosmology (Laboratoire de Physique Subatomique et de Cosmologie) (France)

Mk1 PB-FHR	Mark 1 Pebble Bed Fluoride Salt Cooled High Temperature Reactor
MONICR	molybdenum–nickel–chromium (alloy)
MOSART	Molten Salt Actinide Recycler and Transformer
MOST	Molten Salt Reactor Technology (project)
MSBR	molten salt breeder reactor
MSFR	molten salt fast reactor (such as the one developed by the CNRS, France)
MSR	molten salt reactor
MSRE	Molten Salt Reactor Experiment
NRG	Nuclear Research and Consultancy Group (Netherlands)
ORNL	Oak Ridge National Laboratory
R&D	research and development
SAMOFAR	Safety Assessment of the Molten Salt Fast Reactor
SAMOSAFER	Severe Accident Modelling and Safety Assessment for Fluid-fuel Energy Reactors
SINAP	Shanghai Institute of Applied Physics
SMR	small modular reactor
SOFT	Salt reactor On site reprocessing Fast converter Task
SSR	Stable Salt Reactor
SWATH	salt at wall: thermal exchanges
TFM	transient fission matrix
TMSR	thorium molten salt reactor
TMSR-LF	liquid fuelled thorium molten salt reactor
TRISO	tristructural isotropic
TRU	transuranium (elements) (also known as transuranic (elements))
TU Delft	Delft University of Technology (Netherlands)
VNIINM	A.A. Bochvar High-Technology Research Institute of Inorganic Materials
VVER	water–water energetic reactor (WWER)

CONTRIBUTORS TO DRAFTING AND REVIEW

Abrate, N.	Politecnico di Torino, Italy
Agarwal, K.	International Atomic Energy Agency
Allibert, M.	Laboratoire de Physique Subatomique et de Cosmologie, France
Artisiuk, V.V.	International Atomic Energy Agency
Benes, O.	Joint Research Centre of the European Commission
Bourg, S.	French Alternative Energies and Atomic Energy Commission, France
Boyer, B.	International Atomic Energy Agency
Cammi, A.	Politecnico di Milano, Italy
Carluec, B.	Independent consultant, France
Chen, K.	Moltex Energy, United Kingdom
D'Agata, E	Joint Research Centre of the European Commission
Delpech, S.	IJCLab-CNRS, Orsay, France
Dulla, S.	Politecnico di Torino, Italy
Edwards, L.	Australian Nuclear Science and Technology Organisation, Australia
Fratoni, M.	University of California Berkeley, United States of America
Ghetta, V.	Laboratoire de Physique Subatomique et de Cosmologie, France
Giot, L.	SUBATECH, France
Hania, R.	NRG, Netherlands
Hargraves, R.	ThorCon International, United States of America

Holcomb, D.	Oak Ridge National Laboratory, United States of America
Ignatiev, V.	National Research Centre, Kurchatov Institute, Russian Federation
Kloosterman, J.L.	Delft University of Technology, Netherlands
Konings, R.J.M.	Joint Research Centre of the European Commission
Křepel, J.	Paul Scherrer Institute, Switzerland
Laureau, A.	Laboratoire de Physique Subatomique et de Cosmologie, France
Lauritzen, B.	Technical University of Denmark, Denmark
Leblanc, D.	Terrestrial Energy, Canada
Lecarpentier, D.	Électricité de France, France
Lim, Y.X.S	International Atomic Energy Agency
Lorenzi, S.	Politecnico di Milano, Italy
Martinez-Guridi, G.	International Atomic Energy Agency
Merle, E.	Laboratoire de Physique Subatomique et de Cosmologie, France
Monti, S.	International Atomic Energy Agency
Pater, M.	Seaborg Technologies, Denmark
Peguero, L.	International Atomic Energy Agency
Reitsma, F.	International Atomic Energy Agency
Rickman, R.	Terrestrial Energy, Inc., United States of America
Roelofs, F.	NRG, Netherlands
Saoudi, M.	Canadian Nuclear Laboratories, Canada
Schönfeldt, T.	Seaborg Technologies, Denmark
Silvioli, L.	Seaborg Technologies, Denmark
Soucek, P.	Joint Research Centre of the European Commission

Stubsgaard, A.	Copenhagen Atomics, Denmark
Subki, M.H.	International Atomic Energy Agency
Uggenti, A.C.	Politecnico di Torino, Italy
Uhlíř, J.	Research Centre Řež, Czech Republic
Yoshioka, R.	International Thorium Molten-Salt Forum, Japan
Zou, Y.	Shanghai Institute of Applied Physics, China

Consultants Meetings

Vienna, Austria: 18–20 November 2015; 25–27 September 2017; 29 July – 1 August 2019; 21–23 July 2020

Technical Meeting

Vienna, Austria: 31 October – 3 November 2016

IAEA
International Atomic Energy Agency

No. 26

ORDERING LOCALLY

IAEA priced publications may be purchased from the sources listed below or from major local booksellers.

Orders for unpriced publications should be made directly to the IAEA. The contact details are given at the end of this list.

NORTH AMERICA

Bernan / Rowman & Littlefield
15250 NBN Way, Blue Ridge Summit, PA 17214, USA
Telephone: +1 800 462 6420 • Fax: +1 800 338 4550
Email: orders@rowman.com • Web site: www.rowman.com/bernan

REST OF WORLD

Please contact your preferred local supplier, or our lead distributor:

Eurospan Group
Gray's Inn House
127 Clerkenwell Road
London EC1R 5DB
United Kingdom

Trade orders and enquiries:
Telephone: +44 (0)176 760 4972 • Fax: +44 (0)176 760 1640
Email: eurospan@turpin-distribution.com

Individual orders:
www.eurospanbookstore.com/iaea

For further information:
Telephone: +44 (0)207 240 0856 • Fax: +44 (0)207 379 0609
Email: info@eurospangroup.com • Web site: www.eurospangroup.com

Orders for both priced and unpriced publications may be addressed directly to:

Marketing and Sales Unit
International Atomic Energy Agency
Vienna International Centre, PO Box 100, 1400 Vienna, Austria
Telephone: +43 1 2600 22529 or 22530 • Fax: +43 1 26007 22529
Email: sales.publications@iaea.org • Web site: www.iaea.org/publications